MAMMALS
OF EUROPE

For Ewan, Fiona, and Isobel

Princeton Field Guides

Rooted in field experience and scientific study, Princeton's guides to animals and plants are the authority for professional scientists and amateur naturalists alike. **Princeton Field Guides** present this information in a compact format carefully designed for easy use in the field. The guides illustrate every species in color and provide detailed information on identification, distribution, and biology.

Birds of Kenya and Northern Tanzania: Field Guide Edition, by Dale A. Zimmerman, Donald A. Turner, and David J. Pearson
Birds of India, Pakistan, Nepal, Bangladesh, Bhutan, Sri Lanka, and the Maldives, by Richard Grimmett, Carol Inskipp, and Tim Inskipp
Birds of Australia, by Ken Simpson and Nicolas Day
Birds of Europe, by Killian Mullarney, Lars Svensson, Dan Zetterström, and Peter J. Grant
Birds of Nepal, by Richard Grimmett, Carol Inskipp, and Tim Inskipp
Birds of the Seychelles, by Adrian Skerrett and Ian Bullock
Stars and Planets, by Ian Ridpath and Wil Tirion
Butterflies of Europe, by Tom Tolman and Richard Lewington
Mammals of Europe, by David W. Macdonald and Priscilla Barrett

Princeton University Press
www.birds.princeton.edu

MAMMALS
OF EUROPE

David W. MACDONALD & Priscilla BARRETT

Princeton University Press
Princeton and Oxford

Published in the United States and Philippine Islands by
Princeton University Press, 41 William Street, Princeton, New Jersey, 08540

Originally published by HarperCollins*Publishers* Ltd under the title:

COLLINS FIELD GUIDE TO THE MAMMALS OF BRITAIN & EUROPE

Library of Congress Control Number 2001093007

ISBN 0-691-09160-9

This book has been composed in Times

www.birds.princeton.edu

Printed and bound in Hong Kong

1 3 5 7 9 10 8 6 4 2

Contents

Preface

This book is part of a brace on *European Mammals*: the first volume is a *Field Guide* and the second *Evolution and Behaviour*. While each volume can stand alone, they are intended to function best as a partnership. In the *Field Guide* readers can discover what they have seen and, in condensed form, delve into the vital statistics of each of the 200 species they may encounter in Western Europe. In the *Evolution and Behaviour* they can learn something of the patterns that bring order to the enthralling variation of mammalian lives. To facilitate that link between the specific and the general, we have been at pains to discuss species, in the *Evolution and Behaviour*, that the reader may have identified while using the *Field Guide*. Familiarity, the axiom tells us, breeds contempt, which may be why it is all too common for European naturalists to assume that the really interesting species are confined to far off tropical lands. Hopefully this field guide will demonstrate that fascinating mammals scurry in our own backyard.

Modern mammals have changed a lot since the days of their ancestors, and so too these books have evolved along an unexpected route to an end point that is dramatically different from the one envisaged at the outset. They are born as a complementary pair, but at the outset our sights were not set so high, with the modest aim of a single volume. However, a field guide treats each species in isolation, and thus ignores the patterns that make sense of biological diversity. On the other hand, while one might write about the principles that unite mammals with reference to only a few selected examples, not to introduce all the actors personally runs the risk of making abstract and tidy a story which is actually enthrallingly unfinished. Perversely, one of the most important lessons to be learned from flipping through the pages of the *Field Guide* is just how little is known about most species. All too many general rules in biology rest on all too few particular species. Therefore, realising that we could not achieve both aims within one book, we have collaborated to produce these twin volumes.

Amongst the many debts of gratitude we must record here we should begin with those who got us into this in the first place. The seven year gestation of these two books has had its share of authorial morning sickness, not to mention high blood pressure; but now that the birth pains are over, and the twin off-spring healthy, we thank heartily Sir David Attenborough and Professor Christopher Perrins for their role in the conception.

We are both acutely aware of the impact these books have had on our families. Predictably, but honestly, deep gratitude goes to Jenny Macdonald for her tolerance, and to the children for adapting to the astonishing behaviour of a father who really did spend two family holidays locked in a separate caravan wrestling with this manuscript while they wrestled on the nearby sands. Hopefully Ewan, Fiona and Isobel will one day take some pride in these pages that were traded for too many sunny weekends, but there is no doubt that the swap was tough at the time. Very special thanks also go to Gabriel Horn. He has provided helpful criticism and support throughout a long project, and in addition has stoically borne the disturbed nights, cancelled weekends and truncated holidays that have resulted from being the artist's husband.

In writing these two books primary sources of information were consulted whenever possible. In a more technical book it would be conventional to acknowledge these sources by citing their authors' names in the text. However, that protocol is not accepted in books of this ilk and so the authors of hundreds of technical papers that are synthesised here will go un-named. Their anonymity should not disguise the contribution of their publications to these pages.

Nobody knows enough about the mammalian fauna of a continent to portray it knowledgeably single-handedly. We have relied heavily upon colleagues around Europe who have given their time unstintingly to comment helpfully upon our efforts. So many people have helped over so long a time that we must begin by apologising to those we fail to acknowledge individually. Our general thanks to those not mentioned by name is not diminished by listing the following who have helped us excise errors from the first volume, the *Field Guide*: S. Albon, C.L. Alados, A. Angerbjorn, A. Aquilar, H. Arnold, W. Arnold, M. Artois, R.P.D. Atkinson, S. Baker, D. Balharry, M. Ben-David, R.J. Berry, M.A. Bigg, J.D.S. Birks, L. Boitani, W.N. Bonner, E.L. Bradshaw, S. Broekhuizen, E. Brown, A. Buckle, D. Cantoni, N.G. Chapman, C. Cheeseman, I. Christensen, S. Churchfield, J.R. Clarke, D. Clode, A. Collet, G.B. Corbett, L.K. Corbett, P. Cox, C.P. Doncaster, E.M. Dorsey, M.T. Dowie, M.J. Duchene, N. Dunstone, N. Easterbee, K. Elgmork, H. Ellenberg, S. Engel, S. Erlinge, P.G.H. Evans, J. Fa, C. Faulkes, F.H. Fay, M. Fenn, J. Flowerdew, K.J. Frost, B. Fruzinski, D. Gaskin, J. Gordon, M. Gorman, J.H. Greaves, I. Hanski, P. Hersteinsson, R. Hewson, D. Hirth, J. Holm, I. Hulbert, J. Hurst, A.M. Hutson, H. Ikeda, M. Ireland, P. Jackson, D. Jeffries, G. Jones, R.E. Kenward, C.M. King, K. Kinze, H. Kruuk, F. Larson, D.M. Lavigne, N. Leader-Williams, C.H. Lockyer, A. Loudon, L.F. Lowry, A. Loy, D.K. Ljungblad, S. Lovari, T. Lyrholm, F. Maisels, A.W. Mansfield, R. Mauget, A.R. Martin, M. Martys, M. Masseti, T. McOwat, H. Mendelssohn, J. Morel, P. Morris, E.G. Neal, C.R. Neet, W. Oliver, P.J. O'Sullivan, J.B. Parusel, G. Pigozzi, E. Pulliainen, P.E. Purves, R.J. Putman, P.A. Racey, P.R. Ratcliffe, C.Reece-Engel, R. Reeves, P. Richardson, K. Ronald, G.B.J. Ross, M. Sandell, M.E. dos Santos, J. Searle, B. Saether, D.E. Sergeant, J. da Silva, B. Staines, R.E. Stebbings, I. Stirling, P. Stockley, R.D. Stone, J. Svendsen, S. Tapper, T.E. Tew, M.J.A. Thompson, P.S. Tunkkari, H. Whitehead, B. Wursig, E. Zimen. In addition, T. Clutton-Brock, J. Flowerdew, J. Gordon, J. Harwood, M. Klinowska, and A. Martin all helped by commenting on the artwork. Gordon Corbett kindly delved in his files to unearth improvements to the distribution maps. Judith Mossman helped explore the labyrinthine origins of the scientific names, and Diane Adams, Elizabeth Bradshaw, Lynn Clayton, Dada Gottelli, Clare Hawkins and Laura Handora were amongst those in Oxford University's Wildlife Conservation Research Unit who assiduously helped track down species' statistics.

Special thanks are due to the many people who, between them, offered a wide range of source material for the colour plates. The Norfolk Wildlife Park cheerfully provided access to tracks and scats as well as to live animals. The staff in the Mammal section of the London Natural History Museum gave invaluable guidance with skins and literature. Grateful thanks are due to Ray Symonds and Adrian Friday of the Cambridge University Zoology Museum for cheerfully unearthing, and advising on, so many skulls as well as skins. The bat paintings owe a special debt to Bob Stebbings, who devoted hours to their critical appraisal.

Finally, we have come to regard the *Field Guide* as a staging post in an endless project. Many details of the tracks and signs of European mammals remain undescribed. As we finish this first edition we already have ideas for a later edition, and we hope that it will benefit from feedback from you, the readers.

David W. Macdonald
Oxford

Priscilla Barrett
Cambridge

How to Use This Book

The best way to use this book is as one of a pair. The *Field Guide* will tell you what you have seen, and if it catches your interest you can turn to *Evolution and Behaviour* to discover how that species, its ecology and behaviour, fit into the mammalian scheme of things.

In the *Field Guide* the 64 colour plates illustrate every wild mammal species that you can reasonably expect to find and identify on sight in Western Europe. Characteristic adults are portrayed in every case, and both sexes are shown where their appearance is sufficiently different to warrant it. Juveniles are depicted if they resemble neither parent. Wherever possible the animals are illustrated in characterstic gaits because when watching mammals the observer often has only a glimpse and the style of movement may be the salient memory. The shyness, rarity and nocturnality of many mammals means that signs of their presence are often more commonly seen than the animal itself. For this reason skulls and other field signs are included, together with scenes depicting characteristic dens. If details of the teeth are important for identification they are shown separately from the skull. When details of the grinding surface of cheek teeth are important, as in voles, this is the view shown (the upper tooth row is always drawn above the lower).

Fieldsigns generally include droppings and paw prints. The tracks show the underside view of the pads and claws as they are imprinted in mud. The right hind and right fore prints are shown in each case (the pictures on the left and right respectively). They have been idealised for clarity. Experience is necessary to decipher reliably the blurred, partially registered, overlaid tracks all too commonly encountered in the field. On some plates the field signs extend behind two or more species. An example of this is the hollow tree trunk behind the four pipistrelle bats on Plate 8, indicating that all of them roost in tree cavities. This layout is intended to help the reader identify species as quickly as possible. Similarly, the fact that a cave/mine shaft appears only behind Savi's Pipistrelle indicates that it is the only one of the four thought commonly to roost in such places. The bat plates illustrate the roost sites commonly used: tree holes represent cavities provided by trees, either holes in the trunk or spaces under the bark; a building represents houses or outbuildings; caves represent mineshafts, tunnels and culverts as well as actual caves. Bats are depicted from either dorsal or ventral view, depending on which reveals the greatest array of diagnostic features.

Mammals are grouped on particular plates with attention to both the likelihood of confusing them and their taxonomic relationships. For example, the whales are largely grouped by size, but all members of any one genus are always portrayed consecutively. Identification of whales from brief glimpses at sea is notoriously difficult. The layout of the Cetacean plates aims to provide the observer with clues as to the key points for which to look. Plates 36 and 37, for example, show the shape, height and angle of the snout produced by each species when 'blowing' at the surface, and the amount of the animal's back that is normally visible when it blows. If the back is arched or the tail flukes usually raised before a deep dive this is shown. As whales at the surface may well be feeding with open mouths, the length and colour of their baleen plates are also illustrated.

A really important feature of the layout is exemplified by the rodents, and especially the voles (eg Plate 32). To help the reader pick out the diagnostic features of very similar species, these are generally presented one behind the other, with their special features showing but their shared features hidden. For example, on Plate 32, the Common Vole is presented in front; behind it Gunther's Vole is partly obscured but its hindfoot, tail and rump are all in view, thereby drawing attention to their importance as diagnostic traits to be checked in the species account (p. 00). Carbrera's Vole is yet further behind in the queue, also partly obscured but showing the critically important colour of its rump. We have tried to ensure that every picture tells many stories. For example, not only is there detailed attention to the nuances of fur coloration, but animals are often positioned to show off particular traits (such as the position of scent glands) or characteristic postures.

The reader's attention will be drawn to the intricacies of the plates by the text of the caption pages that face them. These caption pages mention the common and scientific names of the species and cross-refer to the species sections (pages 19–291). For each species the caption texts mention the salient diagnostic traits, along with some less obvious ones that require a careful look at the colour plate. These texts also explain the illustrated fieldsigns. However, for a full account of the animal the reader must cross-refer to the species entry. Great pains have been taken to avoid merely duplicating information in the short captions and the long species accounts. Rather, in the captions you will find an emphasis on such measures as footprint sizes and dimensions of droppings. Measurements of the animal itself are confined to the full species entries. The caption pages also include additional colour illustrations that depict ad-

Fig. 1 General measurements of a mammal used in this book.

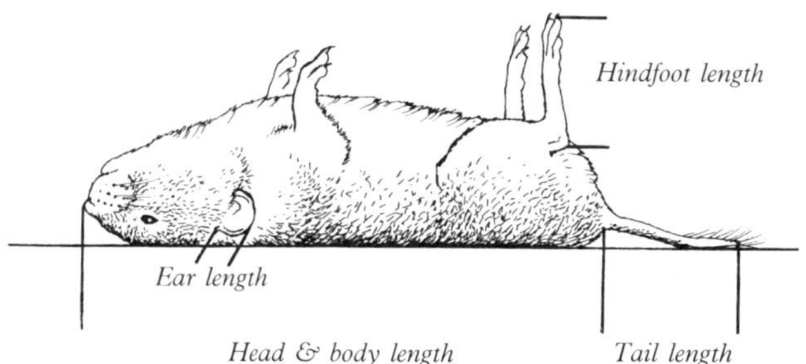

Fig. 2. Names for parts of mammals used in the text (see also the entries in the Glossary).

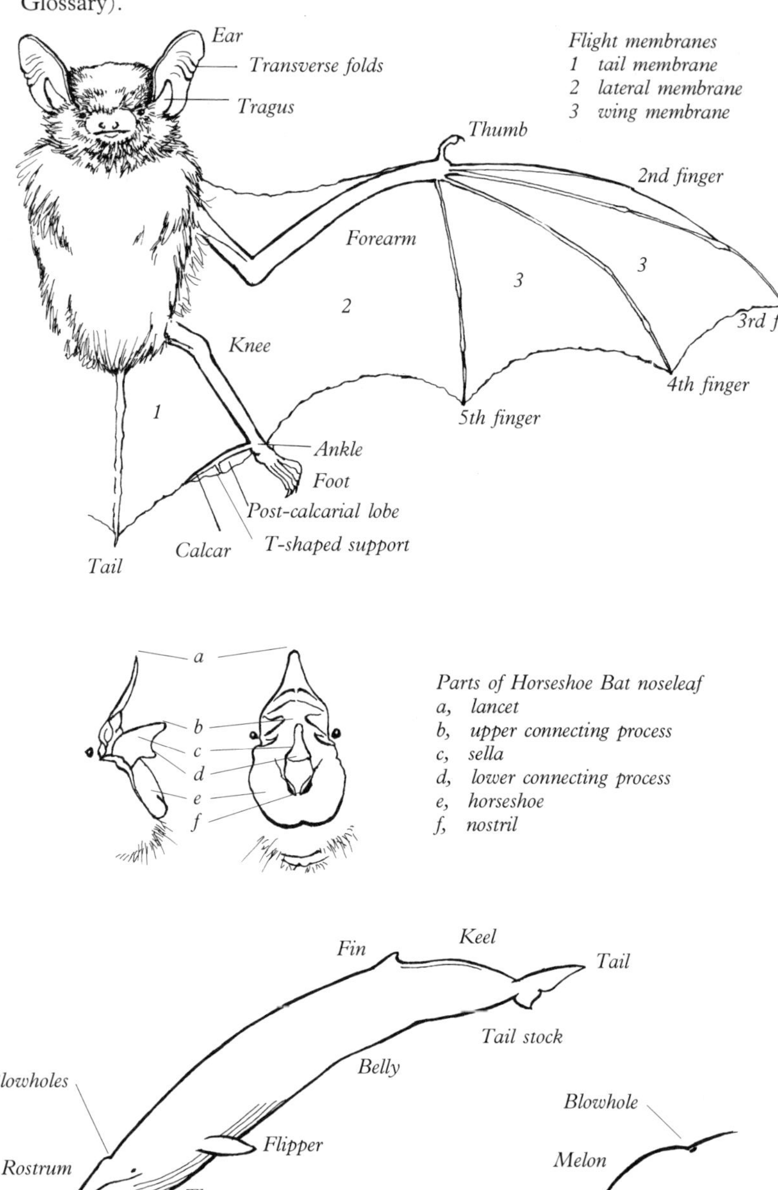

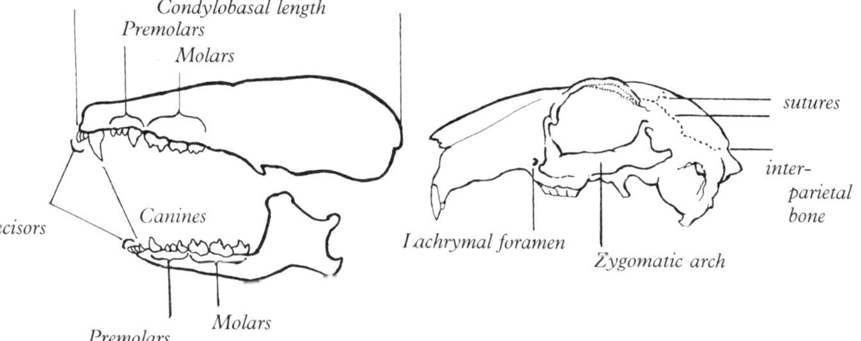

Fig. 3. Names of skull features and teeth used in the text.

ditonal clues to identification (such as regional variations in appearance, or particular behaviour).

The distribution maps are in pairs at the beginning or end of each species account. Predictably, the distributions of many inconspicuous, nocturnal mammals are scantily known as, in particular, are those of bats and whales. In drawing the maps we have consulted not only other field guides but also technical accounts of species' distributions. However, the latter are generally presented in terms that cannot readily be adapted to a simple, European-scale summary, and so the maps remain coarse. For the enthusiast who is not content merely to know that the Pyrenean Desman occurs in the Pyrenees, but requires to know in which valley to search for it, the only recourse is highly specialised texts or, more likely, the records of a local natural history society.

The species entries all follow the same format. However, there is vast inequality in what is known about each species. Some have been the subject of many books, others are known from only a few dried skins or, worse still, their chromosomes. Consequently, in one case it may have been a struggle to decide which of innumerable facts to include, in the next it may have been necessary to scour libraries for anything to say at all. In general, where there is no information under a particular sub-heading for a species this means that either nothing is known of that topic or, if something is known, we failed to unearth it! Each entry is illustrated by a black and white vignette drawing, and these complement the colour plates by portraying aspects of behaviour characteristic of the species. Where the topic of the vignette is mentioned in the species entry attention is drawn to this in the form '(see vignette)'.

The information is presented in the following format:

COMMON NAME *Scientific name* **Plate reference**

Name. Alternative common names and the likely derivation from Latin or Greek words (in italics) of the scientific name. Where several members of one genus are described the derivation of the generic name is given for only the first of them.

Family. Family names are given at the top of each page in the species section. If two or more families are mentioned then the names are mentioned in order. The family indicates which individuals are related to each other within each family; the general features of the order to which the family belongs are given in a short account at the start of the section on that order. The terms genus, family and order represent increasingly broad categories into which it is convenient to classify species; animals in the same category are thought to share a closer common evolutionary ancestry than do those in different categories. For example, the Brown Rat is a member of the genus *Rattus*, the family Muridae and the order Rodentia. It is more closely related to the Black Rat (also a member of the genus *Rattus*) than to the Wood Mouse (genus *Apodemus*), but both are members of the same family and therefore more closely related to each other than either is to the Coypu, which although also a Rodent is classed in the family Capromyidae.

Status. The IUCN, an organisation whose initials puzzlingly stand for the World Conservation Union, has evaluated the conservation status of Mammals in their 1990 list of threatened animals. They define the following terms, used in the *Field Guide*:

Endangered: species in danger of extinction and whose survival is unlikely if the causal factors continue operating.

Vulnerable: species believed likely to move into the endangered category in the near future if the causal factors continue operating.

Rare: species with small world populations that are not at present endangered or vulnerable, but are at risk.

Indeterminate: species known to be endangered, vulnerable or rare, but where there is not enough information to say which of these three categories is appropriate.

Insufficiently known: species that are suspected but not definitely known to belong to any of the above categories, because of lack of information.

Threatened: a general term to denote species which are Endangered, Vulnerable, Rare, Indeterminate or Insufficiently Known.

If no information is given under the Status section, the IUCN has not classified the species in one of the foregoing categories. This does not guarantee that it is secure everywhere, but indicates that it is unlikely to be in immediate danger throughout its entire range. Where there are points to be made about the conservation status of a species on a local scale these are included in the section headed General.

Recognition. This section describes key features in the appearance of the species, and should be read while referring to the colour plates. It also includes some information on moulting, as this can greatly affect appearance. Where recognition depends on the appearance of oddly named structures, these are illustrated in the introduction. The second part of this section describes field signs that may be strong pointers to the species' identity.

Habitat. Describes the broad characteristics of the habitats in which the species is commonly encountered, before describing the circumstances and general appearance of dens and den sites.

Habits. This section begins with a summary of the species' daily rhythms and any seasonal patterns to its activity, before presenting subsections on FOOD: including the main components of the diet, HOME RANGE: the size of area normally travelled by resident animals, their system of spatial organisation, and the distances over which individuals may disperse. Next, there is information on population density. SOCIAL: provides a summary of the species' social behaviour, leading to COMMUNICATION: information is provided on postures, vocalisations and scent glands, together with notes on the acuity of the senses.

Breeding. Information on the likely dates of mating and birth is followed by statistics on age at sexual maturity, gestation times, litter sizes, numbers of litters per annum, and the number of mammae. A sense of variation in these measures is provided, where possible, by average figures qualified by the range of variation commonly reported, and where appropriate these figures are separated for males and females and for a selection of different regions. This section also mentions the YOUNG: describing them (again, cross refer to the colour plate) and giving ages at which they reach critical stages of development, such as weaning. The section finishes with a summary of the type of parental care.

Lifespan. The *maximum* recorded lifespan in the wild and in captivity. Where available, information is presented on population dynamics and the main causes of death.

Measurements. This section condenses an enormous amount of information on the vital statistics of each species. If you happen to have come across a mammal that is prepared to sit still to be measured (a level of cooperation that probably indicates it is dead), then you will be able to use these measurements to identify it. In a more technical book the great variation in sizes between individuals and between regions would necessitate tables of dimensions (and for many species these are not available). Here, in contrast, the aim has been to provide a fair clue as to whether the creature you have just measured fits within the likely size range recorded for a given species. As mentioned with regard to figures provided in the Breeding section, the intention has been to give a sense of variation in these measures by providing, where possible, average figures qualified by the range of variation commonly reported, and where appropriate to separate out these figures for males and females and for a selection of different regions. However, the motley quality of the raw information is such that it has been impossible to be consistent in the style of presentation. As far as possible, the ranges of measurements given are for the sizes of individuals normally encountered, and exclude the confusing influence of giants or midgets (therefore, remember that if your creature does not fit the figures, there is always a small chance that it is one of these extremes of size). Of course all this variation is very annoying, and may mean that you end up being unsure as to whether you have a small something, or a large something else. That, however, is life.

Mammalogists so rarely see their subjects that they tend to be gripped by a desire relentlessly to measure every cranny of their anatomy when the opportunity arises. The most straightforward and generally useful measures are used here. Where information is available, the following vital statistics are given: Head-Body Length, Tail Length, Hind-Foot Length, Ear, Condylo-Basal Length, Weight and Dental Formula. These measurements are taken as illustrated in Fig. 1.

For some species, additional measures are appropriate. For example, in the case of bats there are measures of Wingspan and Fore-finger Length. These measures are illustrated in Fig. 2. The Dental Formula is presented as the number of teeth of a given type in the upper and lower jaws on one side of the skull. The four types of teeth are incisors, canines, premolars and molars. The dental formula for the Roman Mole is 3/3, 1/1, 4/4, 3/3 = 44. Looking at the mole skull illustrated on Plate 7 reveals that on one side of its jaw the mole has 3 upper incisors, 3 lower incisors, one upper canine, one lower canine, four each of upper and lower premolars and three each of upper and lower molars. This gives a total of 22 teeth on each side of the skull or 44 teeth in all. In some species, the distinction between these different teeth is blurred and it is more sensible to refer simply to the number of cheek teeth (premolars and molars combined) or to the length of the post-canine tooth row (e.g. for shrews). The length of the post-canine tooth row is the distance from the front edge of the first premolar to the rear edge of the last molar. In the mysticete whales there are no teeth but their baleen plates are described (and illustrated) instead. For the most part, the dental formulae will be useful in identifying skulls, since intimate inspection has its hazardous aspect with at least larger mammals: if you feel that you can not be certain whether you have a Wolverine or a Badger in sight without probing its dentistry it would be imprudent to pursue the matter without the aid of an anaesthetic for one or other party.

General. This section provides a short summary of points of special interest, including notes on conservation and behaviour patterns you might see in the field. However, for a fuller account of the life histories of these species and the general rules that they follow, readers should refer to the companion volume, *European Mammals: Evolution and Behaviour.*

The accounts in the *Field Guide* inevitably expose only the tips of many icebergs. For example, the reader who discovers that delayed implantation is characteristic of reproduction in Stoats but not in Weasels may guess that this distinction is overwhelmingly important, but the full story could not possibly be compressed into a field guide, and nor would one want it there. This, of course, is what prompted the companion volume, *Evolution and Behaviour*. However, as a short-cut to understanding some technical terms, pages 292–302 of the *Field Guide* include a Glossary. Also, for those wishing to follow up more detail there is an annotated bibliography. The bibliography deals only with books, whereas in fact most of the detailed information can be found only in technical papers published in scientific journals. In compiling the *Field Guide* and *Evolution and Behaviour* we have drawn on the wisdom of a thousand or more original papers; to cite them here would not be appropriate for this type of book, but we cannot repeat too often our gratitude to their authors.

Order Marsupialia – Marsupials

Origins: The Marsupial (pouched) mammals are one of the two divisions (infra-classes) of live-bearing (Therian) mammals. Their ancestors diverged from those of the other infra-class, the placental mammals, 100 mya (million years ago). The first clearly marsupial fossils are found in what is now N. America in sediments dating from the late Cretaceous, about 75 mya, and were mainly like opossums. However, marsupials had died out in N. America by about 15–20 mya, but enjoyed a great evolutionary radiation in S. America until the so-called Great American Interchange, 2 mya, when the linking of north and south via the Panamanian land bridge allowed an influx of competitively superior placental mammals. The earliest marsupials so far found in Australia date from about 23 mya, when they had already diverged into modern families. The parallel evolution of marsupials and placentals means that each infra-class has produced its own attempt at filling a given niche. The Red-necked Wallaby fills a slot between small ungulates and lagomorphs. Since marsupials diverged so early from placental mammals, their closest relatives among the placentals must be the most ancient placentals, the Insectivores.

Main features of the Marsupials: Marsupials today are endemic to the New World and Australia. Their prime distinguishing feature is their method of reproduction. In its early stages, the marsupial egg is like those of reptiles and birds. The embryo is born very early in its development – eg the Eastern Grey Kangaroo (female weighs *c.* 30kg) gives birth after 36 days' gestation to a 0.8g offspring. This is carried in a pouch for about 300 days till it weighs *c.* 5kg, which is about the size, relative to its mother, at which placental mammals would have been born; it continues to suckle till about 18 months old. Marsupials are mostly nocturnal, so their most important senses are hearing and smell. They have a so-called fenestrated palate (i.e. the bones in the roof of the mouth have window-like holes in them). The only place in Europe that a mammalogist is likely to find such a skull is in the Peak District of England: there, a skull with an incomplete palate belongs to a wallaby.

Marsupials in Western Europe: Of 18 marsupial families in the world, just one, the Macropodidae, is found in Europe. The Red-necked Wallaby is one of 266 contemporary marsupial species organised into 76 genera. A special feature of this family (occurring in only a few other marsupial species) is that a pregnancy does not affect the oestrous cycle, and the female is receptive and mates at about the time she gives birth. The resulting embryo begins its development but then becomes quiescent – a situation called embryonic diapause, a phenomena similar to the 'delayed implantation' found in some placental mammals. This diapause is maintained by the hormone prolactin which is produced in response to the sucking stimulus of the young in the pouch. When sucking decreases (ie young leaving the pouch and starting to eat other foods, or if the pouch-young dies) prolactin production decreases and the embryo in suspended animation continues its development.

RED-NECKED WALLABY *Macropus rufogriseus* Pl. 1

Name. *Macros* = long, *pous* = foot (Gk.) *rufus* = red (Lat.), *griseus* = grey (adapted from French).
Status. Introduced exotic in UK.
Recognition. Characteristic bipedal gait. Pelage grizzled grey-brown above, white below. Tail silver, tipped with black. Ears, feet and muzzle tipped with black. Reports exist of fawn and silver individuals. Faeces ovoid pellets *c.* 15 × 20mm, 5–6 pellets in loose string.
Habitat. Woodland and scrub, but feeds more in open (eg heath). No den, lies-up in thick vegetation by day.
Habits. Sometimes diurnal, but especially active towards dusk and throughout night. FOOD: Peak District (UK): largely heather (50% summer; 90% winter), also bracken, bilberry, grasses and pine. Grazer and browser. SOCIAL: Generally solitary, or loose associations of 2–5 females and offspring (mated pairs associate for c. 24 hours). No evidence of territoriality. Usually silent, but faint growling in aggression. Stamp feet when alarmed.
Breeding. Births throughout summer, but peak (in captivity) March-May. Pouch young *c.* 27 days old when next young born. Gestation 30 days (oestrous cycle 32 days). Litter size 1. One litter per year in UK, but second young born if first lost from pouch early in development. YOUNG: 280 days in pouch; female and young separate c. 1 month later. No paternal care.
Lifespan. Max. 18.6 years. Road casualties frequent.
Measurements. Head-body length: 60–70cm (eg male: *c.* 65cm, female: c. 61cm). Tail-length: 60–70cm. Hind-foot length: 20–22cm. Condylobasal length: c. 115mm. Weight: 6.8–22kg (eg male: c. 13.2kg, female: c. 10.5kg). Dental formula: 3/1, 0–1/0, 2/2, 4/4 = 32–34.
General. Accidentally introduced to Britain (escapees) in about 1940.

<*Red-necked Wallaby*. Localised population in Peak District of English Midlands and, since 1975, near Loch Lomond, Scotland.

Order Insectivora – The Insectivores

Origins: The Insectivora originated in the mid-Cretaceous (*c.* 135 mya), and are hence among the oldest surviving mammalian lineages. The first mammals were rather like modern shrews, and many of the modern mammalian orders came from insectivore-like ancestors. The fossils of microchiropteran bats (see p. 43) and ancestral primates (see p. 88) have teeth very like shrews. Modern Insectivores still have much in common with their earliest ancestors, such as *Zalambdolestes*.

Main features of the Insectivora: Their geographical range is worldwide excepting Antarctica and Australia. However, only one species is found in S. America where (as in Australia) they are replaced ecologically by marsupials. All Insectivores are small (none is larger than a Rabbit; the largest is the Otter Shrew weighing about 1kg). The Pygmy White-toothed Shrew is the smallest ground-dwelling mammal. Shrews are also nocturnal, generally solitary, and feed mainly on invertebrates, especially insects. Their sight is poor, but sense of smell is keen as indicated by long, mobile snouts. The brain is small (and smooth), and dominated by large olfactory bulbs. They have continuous rows of teeth (making conventional dental formulae unhelpful but species can be distinguished by the lengths of their tooth rows). Insectivoran characteristics are often described as primitive (meaning conservative); for example shrews have a cloaca (common exit to urinary, digestive and reproductive tracts), and Insectivores walk on the soles of their feet (plantigrade) which each have 5 clawed digits. Nonetheless, there are aspects in which some are highly specialised, eg anti-predator spines in hedgehogs; poisonous secretions from shrew salivary glands; shrew ultrasound – many shrews can echolocate to orientate in the dark beneath dense undergrowth, and to locate prey. The European shrews are all rather similar, but an obvious distinction can be made between members of the genus *Sorex*, which have red tips to their teeth and their ears hidden in their fur, in contrast to members of the genus *Crocidura*, which have white teeth and protuberant ears. Moles are notable for their burrowing habits, and associated anatomy, particularly spade-like paws supported by a stout bone, the humerus, in the forearms. Moles evolved in the Eocene, 54 mya, in Europe. Their semi-aquatic relatives, the desmans, lived in Britain during the Pleistocene 2.5 mya but their continental descendants are now confined to the Pyrenees.

Insectivora in Western Europe: Two out of three insectivoran suborders are found in Europe: Soricomorpha (desmans, moles, shrews) and Erinaceomorpha (hedgehogs). Three out of six families occur in Europe: hedgehogs (Erinaceidae), shrews (Soricidae) and moles (Talpidae), and these are geographically the most widely distributed of the Insectivore families. There are nine *Sorex* species in Europe. Their dental formulae are somewhat arbitrary in that it is hard to distinguish incisors, canines, pre-molars and molars. The fossil record suggests that all modern Insectivores share a common ancestor, but contemporary families diverged 135 mya and have had ample time to develop markedly different characteristics. The Order includes 150 extinct genera, 60 recent genera, and 345 living species. Of these, in Europe today there are 22 species of Insectivore in 8 genera.

WESTERN HEDGEHOG *Erinaceus europaeus* Pl. 2

Name. *Erinaceus* from *ericius* = hedgehog/spiked barrier (Lat.). *Europaeus* from *Europa* = daughter of Agenor (Gk.).

Recognition. Sexes similar. *c.* 6000 erectile spines: pale, creamy-brown with darker band near tip; 2–3cm long on back. Underside and head: coarse, sparse hair. Gait: hesitant, with frequent stops to sniff air. Diagnostic features of skull include short postero-dorsal processes of maxillae, which do not extend behind lachrymal foramena. Moult is not seasonal. Spines long-lasting (eg 18 months), replaced irregularly. Fieldsigns include droppings (eg on lawns) deposited at random. Roughly sausage shaped, variable in size: usually up to 1cm diameter and 4cm long. Normally black, 'tarry', containing beetle elytra and other insect remains. Signs of foraging in dung (especially of cow) with characteristic excavations. Noisy movement at night through vegetation. Dew trails visible in the early morning in long grass; erratic path, grass pushed forward in direction of movement. Often seen as road casualties.

Habitat. Deciduous woodland (or scrub and hedgerows), moist pasture-land/meadows and grassland – particularly favours the borders between these habitats. Suburban gardens, sand dunes with shrubs (up to 2000m in pine zone). Rare in coniferous forest, standing cereal crops, moorland, marshes. Absent above treeline, although in the Alps it occurs in the highest tree zone (*Pinus mugo*). Northern limit determined primarily by climate (and in particular by length of winter) that is suitable for its invertebrate prey. Nest of grass and leaves for hibernation in winter called hibernaculum, and similar for birth in summer. Most animals move nest at least once per winter. In summer shelters by day in vegetation; may move site every few days. Breeding females occupy more permanent nest sites than do males. Sometimes uses abandoned burrows/breeding stops of Rabbits.

Habits. Nocturnal. FOOD: Largely invertebrates at ground level e.g. earthworms, carabid beetles, caterpillars, spiders, slugs. Occasionally vertebrates eg frogs, lizards, young rodents, and nestlings as well as bird's eggs and carrion including fish. Some plants and fungi eg fruits, mushrooms. Nightly intake *c.* 70g. Food caught, and dealt with, entirely by mouth. HOME RANGE: Normal annual range for adults (probably conservative): male: 15–40ha but up to 100 ha, female: 5–12 ha. Nightly range can vary from 0.5–25ha (male), 0.5–10ha (female). Depends on season and habitat. Population densities – approx. 1 per 0.5ha to 1 per 3 ha; in mixed farmland (UK), normally about 1 per 2.5 ha. Range overlap, but possibly temporal avoidance reduces chance of encounters. Average speed: *c.* 3m per min; occasional sprints of 30–40 m. Young disperse after weaning. SOCIAL: Solitary in wild. Social hierarchy in captivity; males especially aggressive. Very occasionally male and female share nest for short time. Suggested scent based mutual avoidance system (temporal). Non-territorial. Few aggressive encounters observed in the wild. COMMUNICATION: Snuffles and snorts when feeding; pig-like squeals when alarmed. Notable absence of territorial demarcation and associated behaviour patterns such as scent marking. Both sexes possess relatively poorly developed anal glands and prominent proctodeal glands. Chemical signals probably of considerable significance in reproduction – accessory sex glands well developed – male said to discharge secretion in urine behind female (or on her during mounting). During courtship behaviour, female repeatedly passes over scent marks – possible releaser pheromone acts to promote receptivity to mating in female. Vision poor, but senses of smell and hearing are acute.

Breeding. Most matings in spring after hibernation (males are fecund from April–August). Pregnancies found from May–October, but peaks in May/July and September. Sexual maturity at 12 months. Gestation 31–35 days. Litter size averages 4–6 (2–10) eg Britain 3.7, Sweden 5.2. (there is some indication that Hedgehogs show geographic variation in litter size, with litters increasing in size at northern latitudes, or higher altitudes). Litters per year 1 (sometimes 2). Mammae 10. YOUNG: Blind at birth, coat of white spines appears soon after birth. 2nd coat of dark spines with white tips visible by 36 hrs. Can roll up at 11 days. Begins to leave nest at 22 days. Weaning age 4–6 weeks. Only female tends young, but radio-tracking data suggest that she leaves them alone all night.

Lifespan. Max. 7–10 yrs, averaging 3. Pre-dispersal mortality estimated at 20%, 1st yr mortality *c.* 60–70%, with subsequent good survival until 4th-5th yrs. Principal cause of mortality probably starvation during hibernation, although remains have been found in the stomachs of Badgers, Red Foxes, Pine Martens and most other Carnivores and in pellets of raptors such as Golden Eagles and Eagle Owls: however, some of these remains may have resulted from the predators feeding on carrion (eg from road kills). Hedgehogs are generally absent where badgers are numerous.

Measurements. Head-body length: 225–275mm (160mm at weaning).

< *Western Hedgehog.* Widespread throughout W Europe and N Russia (replaced to SE by Eastern Hedgehog). Overlaps with Algerian Hedgehog in S Spain and S France.

Tail-length: 15–30mm. Hind-foot length: 40–45mm. Shoulder-height: 120–150mm. Condylo-basal length: adult 55–60mm (weaned: 40mm) – depends on age. Weight: 400–1200g (birth 11–25g; weaned 120g; max. 2000g (UK usually smaller)). 30% reduction due to fat loss during hibernation. Dental formula: 3/2, 1/1, 3/2, 3/3, = 36 (deciduous teeth: 3/2, 1/1, 3/2, 0/0 = 24.).

General. Common and widespread, with occasional unexplained areas of absence. Rolls-up when in danger (see vignette this page). Climbs and swims well (see vignette previous page). Little evidence for belief that hedgehogs kill and eat snakes. Puzzling self-anointing behaviour, when grooms itself with foamy saliva which may be repellant to predators (but may have different explanation, e.g. anti-parasitic). Hibernation: South: UK adults October/November–March/April; UK young December–April. Hibernation in tidy nest of leaves and grass under cover or in rabbit burrows. Adult males go into hibernation first. Metabolism slows (breaths per min: summer active 40–50, hibernating 9). Temperature drops (summer 34°C, winter *c.* 4–6°C, fluctuating with environment). Heart beat (summer 190 per min, winter 20 per min). Young from later litters hibernate as sub-adults, unlikely to survive if weigh less than 400g.

EASTERN HEDGEHOG *Erinaceus concolor* — Pl. 2

Name. *Erinaceus* from *ericius* = hedgehog/spiked barrier (Lat.). Concolor = having the same colour (Lat.).

Recognition. Distinguished from Western Hedgehog only by throat and breast being lighter colour than belly, and certain skull features (postero-dorsal processes of maxillae extending behind lachrymal foramena).

Habitat. In Europe, as for Western Hedgehog.

Habits. Nocturnal. As for Western Hedgehog.

Lifespan. Causes of death reputedly same as for Western Hedgehog, except those associated with hibernation, because the Eastern form does not hibernate near the southern limit of its geographical range. Pine Marten reputed to prey on this species.

Measurements. As for Western Hedgehog. Dental formula: 3/2, 1/1, 3/2, 3/3 = 36

General. All details as for Western Hedgehog. Zone of sympatry with

< *Eastern Hegehog.* Widespread in E Europe.

> *Algerian Hedgehog.* Scattered locations in Iberian peninsula and Balearic Islands.

E. europaeus at eastern limit of *E. europaeus* and western limit of *E. concolor* – a band about 200km wide. Variability of animals in this zone suggests some hybridisation takes place.

ALGERIAN HEDGEHOG *Atelerix* (*Erinaceus*) *algirus* Pl. 2

Name. *Aethechinus* from *aithon* = fiery; *-echinos* = hedgehog (Gk.).

Recognition. Spine-free 'parting' on crown of head wide enough to insert pencil. Longer legs than European hedgehogs, has larger external ears, and generally paler spines with off-white ventral pelage, and white mask on face (except in S. Spain where Western Hedgehog equally pale). Also, said to have more noticeable neck than other hedgehogs. Moult as for *Erinaceus europaeus.*

Habitat. Mediterranean scrub zone, plateau grasslands and around cultivation. Sometimes nests in burrows.

Habits. Nocturnal. FOOD: Snails, centipedes, insects, small vertebrates (lizards, frogs etc), carrion, fungus. Good senses of smell and hearing.

Measurements. Head-body length: 200–250mm. Tail-length: 25–40mm. Hind-foot length: 35mm. Condylo-basal length: 50–58mm. Weight: 70–100 gm. Dental formula: 3/2, 1/1, 3/2, 3/3 = 36. Third upper premolar has 3 roots.

General. Does not hibernate. Basically a N.W. African species, but records of isolated individuals from mainland France and Spain. Probably these stray sightings, and the populations in the Balearic and Canary Islands, have been introduced by man. Whereas the latter island populations are well established, it is uncertain if this is true for any population on mainland Europe.

LEAST SHREW *Sorex minutissimus* Pl. 4

Name. *Sorex* = shrew (Lat.) from *hurax* = shrew (Gk.); *minutissimus* = most tiny (Lat.).

Recognition. Extremely small. Tail short and narrow. Clear cut division between dark brown back and lighter flanks and underside. Short tail in comparison to Pygmy Shrew.

Habitat. Wet coniferous forest with moss cover, damp swamp edges, forest edge with lush grass (also open sandy and dry pine heaths). Resting period: 10–15 mins. Has been observed to have up to 70 rests in 24 hrs. Longer activity periods than in larger species like *S. araneus* and *S. caecutiens*. FOOD: small insects, insect larvae, spiders and snails. Eats up to 2–5 times its own weight within 24 hrs. COMMUNICATION: responds only to very high frequency sounds and probably produces ultrasound. SOCIAL: unknown.

<*Least Shrew.* Distributed in Scandinavia and eastwards.

>*Apennine Shrew.* Recorded from scattered locations in Italy.

Breeding. Season: May-June; litter size: 3–6, 2 litters per year.
Measurements. Head-body length: 33–53mm. Tail- length: 20–31mm. Hind-foot length: 7–9mm. Condylo-basal length: 11.7–13.6mm. Weight: 1.2–4g. Dental formula: 3/1, 1/1, 3/1, 3/3 = 32 (tooth row less than 6mm).
General. Extremely uncommon and never abundant, but found in northern Europe.

APPENINE SHREW *Sorex samniticus* Pl. 4

Name. Italian Shrew. *Samniticus* from Samnium, a region of the S. Appenines.
Recognition. Visually indistinguishable from Common Shrew, though it tends to be slightly smaller and with a flatter skull. Upper incisor cusps divided by rounded notch (cf V-shaped notch of Common Shrew). Ecologically almost identical to Common Shrew, but biochemical karyotypic analyses and lower jaw and skull measurements indicate it to be taxonomically distinct. Reputed not to burrow.
Habitat. Above 1160m in areas of overlap with Common Shrew, and at lower altitudes (300–1000m) where it seems to replace the Common Shrew. Tends to prefer humid habitats.
Breeding. Season starts in May.
Measurements. Head-body length: 68–78mm. Tail-length: 33–45mm. Hind-foot length: 11.0–12.5mm. Condylo-basal length: 17.7–19.2mm. Weight: 6.5–10g. Dental formula: 3/1, 1/1, 3/1, 3/3 = 32.
General. Similar to Common Shrew in all respects, but confined to continental Italy.

COMMON SHREW *Sorex araneus*  Pl. 4

Name. *Araneus* = spider (Lat.).

Recognition. Pelage dark brown to almost black on back, lighter brown flanks and greyish-white underside. Occasional white spotting on ears/tail tip. Reddish-brown tips to teeth. Tail of young covered with short, bristly hair with distinct 'pencil' of hairs at tip. Older animals (8 months) lose hair on tail, which can also appear heavily scarred;

males lose hair around testes when sexually mature, producing distinct 'bald' patches. Juvenile pelage lighter above and less contrast between the three zones. Gait – bustling trot. Lateral scent glands on flanks, particularly obvious in adult male. No protruding tactile hairs on tail as in *Crocidura* shrews. Moult: one in autumn to thicker, longer, denser winter pelage; one (some authorities say two, in quick succession) in spring (April/May) into thinner summer coat. Fieldsigns include very small runways through leaf litter and tunnels in soil. Faeces blackish and granular, containing remains of chitinous exoskeleton of insects.

Habitat. Widespread. Thick grass, hedgerows, bushy scrub, bracken, up to 2480m in the Alps. Mainly in deciduous woodland and wetlands in Central Europe. Particularly numerous in grassy 'edge' habitat, eg especially road verges in UK. Nest under logs or grass tussocks, using burrows of other species. Ordinary nests small, but breeding nests more substantial, balls of interwoven dried grass and leaves.

Habits. Frequent (eg 10 per day) periods of rest alternate with activity but especially active at night, dawn and dusk. More active below ground than *Sorex minutus*. FOOD: Largely invertebrates, especially earthworms, slugs, snails, insect larvae, beetles, spiders, woodlice. Occasionally vertebrates. Daily intake: 80–90% body wt. (but depends dramatically on ambient temperature (lactating female intake 150–200% average body wt.). Can locate prey hidden 12cm deep in soil, probing with snout and digging. HOME RANGE: 100–1000m^2, mostly 370–630m^2, occupied by single individual. During autumn/winter, partially exclusive home range, maintained by females in breeding season. In a high density grassland site, most animals may breed within 100m of birth place. Major dispersal phase during summer of birth. In the breeding season some adults appear to be 'nomadic'. SOCIAL: Solitary and probably territorial except during breeding season: highly aggressive at all times except when with young. COMMUNICATION: Vocal: twitters during normal activity, shrill screamduring aggression. Scent: flank glands. Sight less developed than smell, touch and hearing. Produces ultra-sound and uses echolocation to find burrows.

Breeding. Season: April-August. Mating: March onwards (first oestrus in Britain mid-late April) with successive litters until autumn. Early litters followed by post-partum oestrus, latterly intervals between successive oestrus periods become longer. Adults uncommon by July. Sexual maturity: usually in Spring following their birth, young of early litters may very rarely breed in July-September. Gestation: 20 days, up to 24–25. Litter size: 1–10 (average 6–7). Litters per year: probably 2, rarely more, often one. Mammae 6. Testes and accessory sexual organs constitute 10% of adult male body weight. Females receptive for only 24 hours during oestrus. Male bites female's neck while mating, so pregnant females generally identifiable by nape scars and by weighing more than 10g. YOUNG: Born hairless and blind, pelage developing at 9 days (see vignette), eyes open at 16 days, first invertebrate prey at *c.* 21 days. After leaving nest lose weight over winter until *c.* March of

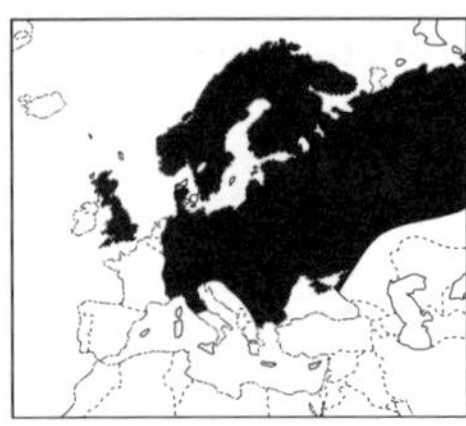

< *Common Shrew.* Widespread in C&N, largely absent W Europe except widespread in mainland Britain.

> *Pygmy Shrew.* Widespread in Europe, but missing from S Iberian peninsula.

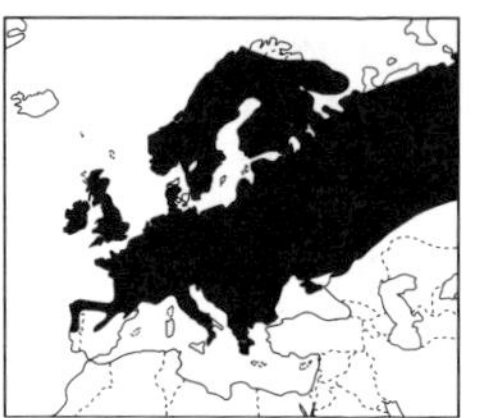

subsequent year, when rapid surge in growth of body and reproductive system. Weaning age: 22–25 days. Parental care: female only.

Lifespan. Max. 23 months in the wild. Mortality: 50% die before 2 months; 20–30% survive to breed.

Measurements. Head-body length: 54–87mm. Tail-length: 32–56mm. Hind-foot length: 10–15mm. Condylo-basal length: 18–20mm. Weight: 6–12g (birth, 0.47g; 14 days, 6.7g). Dental formula: 3/1, 1/1, 3/1, 3/3 = 32.

General. Swims well. Swift, bustling movement, exploring busily with snout and vibrissae. Tight fit of body in burrows keeps fur clean. Individual recorded covering one kilometre in one hour. Rarely leaves thick cover. Activity less in winter when spends *c.* 80% of time underground. Jersey: *Sorex coronatus* has blunter snout, and rostrum of skull is shorter, broader and deeper than mainland British shrews. Population densities (peak): 42–100 per ha in grassland, 7–21 per ha in dunes; peaks summer, lower numbers in winter.

PYGMY SHREW *Sorex minutus* Pl. 4

Name. *Minutus* = tiny (Lat.).

Recognition. Bi-coloured, with gradation between flanks and underside (cf dividing line of tri-coloured Common Shrew). 3rd unicuspid tooth as large as (or larger than) 2nd (opposite for Common Shrew). No seasonal variation in colour, lighter in colour than adult Common Shrew. Ears almost (or completely) hidden in fur; tail hairs denser and longer than those of Common Shrew and so tail appears thicker. Tail 65–70% of head-body length. Tips of teeth red. Moult as for *Sorex araneus.*

Habitat. Areas with good ground cover e.g. heaths, grasslands, sand dunes, woodland edge, but uncommon in woodland. Higher altitudes in South. Nest is small ball of grass in dense cover.

Habits. Alternation of activity and rest in short periods. Equally active by day and night, predominantly on ground surface. Relatively more diurnal than Common Shrew. FOOD: Invertebrates of leaf litter (mostly beetles, spiders, woodlice, not earthworms). Intake (in captivity): 100% body weight per day. Unlike Common Shrews, does not dig for prey. HOME RANGE: British grassland 1900–1700m^2; Dutch dunes 530–

1860m^2. Juveniles disperse quickly from nest area at weaning. Population density approximately 12 per ha maximum in English grassland, 4–11 per ha on dune grassland-scrub in Netherlands. Population peaks in summer, with troughs in winter. SOCIAL: Territorial. Solitary and aggressive except in breeding season. Strict territoriality abandoned by sexually mature males as they travel in search of females. COMMUNICATION: High-pitched abrupt 'chit' when threatened or alarmed, almost inaudible squeaking whilst foraging.

Breeding. Births in April–August (peak in June in UK). In Netherlands mean annual production of young = 15 per ha (cf 58 per ha for Common Shrew). Sexual maturity usually in year following birth but, eg in Poland, 22% of females breed in year of their birth. Gestation *c.* 22 days. Litter size 4–7. Litters per year probably 2. Mammae 6. YOUNG: Tail hairs protrude to give middle of tail bushy appearance. Weaning age 22 days. Parental care by female only.

Lifespan. Max. 16 months, usually 13 in wild. Reputedly higher mortality throughout life than *Sorex araneus*, doubtless because small size makes more prone to adversity caused by environmental unpredictability. Estimated 50% mortality as juveniles, *c.* 20% of those born will survive to breed.

Measurements. Head-body length: 40–64mm. Tail- length: 30–46mm. Hind-foot length: 9–12mm. Condylo-basal length: 14.8–16.6mm (upper tooth row: 6.2–6.6mm). Weight: 2.5–7.5g (Birth: 0.25g; 14 days *c.* 2.5g). Dental formula: 3/1, 1/1, 3/1, 3/3 = 32.

General. Less subterranean, and usually lower population densities than Common Shrew. Does not burrow, but uses runways of other species. Scuttles faster, and is a better swimmer and climber than Common Shrew.

DUSKY SHREW *Sorex isodon* (*sinalis*) **Pl. 5**

Name. *Isodon* = equal teeth, *Sinalis* from *sinis* = plunderer(Gk.).

Recognition. Bi-coloured but underside almost as dark as upper body. Broad front feet. Unicuspid teeth decrease evenly in size from front to back.

Habitat. Various, but especially in wet coniferous forest, and productive habitats where food availability is high. FOOD: Insects, earthworms, spiders, myriapods, small vertebrates (frogs).

Breeding. Season: June–August, 2–3 litters per year. Litter size: 1–10, mean 6–7.

Measurements. Head-body length: 55–82mm. Tail-length: 41–55mm. Hind-foot length: 13–15mm. Condylo-basal length: 16–20.3mm. Weight: 6.5–14.5g. Dental formula: 3/1, 1/1, 3/1, 3/3 = 32.

General. Previously considered conspecific with Chinese *S. sinalis.*

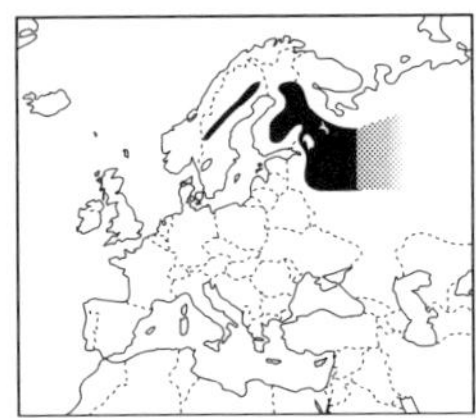

<*Dusky Shrew.* Narrow band of Sweden, through Finland to Russia.

>*Masked Shrew.* N Sweden, Finland and eastwards.

MASKED SHREW *Sorex caecutiens* Pl. 5

Name. Laxmann's Shrew. *Caecutius* = seeing badly (Lat.), *caecus* = blind).

Recognition. White feet with silvery, shining hair. Bi-coloured: sharp demarcation of upper body and underside. Tail tipped with bushy tuft. Unicuspid teeth rather uniform size (cf. Pygmy Shrew). Size intermediate between Common and Pygmy Shrews.

Habitat. Coniferous forests, tundra and moorland. Presumably more active at night than during day time. FOOD: mainly insects, especially Coleoptera, a few spiders, myriapods and earthworms, larch seeds; may feed on other small mammals.

Breeding. Season: June-August, up to 4 litters per yr. Litter size: 2–11, mean 7–8. In years of low spring density, young females of the first litters take part in reproduction greatly contributing to rapid population recovery.

Measurements. Head-body length: 44–70mm. Tail-length: 31–45mm. Hind-foot length: 10.5–12.0mm. Condylo-basal length: 16.1–17.3mm. Weight: 3–8g. Dental formula: 3/1, 1/1, 3/1, 3/3 = 32.

General. Assumed to be similar to Common and Pygmy Shrews. Patchy distribution across Europe – possibly relict of previously widespread species.

ALPINE SHREW *Sorex alpinus* Pl. 5

Name. *Alpinus* = Alpine (Lat.).

Recognition. Body colour uniformly dark, feet and underside of tail pale. Lower canine and premolar clearly bicuspid; 4th and 5th unicuspid teeth same size. Tail as long as head and body.

Habitat. Alpine meadows and moors from 200–3335m (Alps: 600–2500m, Germany above 180 m). Often in rocky habitats, frequenting the stony banks of mountain streams.

Habits. Climbs well, better than *S. araneus.* Uses its tail for balance and

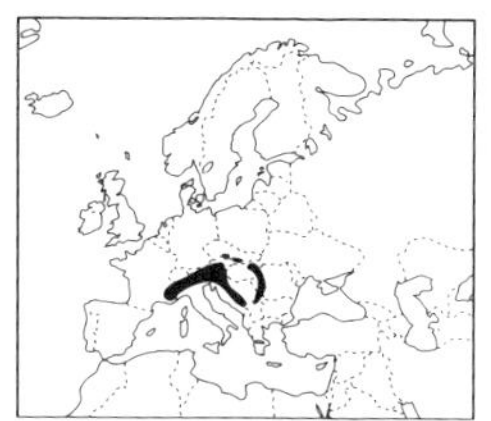

<*Alpine Shrew.* Mountain habitats in S Europe, principally the Alps, Pyrenees and areas of Yugoslavia, Romania, Hungary and Czeckoslavakia.
>*Millet's Shrew.* France, with reports from neighbouring regions.

support when climbing. FOOD: snails, earthworms, spiders, isopods, chilopods, insects and insect larvae.
Breeding. Breeding season May-October. Litters per year 2–3. Litter size 3–9 (average 5–6).
Measurements. Head-body length: 60–75mm. Tail-length: 60–75mm. Hind-foot length: 13–16mm. Condylo-basal length: 17.9–21mm. Weight: 5.5–11.5g. Dental formula: 3/1, 1/1, 3/1, 1/3 = 28.

MILLET'S SHREW *Sorex coronatus* Pl. 5

Name. French Shrew. *coronatus* = crowned (Lat.).

Recognition. Visually indistinguishable from Common Shrew although tends to be slightly smaller, except on Jersey, where it is larger than mainland British *Sorex araneus.* Identification depends on detailed skull and lower jaw measurements, chromosome counts and detection of biochemical differences. Only reliable fieldsign is presence of bones in pellets of owls.
Habitat. As for *Sorex araneus* but tends to be found in drier and warmer places. On Jersey found in coastal habitats of sand dunes, heath and coastal scrub, deciduous woodland, hedgebanks, gardens.
Breeding. Breeding season May-September. Gestation 20 days. Litter size 3–7 (average 5–6), cf Common Shrew where average 4–6 litters per year. Sexual maturity in spring of year following birth.
Measurements. Head-body length: 68–80mm. Tail-length: 37–47mm. Hind-foot length: 12–13mm. Condylo-basal length: 18.5–20mm. Weight: 6.5–11.5g. Dental formula: 3/1, 1/1, 3/1, 3/3 = 32.
General. Distributed throughout France (except N.E.), in N. Spain, Austria, Boun region of Germany, lower altitudes in Switzerland. Probably introduced on Jersey. Resembles *Sorex araneus* in all habits, and the two species are in strict competition and do not occur together.

SPANISH SHREW *Sorex granarius* Pl. 5

Name. *Granarium* = granary (Lat.).
Recognition. Visually indistinguishable from Common Shrew, but tends

to be slightly smaller and has shorter skull and broader muzzle. Identification depends on detailed skull measurements and chromosome counts.

Habitat. Close to streams and wet places, above 1000m in mountains of central Spain. Towards the Atlantic may be found at lower altitudes and in less characteristic habitats.

Measurements. Head-body length: 62.5–72mm. Tail-length: 38–45mm. Hind-foot length: 11–12.5mm. Condylo-basal length: 17.3–18.3mm. Weight: 4.5–8g. Dental formula: 3/1, 1/1, 3/1, 3/3 = 32.

WATER SHREW *Neomys fodiens* Pl. 6

Name. *Neos* = new (Gk.); *mys* = mouse (Gk.); *fodiens* = digging (Lat.).

Recognition. Fringe of bristly silvery-white hairs on hind feet, double row of hairs act as keel on tail (for swimming). Very dark above, but underside varies from white to grey. Sometimes occurs as very dark morph. Ear openings can be closed. 4 unicuspid teeth (cf 3 in *Sorex* and 5 in *Crocidura*). Moults in spring and autumn. Fieldsigns include caches of partially eaten snails, amphibia or fish. Uses runways in dense vegetation. Remains of prey, eg snail shells, caddis fly larvae cases, left at habitual feeding sites on stream banks or rocks beside streams. Faeces in middens established in 'runways' and near burrow entrances.

Habitat. Generally close to water (eg mostly banks of swiftly flowing streams and weirs). Especially abundant in watercress beds, occasionally near ditches, ponds and seashore. Up to 2500m in Alps. Recorded 3km from water. May occur as temporary visitor in woodlands and grasslands. Nest is ball of vegetation in extensive burrow system, excavated by shrew, using front feet and nose. Entrance may be above or below water level. May re-use mole tunnels.

Habits. Mostly nocturnal, especially just before dawn. Least active in late morning. FOOD: Mostly aquatic crustaceans and insect larvae, and terrestrial beetles, molluscs, worms, occasionally small fish, amphibia and

< *Spanish Shrew.* Recorded from scattered locations in Iberian peninsula.

> *Water Shrew.* Widespread in continental Europe and mainland Britain, but absent from Iberian peninsula and Ireland.

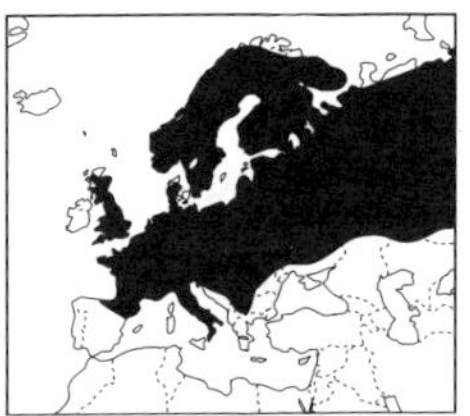

mammals. Prey caught underwater and on land. Captive shrew consumed 50% body weight daily. Accumulates remains of partially consumed meals in specific places. HOME RANGE: *c.* 160m long. Solitary individuals occupy peripherally overlapping ranges (defends nest in captivity). 20–30m^2 on land. 60–80m^2 of water surface along brooks in Germany. Most activities restricted to a small stretch of bank and nearby water. Dispersal distances of 28–162m. Population density very variable but lower than *Sorex araneus*, eg 3.2 per ha in watercress beds in England, where often seem to occur in clusters. Peak numbers in summer. SOCIAL: Mostly solitary, but reputedly more socially tolerant than other red-toothed shrews; often live in small groups in quite close proximity. COMMUNICATION: Vocalizations include rolling churrs and squeaks tends to be noisy. Highly developed flank glands fringed with white hairs in adult males only. Olfaction and hearing well developed. Sensitive, mobile whiskers for detecting prey while swimming.

Breeding. Breeding season April-September, with peak of births in May-June. Sexual maturity 2nd year (some females in first year). Gestation: 14–21 days. Litter size: 3–15, mean 6. Litters per year: at least 2 possible (post-partum oestrus). Mammae 10. YOUNG: May remain with mother until 40 days old. Weaning age 27–28 days. Parental care by female only.

Lifespan. Max. 14–19 months. High juvenile mortality, and also high adult mortality during breeding season.

Measurements. Head-body length: 63–96mm. Tail-length: 47–82mm. Hind-foot length: 16–20mm. Condylo-basal length: 19–22mm. Weight: 8–23g (birth: 1g). Dental formula: 3/1, 1/1, 2/1, 3/3 = 30 (tips of teeth red as in *Sorex* species).

General. Venom in saliva, produced by submaxillary gland (effective in small mammals). Fast, adept swimmer, good diver, floats like cork. Buoyant due to air trapped in coat, so starts dive with leap to break water surface. Propulsive power of feet and tail increased by fringing erectile hairs. Melanism rather common (albinism rare). Ears may or may not have white tufts. Fur very water-repellent.

MILLER'S WATER SHREW *Neomys anomalus* Pl. 6

Name. *Anomalos* = abnormal (Gk.).

Recognition. Distinguished from Water Shrew by greatly reduced or absent keel on tail, and reduced fringe on feet. Underside pale, often white.

Habitat. Lush grassland close to water and swamps, but less adapted to, or dependent upon, water than Water Shrew. In south and west Europe often limited to mountainous terrain; further east occurs at lower altitudes. Otherwise, apparently much as *Neomys fodiens*.

Habits. FOOD: Much as *Neomys fodiens*, but more terrestrial.

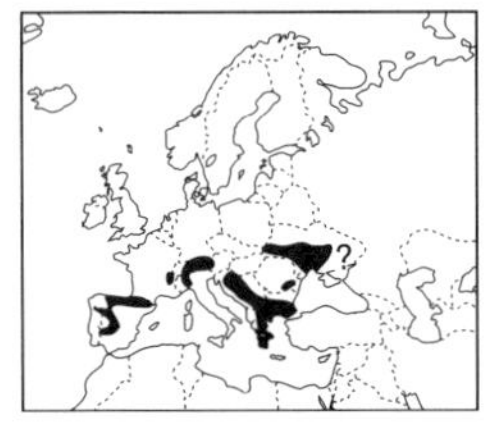

<*Miller's Water Shrew* Scattered through S Europe, notably Spain, the Alps, Yugoslavia and Greece.
>*Bi-coloured White-toothed Shrew*. Throughout C Europe, westwards through N France, absent from N Europe and Iberian peninsula.

Breeding. Mammae 10.
Measurements. Head-body length: 64–88mm. Tail-length: 42–67mm. Hind-foot length: 14–20mm. Condylo-basal length: 19–21mm. Weight: 7.5–16.5g. Dental formula: 3/1, 1/1, 2/1, 3/3 = 30.
General. Saliva less venomous, but all other details similar to Water Shrew.

BI-COLOURED WHITE-TOOTHED SHREW *Crocidura leucodon* Pl. 3

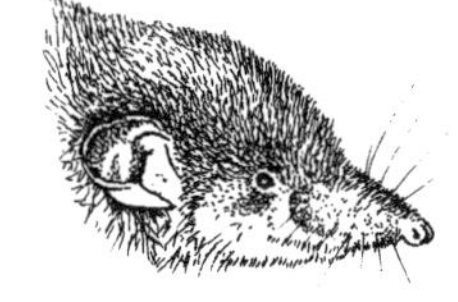

Name. *Krokos* = yellow (Gk.); *oura* = tail (Gk.); *leucos* = white (Gk.); *odous, odontos* = tooth (Gk.).
Recognition. Clear delineation of greyish-brown upper body and yellowish-white underside. Tail clearly bicoloured. Isolated, pale long hairs on tail and rear of body. Rostrum of skull shorter and deeper than that of Greater White-toothed Shrew, and unicuspid teeth more crowded. Large protuberant ears. 2 moults: one in autumn and one in spring. Fieldsigns as for all Crocidurinae, footprints hard to distinguish from those of small rodents. Size of fresh droppings 6 × 3mm, usually dark, ovoid, rather soft and often irregular shape – if found (eg in traps), easy to distinguish from those of small rodents (as for all Crocidurinae).
Habitat. Low, dry habitats, shrubby undergrowth, fringes of woods and gardens, compost heaps. Nest of fresh or dry grass in sheltered undergrowth, avoiding damp.
Habits. Largely nocturnal, but also active by day. FOOD: Invertebrates (insects, insect larvae, worms etc). COMMUNICATION: shrill cry (single or repeated); soft twittering. Scent from lateral flank glands, especially from the male during the breeding season.
Breeding. Breeding season: March-October. Gestation 31 days. Litter size 2–6 (average 4.2). Mammae 6. YOUNG: Young born naked. Weaning age *c.* 26 days. Noted for caravanning behaviour.
Lifespan. Max. 2–5 years in captivity. Principal predators are probably owls.
Measurements. Head-body length: 60–90mm. Tail- length: 27–43mm. Hind-foot length: 11–13mm. Condylo-basal length: 17.5–21mm. Weight: 6–15g. Dental formula: 3/1, 1/1, 2/1, 3/3 = 30 (tooth row 5–6mm).
General. Torpid in periods of scarce resources and colder days. Habits thought to be similar to Greater White-toothed Shrew.

LESSER WHITE-TOOTHED SHREW *Crocidura suaveolens* Pl. 3

Name. *Suavis* = sweet (Lat.); *olens* = smelling (Lat.).
Recognition. White teeth (cf *Sorex* spp.), 2nd unicuspid tooth markedly smaller than 3rd in crown view. Teeth appear more variable in size than those of other white-toothed shrews. Similar to Greater White-toothed Shrew but smaller and more yellowish underside. 2 moults: one in autumn and one in spring. Fieldsigns include extensive runs in litter zone, and remains of crustacean prey under rocks and in seaweed.
Habitat. Dry bracken, tall vegetation and other habitats offering good cover and litter zone (eg all habitats on Scilly Islands but especially heathland). Nests of soft vegetation in sheltered places on heathland.
Habits. Several phases of activity throughout day and night, but predominantly nocturnal. FOOD: Insects and other invertebrates (eg sand hoppers). SOCIAL: Generally solitary, but not as aggressive as *Sorex* shrews. Home ranges overlap. COMMUNICATION: Sharp squeaks when disturbed. Twitters almost continuously. Lateral flank glands especially developed in the male during the breeding season. So-called bellymarking involves spreading hind limbs and dragging underside and anal region on ground while pulling the body forward with the forelimbs. Both lateral flank glands and anal area touch the ground during marking.
Breeding. Breeding season is spring to autumn. May breed in first year. Sexual maturity 45–50 days old. Gestation 28 days. Litter size 1–6. Up to 4 litters per year. Mammae 6. YOUNG: Born naked. Eyes open on the 10th day. Weaning age *c.* 22 days. Young 'caravan' from *c.* 8 days old.
Lifespan. Max. less than 2 years in wild; in captivity: 792 days. Principal predators are probably owls.
Measurements. Head-body length: 50–82mm. Tail-length: 24–44mm. Hind-foot length: 9.8–13.5mm. Condylo-basal length: 15.6–18.3mm. Weight: 3.5–6g (at birth, 0.63g; adult females during the breeding season, 7–8g; males, 7–10g). In Corsica individuals are notably larger, weighing 10–13g. Dental formula: 3/1, 1/1, 2/1, 3/3 = 30 (upper tooth row: 7.4–8mm).
General. Numbers declining in Europe. Undergoes torpor. Tail partly prehensile.

<*Lesser White-toothed Shrew.* Widespread in S Europe, excluding most of Spain.
>*Greater White-toothed Shrew.* Throughout W Europe, including Sardinia; absent from Corsica where replaced by Lesser White-toothed Shrew.

GREATER WHITE-TOOTHED SHREW *Crocidura russula* Pl. 3

Name. House Shrew. *Russula* = small, red-headed (Lat.; *russus* = red headed).

Recognition. Teeth white (cf. *Sorex* spp.). Skull rather narrow and elongated in the unicuspid tooth region. Three unicuspid teeth (3rd relatively large). Gradual, blurred division between reddish upper and paler underside. Isolated long sparse hairs on tail. 2 moults: one in autumn and one in spring.

Habitat. Dry ground, grassland, fringes of woods, gardens and hedges. Commonly enters houses, farm outbuildings etc. Generally below 1000m (occasionally 1600m in Alps). Nest of dry grass (or man-made materials in houses) lined with leaves, situated under cover. Nest sometimes open, saucer-shaped. Den sites sometimes found when turning over compost heaps.

Habits. Active in bouts for *c.* a third of each 24 hrs (polyphasic), but predominantly during the night. More diurnal than Lesser White-toothed Shrew. FOOD: Insects and other invertebrates, very occasionally takes small vertebrates (lizards and young rodents), probably scavenges dead vertebrates. HOME RANGE: Winter home range: 35–100m^2, breeding home range: 80–320m^2. Some overlap between neighbouring home ranges. Juveniles of the first litter disperse rapidly. SOCIAL: Pairs form in breeding season and are aggressive to other individuals. 'Caravanning' reported for female and young. During winter, communal nests form, where up to six individuals sleep together. COMMUNICATION: Shrill cries (11.4 KHz). Males have conspicuous subcaudal gland on ventral surface of tail root, different from those of other soricids (growth of this gland coincides with sexual maturity). Both sexes possess lateral flank glands. These enlarge in the male during the breeding season and have a musky odour. Defecate on prominent sites.

Breeding. Breeding season February-November (Channel Islands: February-October). Sexual maturity *c.* 3 months. 15% of juveniles mature in the year of their birth. Gestation 28–33 days. Litter size 2–10 (averages 3.5–4.0). Litters per year: 4–5. In captivity up to 12 litters per year. Mammae 6. YOUNG: Born naked. Eyes open 7–14th day. Weaning age 20–22 days. Males may show parental care. On and after the 7th day, young may be moved by the female by caravanning.

Lifespan. Max. at least 18 months in wild (48 months in captivity). Principal predators probably owls.

Measurements. Head-body length: 51–86mm. Tail-length: 24–46mm. Hind-foot length: 10.5–13.5mm. Condylo-basal length: 18–20.4mm. Weight: 9–14g (juveniles at weaning 5–7g). Dental formula: 3/1, 1/1, 2/1, 3/3 = 30 (upper tooth row: 7.7–8.5mm).

General. *Crocidura russula* present on Alderney, Guernsey and Herm Islands, *Crocidura suaveolens* on Sark and Jersey. *Sorex coronatus* only on Jersey. *Crocidura russula* becomes torpid when food is limited or less accessible (snow), especially in winter. *Crocidura sicul* on Sicily. *Crocidura zimmermanni* on Crete.

PYGMY WHITE-TOOTHED SHREW *Suncus etruscus* **Pl. 3**

Name. Savi's pygmy shrew, Etruscan shrew
Recognition. Minute – one of smallest terrestrial mammals in the world. Four unicuspid teeth. Teeth white (cf *Sorex* shrews). Large, protuberant ears. Long hairs scattered through pelage, especially on tail and snout (gives slightly frosted appearance). Colour: back greyish brown; underside, from the neck to abdomen, pale grey. 2 moults: the first in autumn when individuals develop darker, thicker fur; second in spring.
Habitat. Open terrain, grassland, scrub, gardens, deciduous woodland. Under logs and boulders. Recorded up to 630m above sea level in France, 1000m in Italy. Found frequently in dry stone walls and ruins. Resting place between roots or similar crevices.
Habits. Alternates periods of activity and rest, but especially active at night. FOOD: Invertebrates up to size of grasshoppers. SOCIAL: Pairs form during the breeding season and tolerate juveniles for a long time in the nest. During winter *Suncus etruscus* seems to be aggressive towards any conspecific. COMMUNICATION: Vocalisation includes shrill cry (torpid shrews also make a shrieking sound). Lateral flank glands give off strong musky odour, especially developed in males during the breeding season.
Breeding. Births between March/April and September/October. Apparently all sub-adults reach sexual maturity after passing the winter. No reproduction has been observed in the year of birth, even for juveniles born at the beginning of spring. Gestation 27–28 days. Litter size 2–5. Up to 6 litters per year (in captivity). Mammae 6. YOUNG: Naked young 0.2g at birth. Eyes open on the 14–16th day. Weaning age 20 days. On and after the 9–10th day, the young are moved; if disturbed the female leads them by caravanning.
Lifespan. Max. 26 months in captivity. In the field, probably about 18 months. Owls probably important cause of death.
Measurements. Head-body length: 35–52mm. Tail-length: 24–30mm. Hind-foot length: 7–8mm. Condylo-basal length: 11.8–13mm. Weight: 1.5–2.5g. Dental formula: 3/1, 1/1, 2/1, 3/3 = 30 (tooth row 5–6mm)

<*Pygmy White-toothed Shrew.* Found in S Europe, including Balearic Islands, Corsica, Sardinia, Sicily and Crete.

>*Pyrenean Desman.* Two isolated populations in the Pyrenees.

General. This species becomes torpid, especially in periods of scarce resources and colder days.

PYRENEAN DESMAN *Galemys pyrenaicus* Pl. 6

Name. *Gale* = weasel (Gk.); *mys* = mouse (Gk.).
Status. Vulnerable.
Recognition. Thick-set, rotund body, neck very short. Long (20mm) proboscis-like snout in continual movement. Large hind-feet edged with stiff bristles, forefeet fringed with bristles. Forefeet partly webbed, hindfeet webbed. Long tail laterally flattened at end as rudder. Small eyes. Soft, dense, water-proof fur gives off metallic glints. Feet (especially forefeet) have very sharp claws for gripping slippery surfaces. Fur dark brown/black above, grey/tan below. Incessant grooming. Moult not known. Very difficult to observe. Droppings small and twisted; black and glistening when moist. They are deposited systematically on any object protruding above water level. Individuals have a strong odour.
Habitat. Cold, swift-flowing mountain streams, and canals between 65–1200m. Rarely found in reservoirs – very sensitive to human disturbance, vacating areas disturbed by hydroelectric plants or which are polluted. Nest site at riverside. Lives in natural holes or burrows of water voles, or enlarges existing crevices. Nest of dry grass and leaves. Nest site slopes upwards and away from stream.
Habits. Two activity periods daily, nocturnal but active for about 2 hours during day. FOOD: Aquatic and waterside invertebrates eg nymphs of stone fly larvae (and dragon fly nymphs when found, but water generally too cold), crustacea, worms, plus occasionally some small vertebrates eg fish. Brings prey to land to eat. Daily consumption 30–50% of its body weight. HOME RANGE: 200–400m of stream length, but depends on the season and on food availability in the area. Dispersal predominantly by juveniles (males dispersing further than females). SOCIAL: Intolerant of conspecifics. Adult males and females form stable pairs and together defend territory. However, within each pair, the male and female are usually highly aggressive. COMMUNICATION: High-pitched shriek used in defence. Large musk gland at base of tail. Proctodaeal glands similar to those of other insectivores. Subcaudal organ with ducts leading out between in dividual scales on ventral surface of tail – emits 'musky' odour. In captivity, faeces are deposited in specific areas when two sexes are placed together, and frequently investigated. Black, glistening streaks deposited on substrate by subcaudal gland, particularly by males in breeding season; the 'gassy' odour of the secretion is detectable by humans. Hearing and sense of smell acute (prey detected by latter). Poor vision, possibly restricted to detecting changes in light intensity from bright to shade.
Breeding. Protracted breeding season. Matings in January-May, with

pregnant females found January-June. Sexual maturity in first year after birth. Gestation *c.* 4–5 weeks. Litter size 1–4. Litters per year: 1–2. Mammae 8. YOUNG: Furred when new born. Weaning age 4–5 weeks. Parental care by female only.

Lifespan. Average lifespan at least 3–4 years, max. 5yrs. Seriously threatened by habitat destruction such as cutting fringe vegetation along streams, dams, electro-fishing and pollution. Mink blamed for disappearance in many areas in Spain, but no direct evidence.

Measurements. Head-body length: 110–150mm. Tail-length: 120–155mm. Hind-foot length: 31–38mm. Condylo-basal length: 33–35.5mm. Weight: 50–80g (birth: 4g). Dental formula: 3/3, 1/1, 4/4, 3/3 = 44.

General. Adapted to aquatic life. Swims rapidly, propelled by webbed hind feet. Climbs nimbly, but ungainly walk. First discovered in 1811, IUCN classified as vulnerable in 1988, due to pollution of mountain streams. Only found in restricted areas of N. Spain, N. Portugal and central French Pyrenes.

BLIND MOLE *Talpa caeca* **Pl. 7**

Name. *Talpa* = a mole; *Caeca* = blind (Lat.).

Recognition. As Common Mole, but smaller, with longer, slender nose, whitish hair on lips, legs, tail and particularly forelegs. Eyes permanently closed by thin membrane which cannot be opened without damage. Upper incisor teeth form a V-shape with largest, central (1st) incisors being more than twice the size of the smallest, peripheral (3rd) ones.

Habitat. Confined to mountainous areas, in contrast to Roman and Common Moles which are both found in various habitats from the Appenine Mountains and Alps through to coastal areas. In contrast to the Roman Mole (which does not overlap with the Common Mole), the Blind Mole is generally thought to overlap in distribution (i.e. be sympatric with) the Common Mole (both species have been caught in the same tunnel in Gran Sasso in the Abruzzi region of Italy).

Measurements. Head-body length: 95–140mm. Tail-length: 20–30mm. Hind-foot length: 15–17mm. Condylo-basal length: 29–32mm. Weight: 65–120g. Dental formula: 3/3, 1/1, 4/4, 3/3 = 44.

General. Habits probably similar to Common Mole. Despite name, vi-

<*Blind Mole.* Iberian peninsula, N Italy (the Appenines) Adriatic coast and Greece.

>*Common Mole.* Missing from southern Italy, Albania, Greece, Portugal, Iceland and Ireland; present on most small British islands.

sion less poor than that of Roman Mole. One subspecies, *T. c. occidentalis* from the Iberian Peninsula, is often considered a separate species, and a second subspecies, *T. c. hercegoviniensis*, occurs in Yugoslavia.

COMMON MOLE *Talpa europaea* Pl. 7

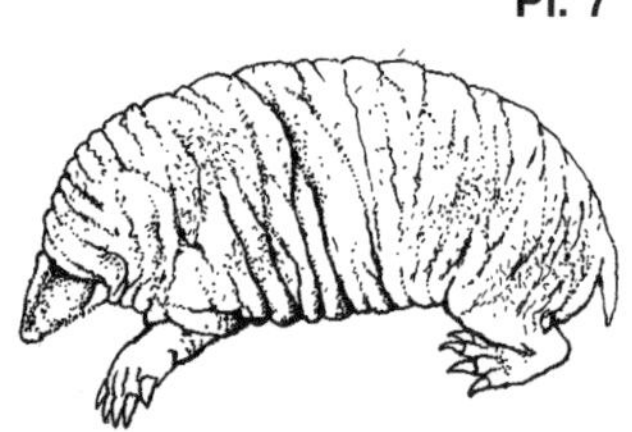

Name. Northern Mole. *Talpa* = mole (Lat.).

Recognition. Short, black, velvety fur. Cylindrical body and broad, heavily-clawed feet. Spade-like forefeet are unmistakable diagnostic character. Tiny eyes (1mm diameter), largely obscured by fur but fully functional. Tail carried erect. Generally has bright pink nose. External genitalia similar in both sexes, except during the breeding season, when females are perforate. Non-virgin females have a vaginal scar. Out of breeding season the distance between genital and anal papillae can generally be used to distinguish the sexes: females = less than 4mm; males = more than 5mm. Penis *c.* 7.5mm and slightly larger than female papilla but in all these measurements there can be considerable overlap between the sexes. Moults three times annually: February-March/April, July-September, October–December – dependent on latitude. Molehills are conspicuous fieldsigns (including the larger, more permanent fortresses). Surface runs leading from mole hills – watch for shaking grass stems: the mole can often be 'seen' pushing up under the grass, making a tearing noise as it breaks the roots. Occasionally the mole's exit tunnel is visible in the middle of a mole hill: approx. 35mm diam., used for exit to collect nest material or surface food.

Habitat. Abundant in deciduous woodland, arable fields and permanent pasture. Avoids stony, sandy, waterlogged soil because cannot construct a proper burrow system, and acid (less than pH 4.4) soil because of lack of suitable earthworm species. Found up to 1000m in Scotland, up to 2000m in Alps. Semi-permanent burrows (shallow to more than 100cm deep) plus one or more sleeping chamber (lined with moss, grass and leaves – other materials have included cigarette papers, fertiliser bags, fish and chip bags etc. woven into a tight ball – animal climbs in and re-seals) only rarely under an especially large mole hill. Nests may form hub from which burrows radiate – sometimes large fortress built over nest. Mounds may contain stores of worms. Fortresses occur in winter (possibly act to retain heat).

Habits. Active day and night (in some cases a cycle of 4.5 hr activity, 3.5 hr resting in nest). Moles spend most of their lives in an elaborate burrow system, dug exclusively with their forelimbs. Sleeps in upright position, head between forelegs. Synchronized activity rhythm between neighbours in some cases. Activity seasonal – males more active when searching for females, females more active when suckling young. FOOD:

Bulk of diet is earthworms (winter: more than 90%; summer: *c.* 50%), plus beetle and fly larvae, myriapods (centipedes, millipedes etc.) and molluscs (especially slugs). Occasionally vertebrates (e.g. amphibia). Daily intake: 40–50g. Tunnel acts as a food-trap as soil invertebrates fall in. Also constructs shallow, short-term feeding runs – possibly a compromise between high energy used and high food density for surface runs, and low energy used but low food density permanent runs. HOME RANGE: Highly seasonal (for males) and dependent on habitat quality. 200–2000m^2 (and three-dimensional). Home ranges very rarely regular in shape. Population density: winter 8 per ha, summer 16 per ha. Dispersal by juveniles in July/August (S. Britain), June/July (Scotland). Dispersal largely on surface, or just beneath. Until find unoccupied living space, very vulnerable and there is a high mortality at this time to predators; also if burrow flooded, adults leave and recolonise elsewhere. SOCIAL: Slightly overlapping territories occupied by one mole, but neighbours may coordinate movements to avoid each other in shared zone – via olfactory recognition and perhaps some acoustics. During breeding season males abandon strict territoriality and tunnel over extensive areas in search of females. COMMUNICATION: aggression accompanied by shrill twittering and squeaking. Preputial glands important in territorial demarcation. Can scream when scared. Tactile senses very well developed; erectile hairs with specialised receptors on tail, feet and tip of snout. Eimer's organs in nose sense touch and changes in humidity and temperature and are very sensitive and easily damaged. Vision is poor and hearing moderate, although no external ear.

Breeding. Mating generally in April in the North, earlier further south. Births in May-June, but huge seasonal variation depending on latitude. Sexual maturity 11 months. Gestation 28 days. Litter size av. 3–4 (range 2–7). Litters per year 1 (probably 2 in S. Europe). Mammae 8. YOUNG: Born naked. Fur grows at 14 days (see vignette); eyes open at 22 days; venture from nest at 33 days. Weaning age 4–5 weeks. Parental care by female only.

Lifespan. Max. 7 years (most die at less than 3 yrs). Mortality very heavy when young dispersing due to predation (foxes, buzzards, herons etc). Otherwise, causes of death include starvation (drought), flooding and Man, but very little is known about precise numbers in adults.

Measurements. Head-body length: 110–160mm (av: male = 143, female = 135mm). Tail-length: 20 (subadult)-40mm. Hind-foot length: 17–19.5mm. Depth of chest (on which it rests) *c.* 30–40mm. Condylo-basal length: 32.5–37.00mm. Weight: 65–130g (av. male = 85–95g, female = 70–75g) (Birthweight *c.* 3.5g). Dental formula: 3/3, 1/1, 4/4, 3/3 = 44.

General. Highly adapted to subterranean (fossorial) life, although swims well using all four limbs. When running forefeet take two steps to each hindfoot step. High incidence (0.1%) of colour variants eg pale (cream to rust coloured), piebald or silver-grey. Often food caches in or near nest chamber in autumn and winter (largest cache = 1200 worms, weighing 2kg, paralysed by decapitation). A mole's nose is reportedly so sensitive that striking it will kill the mole but this may be an Old Wife's Tale. Viewed as a pest when mole hills cause soil to contaminate silage with soil bacteria when grass being cut, but this problem readily solved by silage additives, so pest status probably exaggerated. Mole burrows can be readily detected by poking with a stick (e.g. bamboo cane c. 1 cm diam) – there is a sudden change in resistance against the stick as it breaks through compacted soil into the burrow).

ROMAN MOLE *Talpa romana* Pl. 7

Recognition. As Common Mole, but a little larger, eyes covered by a membrane (as for Blind Mole) and a wider muzzle (13.5–15.0mm across cheek bones). Longer body and hind feet than Blind Mole. The Roman and Common Moles can be distinguished by the shape of the three cusps on each of the three upper molar teeth of each jaw: in the Common Mole these cusps are simple cones, but in the Roman Mole the central cusp of each molar has an additional notch (called the mesostyle).

Habitat. Southern and central Italy. The distributions of the Roman Mole and Common Mole do not overlap, and the boundary between them runs from Rome to Ancona.

Measurements. Head-body length: 126–165mm. Tail-length: 20–32mm. Hind-foot length: 15–20mm. Condylo-basal length: 34.2–37.6mm. Weight: 65–120g. Dental formula: 3/3, 1/1, 4/4, 3/3 = 44.

General. As for Common Mole in habits, but smaller and possibly less active at digging. Poorest vision of European moles. A nominal subspecies of *T. romana* occurs along the Tyrrhenian coast of Italy with the exception of central and southern Calabria. The population of Mount Gargano is described as a new subspecies, *T. romana wittei*, and *T.r. adomei* from central/southern Calabria and Apulia is the smallest Italian subspecies. The moles found in the Balkans (often included under *T. romana*) are most probably an entirely separate species (*T. stankovici*).

<*Roman Mole.* Confined to S Italy. Some authorities consider the variety of Comon Mole found in Greece to be a separate species, *Talpa stankovici.*

Order Chiroptera – Bats

Origins: The bats are divided into the Megachiroptera and Microchiroptera. These 'megabats' and 'microbats' probably arose from different ancestors. The first known microbat was *Icaronycteris* whose remains were found in early Eocene sediments (54 mya); the first known megabat was *Archaeopteropus* from the early Oligocene (37 mya). Early microbats had very shrew-like teeth. Opinion is mounting that megabats are evolutionarily close to the primates.

Main features of the Chiroptera: Aside from their dramatic conquest of the air, bats are structurally not very different from typical mammals except for the extension of their finger bones to carry flight membranes, and their backward bending knees. They are nocturnal. Echolocation in microbats is associated with large ears and in some cases facial skin growths, and the ability to catch flying insects at night. This food source is seasonal which, combined with their small size, has favoured the evolution of hibernation. Daily torpor (body temperature drops to ambient temperature) is common among temperate bats, and is another energy-saving adaptation necessitated by small bodies engaged in such a high energy activity as flying. Short migrations to roosts of appropriate temperatures occur throughout the year. Mating begins in the autumn, and continues at intervals throughout hibernation. Sperm is stored and ovulation and fertilisation take place in the spring.

Bats are a very vulnerable order because they live in huge colonies which can be destroyed at one blow. This occurs when caves flood or collapse, trees blow down or buildings catch fire. A major danger to bats is remedial timber treatment to buildings, using lethal chemicals such as Lindane and Dieldrin, which they take in via the lungs, mouth and skin. Habitat destruction is also a big problem and it is all too easy to annihilate a roost site just by blocking up small holes in cellars, tunnels and ancient monuments; also by blocking or commmercialising caves. All these threats are added to those, such as intensifying agriculture, faced by all European mammals. Therefore all bat species are legally protected in all EEC countries, and to varying extents in all other European countries as well. A high priority is to formulate better timber treatments, eg Permethrin. An inherent difficulty for bat conservation is their slow reproduction: they have small litters (generally one infant), some do not breed until they are several years old and then may not breed each year. The result is that any recovery of a protected population may, at best, be slow.

Chiroptera in Western Europe: Nearly a quarter of all mammalian species are bats (951 species). 780 are microbats, in 143 genera. There are 32 European species, none of which is a member of the Megachiroptera, but three of the 18 Microchiropteran families occur in Europe: the Rhinolophidae, the Vespertilionidae and the Molossidae.

Rhinolophids, the horseshoe bats, possess a noseleaf consisting of a horseshoe-shaped process around the nostrils surmounted by a vertical sella and lancet (see p.12). Their echolocation signals are emitted through

A Key to Bat Indentification using a Bat Detector

This information is based on the Queen Mary College Mini Bat detector, but also applies to other tuneable detectors. Initially the detector should be aimed at the bat and scanned across the entire range, but concentrating on the 40–50kHz range. Notes should be taken of: frequency and range of loudest sound, spacing and sharpness of call, distance from bat to detector, volume of call, size of bat and manner and height of flight. This information can then be used in conjunction with Table 1 to identify which group of bats the individual belongs to. Tables 2,3 & 4 can then be used to identify the species.

Table 1. Call characteristics of Bat groups

	Horseshoe Bats; see Table 2	***Myotis* sp., Pipistrelles & Long-eared Bats; see Table 4**	**Noctule & Leisler's Bat; see Table 3**	**Serotine; see Table 3**
Peak frequency	85 or 110kHz	40–50kHz	20–35kHz	
Sound	Continuous warble	Rapid clicks	Loud 'chip shop' frying sound, some clicks	Like hand clapping

Table 2. Horseshoe Bat call characteristics

	Greater Horseshoe Bat	**Lesser Horseshoe Bat**
Peak	85kHz	110kHz
Frequency range	Very narrow	Very narrow
Size of bat	Big	Small

Table 3. Noctule, Leisler's Bat and Serotine call characteristics

	Noctule	**Leisler's Bat**	**Serotine**
Peak	25kHz	25kHz	25–30kHz
Range	20–45kHz	15–45kHz	15–65kHz
Volume	Very loud	Not as loud as Noctule	Quite loud
Sound	'Chip shop' over call range, occasional clicks when feeding	'Chip shop', but clicks at top of range	Irregular hand-clapping
Flight	Very high (above tree tops, with sudden dives to ground level	Quite high, only shallow dives less frequently than Noctule	Roof height or lower
Size	Big; broad wings	Smaller than other two	Big, narrow wings

Table 4. Myotis spp., Pipistrelles, Long-eared, Daubenton's, Natterer's, Whiskered and Brandt's call characteristics

	Myotis spp.	Pipistrelles[1,2]	Long-eared Bats[3]	Daubenton's Bat	Natterer's Bat[3]	Whiskered/ Brandt's Bat
Peak	50 or 60 kHz	50kHz	35–40kHz	45–50kHz	30 and 50kHz	60kHz
Frequency range	25–80kHz	40–60kHz	25–50kHz	35–85kHz	20–50kHz	40–80kHz (or lower?)
Volume	Usually loud	Loud	Very quiet. Rarely detectable	Strident	Very quiet	Not as loud as Daubenton's
Sound	Clicks over whole of range	Clicks turning to 'wetter' slaps at bottom of range	Clicks like Geiger counter	Regular machine gun like sound, audible for 5–10 seconds	Irregular soft clicks, like stubble burning	Regular, can sound distorted when turning
Flight	Usually level, often amongst trees or over water	Twisting and turning in open areas, often at head height	Slow, amongst foliage. Can hover	Skims water with quivering wing beat	Around tree canopy	Regular circling/gliding at fixed height: looks like Peacock butterfly in sailing flight
Size				Medium	Medium	Small/medium, dumpy shape

1 Pipistrelle usually only heard for short bursts. Also give a 'chonk' ('social' call) at 20–30kHz
2 Nathusius, similar to Pipistrelles but peaks at 30kHz
3 Barbastelle Bat like Long-eared Bats but quiet, as is Natterer's Bat

the nose with the mouth shut. Horseshoe bats always hang free by their toes; they may hang by one leg, spinning and echolocating simultaneously. When torpid the wings envelop the body almost completely (and with it the baby in the first few days after its birth), with the tail downwards over the back (so dripping water meets the membranes rather than soaking into the fur). Movement on all fours (quadrapedal) is poor. The calcar extends about one third of the length along the tail membrane; rhinolophids have no postcalcarial lobe. Females of this family have 2 mammary glands in the pectoral region and two 'false nipples', which develop at the first pregnancy, located slightly above and to the side of the genital opening (= the inguinal area); young cling to these false nipples so that when the mother hangs head downwards, the young has its head upwards. Ears move independently of one another. Horseshoe bats have no tragus, but have a well developed anti-tragus.They become only slightly torpid by day in summer, so they can take off quickly. Prey are often caught in the wing membrane and can be stored briefly in cheek pouches. Larger prey are eaten at established feeding sites. Due to the horseshoe bats' short tail length, the tail membrane always forms a concave arch during flight.

Vespertilionids, in contrast, have a pointed or convex arch formed by their tail membranes. They all have a prominent tragus, and a post-calcarial lobe. They sometimes hang free but more often are found in crevices or in tight corners. Most of them are very agile quadrapedally. Those which do hang free hold their wings at their sides (except sometimes Long-eared Bats). They do not possess false nipples. Vespertilionid bats emit echolocation signals through open mouths.

Molossidae are the 'free-tailed bats', so-called because their thick tails always project considerably past the edge of their tail membranes. The only European species is *Tadarida teniotis*. They have distinctive heads and faces, and often wrinkled lips; very long, narrow wings, strong legs, generally short velvety fur. They are strong, fast-flying bats that live on insects caught chiefly on the wing, as do horseshoe and vesper bats. Free-tailed bats roost in a wide variety of places: in caves, under bark, under eaves, palm fronds etc.

EGYPTIAN SLIT FACED BAT *Nycteris thebaica* Pl. 17

Status. A vagrant in Europe, widely distributed in Africa, and recorded in Corfu (west coast of Greece). Recognition. Conspicuous long ears. Upper parts buffy-brown, the base of the hair slate grey (but buff on sides of neck and head). Ears and wing membranes are light brown. Long, soft hair extends onto both surfaces of the wing membranes near the body. Underparts lighter colour. The upper incisors have two lobes on their cutting edges. The tragus is pear-shaped with an indentation on the lower part of its outer edge.

Habitat. Open savannah woodland, in association with buildings, from sea level to over 1500m. Roost in caves, culverts under roads, mines, hollow trees and rock fissures, also in roofs.
Habits. Nocturnal. Leave roost well after dusk, returning well before dawn. Individuals forage separately. SOCIAL: Gregarious, occurring in roosts with up to hundreds of individuals. Insectivorous gleaners, with echolocation pulse repetition increasing as they close on the target. Feeds mainly on arthropods, especially crickets and grasshoppers, with Lepidoptera, Isoptera, Homoptera and Coleoptera also taken.
Breeding. Copulation and fertilisation in early June, gestation 5 months, births early November. Lactation 2 months. 2 mammae. Litter size 1, carried around by the mother attached to nipple.
Measurements. Head-body length: male, mean 49mm; female, mean 55mm. Tail-length: male, mean 53mm; female, mean 55mm. Hind-foot length: male, mean 11mm; female, mean 12mm. Ear: male, mean 33mm; female, mean 34mm. Weight: male, 9.0–11.5g; female, 10.0–13.7g.
General. Slow fliers, often foraging within a metre of the ground. They can hover, to glean prey from vegetation and the ground.

LESSER HORSESHOE BAT *Rhinolophus hipposideros* Pl. 8

Name. *Rhis, rhinos* = nose (Gk.); *lophos* = crest (Gk.) *Hippos* = horse (Gk.); *sideros* = iron (Gk.).
Status. Endangered. Has become extinct in northern England within the last 50 years. Threatened with extinction throughout northern Europe.
Recognition. Smallest European horseshoe bat. Soft, fluffy fur; base of hairs light grey; back smoke-brown without reddish tint, underside grey to grey-white. Ears and wing membranes light grey-brown. Noseleaf very delicate: upper connecting process short and rounded, lower connecting process distinctly longer, and pointed in profile. Distinguished from other species by small size and rounded upper connecting process of noseleaf. Noseleaf lacks prominent upper projection on central, wedge-shaped sella. Ears fold to within 5mm in front of nose tip, and as with all horseshoe bats they lack a tragus. Maternity colonies sometimes noisy with continual chattering. Flight fairly fast, wing movements almost whispering. Hunts in open woodland and parks, among bushes and shrubs, close to ground (up to 5 m). Much faster wingbeat than Greater Horseshoe Bat. Juvenile Greater and Lesser Horseshoe Bats can be distinguished from adults by tapering rather than knobbly finger joints, and by grey rather than brown fur.
Habitat. Warmer regions in foothills and highlands, preferring partially wooded areas, areas of limestone. In the more northern latitudes summer roosts (nurseries) are in warm attics, buildings; in the south summer roosts are in caves, mine tunnels. Usually requires direct flight path into roost without obstruction but can crawl backwards into crevices. Winter

roosts in caves, mines, cellars; temperatures 6–9°C, high humidity, always hangs separately without touching neighbour, up to 500 animals per roost. If disturbed during hibernation draws body up by flexing legs; thus an indication that bat undisturbed.

Habits. Emerges in evening, active throughout night. FOOD: A 'gleaner', picking prey off stones and shrubbery, sometimes also hunts by aerial hawking and by pouncing on ground. Prey craneflies, small moths, mosquitos, gnats, beetles and spiders. HOME RANGE: Sedentary: distances of 0–100km recorded between summer and winter roosts, maximum 153km. SOCIAL: Summer: females form nursery colonies of 10–500; immature males also may be present. Winter: generally solitary, but occasional loose aggregations with 25–50cm between individuals. COMMUNICATION: Vocalisations include chirping or scolding. Echolocation 108–110 khz.

Breeding. Mating generally in autumn (eg September–November), but sometimes during winter in hibernaculum. Mating chases prior to brief copulation during which the male hangs himself up behind and over the female. Females store the sperm until ovulation and fertilisation in March/April. Vaginal plug formed inside female after mating (as in all horseshoe bats). Move to nurseries from April onwards (initially to 20% of bats at roost may be male, but most depart when young born). Approx 50–60% of females give birth between mid-June and beginning of July, but some births until early Aug. Nursery roosts have 10–500 adults. Nurseries disperse between August–October. Roosts often shared with Greater Mouse-eared Bats but no mixing occurs. Sexual maturity of both sexes at 1 year old. Litter size 1. Litters per year 1. Mammae 2 (+ 2 false). YOUNG: Weight at birth *c.* 1.8g, covered in fine hair except on abdomen, sensory hairs around horseshoe. Eyes open *c.* 3 days; fully weaned at 4–5 weeks; completely independent at 6–7 weeks. Juveniles dark grey. Parental care by female only.

Lifespan. Max. recorded 18 years.

Measurements. Head-body length: 37–45mm. Tail-length: 23–33mm. Hind-foot length: 7.5–10.5mm. Forearm length: 32–42.5mm. Wingspan 192–254mm. Ear: 15–19mm. Condylo-basal length: 13.4–16mm. Weight: 5–9g. Dental formula: 1/2, 1/1, 2/3, 3/3 = 32.

General. Cline in size from W. France eastwards, so smallest bats in France and UK, largest in Romania. Horseshoe bats have false nipples.

< *Lesser Horseshoe Bat.* From SW Europe and NW Africa, E to the Himalayas.

> *Mediterranean Horseshoe Bat.* S Europe E to the Black Sea.

MEDITERRANEAN HORSESHOE BAT *Rhinolophus euryale* **Pl. 8**

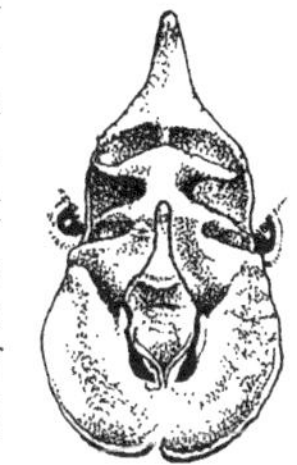

Recognition. Medium-sized. Fur fluffy, base of hairs light grey. Back grey-brown, slight reddish or lilac tinge; underside grey-white to yellowish-white; blurred border between back and underside. Often darker hairs around eyes. Noseleaf and lips light brown; ears and wing membranes light grey. Noseleaf: upper connecting process of sella pointed, slightly bent downwards; long, connecting process, which looks rounded when seen from below; lancet tapers off evenly towards blunt, wedge-shaped tip; sella with parallel sides. Wings broad, 2nd phalanx of 4th finger (17.9–19.1mm) more than twice as long as 1st phalanx (6.6–8.5mm). At rest 3rd to 5th fingers are slightly bent at the joint between 1st and 2nd phalanx, so bat not usually wrapped up completely in wing membranes (even during hibernation). Similar species are Greater, Lesser, Mehely's and Blasius' Horseshoe Bats. Slow, fluttering flight, able to hover: very agile. Hunts low over ground on warm hillsides, or in relatively dense tree or scrub cover.

Habitat. Warm wooded areas in foothills and mountains. Prefers limestone areas with numerous caves and access to water. Essentially a cave dweller; although nurseries are usually in caves, may use warm attics in north of range. Winter roosts in caves and mine tunnels; temperatures above 10°C; hang free from ceiling, bodies sometimes in contact with one another.

Habits. Nocturnal. FOOD: Moths and other insects. Often eats prey at feeding sites. HOME RANGE: Usually sedentary: greatest recorded movement 134km. SOCIAL: Much more sociable than Lesser or Greater Horseshoe Bats. Summer: females in nurseries of 50–400, males sometimes also present. When active in roosts often hang with bodies in contact, embracing each other with wing membranes and licking each other's faces and heads. COMMUNICATION: Deep chirping, squeaking, scolding.

Breeding. In some areas (eg Bulgaria) pregnant females found in mid-July, alongside juveniles that have learnt to fly. Litter size 1. Mammae 2 (+ 2 false). YOUNG: Weigh *c.* 4g at birth. Can fly mid-July to mid-August, depending on area. Generally grey in colour.

Measurements. 1st phalanx, 4th finger 6.6–8.5mm. 2nd phalanx, 4th finger 17.9–19.1mm. Head-body length: 43–58mm. Tail-length: 22–30mm. Hind-foot length: 9–11mm. Forearm length: 43–51mm. Wingspan: 300–320mm. Ear: 18–24mm. Condylo-basal length: 16.4–17.8mm. Weight: 8–17.5g. Dental formula: 1/2, 1/1, 2/3, 3/3 = 32.

General. Decline in numbers observed in north of range (particularly France and Czechoslovakia). Habits generally supposed to be similar to Greater and Lesser Horseshoe Bats. Frequently share roosts with other horseshoe bat species, as well as Geoffroy's and Long-Fingered Bats. Decline in this species is associated with use of pesticides.

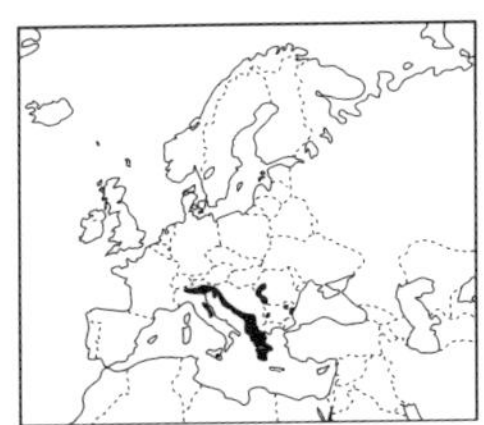

< *Blasius' Horseshoe Bat.* Adriatic coast of Italy, Yugoslavia, Albania and Greece.

> *Greater Horseshoe Bat.* S continental Europe and SW England.

BLASIUS' HORSESHOE BAT *Rhinolophus blasii* Pl. 8

Status. Endangered in Italy, rare in Greece. Distribution only partially known.
Recognition. Medium-sized. Fur fluffy, base of hairs very pale, almost white. Dorsal side grey-brown, sometimes vague lilac tinge; underside pale – almost white or with yellow tinge; sharp dorso-ventral border. Rarely dark 'spectacles' around eyes, and then very faint. Horseshoe broad, flesh-coloured, ears and wing membranes light grey. Noseleaf: upper connecting process of pointed, blunt, wedge-shaped sella is straight, not bent downwards, longer than lower connecting process, which is narrow and rounded when seen from the front; horizontal fold slightly indented in middle; lancet has concave sides and tapers off evenly towards the top. Wings broad. 2nd phalanx of 4th finger at most twice as long as 1st phalanx. Similar species are the other four horseshoe bat species described.
Habitat. Warm limestone areas with fairly open cover of shrubs and trees. Cave dweller, both for summer and winter roosts. Hangs freely, with no body contact with other bats.
Habits. Nocturnal. FOOD: Hunting and feeding probably similar to Mediterranean Horseshoe Bat. HOME RANGE: Probably sedentary. SOCIAL: Up to 200 females in a nursery.
Breeding. Litter size 1. Mammae 2 (+ 2 false).
Measurements. 1st phalanx, 4th finger, *c.* 8.2mm; 2nd phalanx, 4th finger *c.* 14–15mm. Head-body length: 44.5–54mm. Tail-length: 25–30mm. Hind-foot length: 9.5–10mm. Forearm length: 45–48mm. Wingspan: *c.* 280mm. Ear: 16.5–21mm. Condylo-basal length: 15.8–17.5mm. Weight: 12–15g. Dental formula: 1/2, 1/1, 2/3, 3/3 = 32.

GREATER HORSESHOE BAT *Rhinolophus ferrumequinum* Pl. 9

Name. *Ferrum* = iron (Lat.); *equinus* = of a horse (Lat.).
Status. Endangered. Has disappeared from half former UK range: only 1% survive. Also threatened with extinction in northern Europe.
Recognition. Largest European horseshoe bat. Fur soft, fluffy: base of hairs dark grey; grey-brown or smoke grey with reddish tinge on back. Underside grey-white to yellow-white. Wing membranes and ears light grey-brown. Noseleaf: upper connecting process of sella short, rounded;

lower connecting process pointed in profile. Not easily confused with other species because of size and blunt upper connecting process. Depending on temperature may wrap itself completely in its wing membranes during hibernation and usually when torpid during the day. False nipples in females not fully developed until after first birth (usually at 3 yrs old). Fieldsigns include insect remains and droppings beneath feeding perches in trees or cave entrances, these signs being most evident in spring/summer. Maternity colonies can be noisy with continual chattering. Slow, fluttering flight with short glides, usually low (0.3–6m). Prefers dense tree cover for hunting but also hunts in open if prey available there; hillsides, cliff faces, gardens.

Habitat. Sheltered wooded valleys with areas of open trees and scrub, access to pasture, close to suitable roost sites, and to areas of flowing or standing water. Semi-natural woodland important in early spring, meadow and pasture in late spring, pasture in autumn. Also limestone areas and places of human settlement. Traditionally cave dwellers, especially in the south, where they breed in caves; have a tendency to be house dwellers in north. Summer: roof spaces of barns, cathedrals, warm attics; in south also in caves and mine tunnels. Winter: hibernation deep in mine shafts and caves, also cellars and roofs; temperatures 7–10°C, rarely below; hangs free from ceiling, singly or in dense clusters (shared body heat maybe important).

Habits. Emerge *c.* 50 mins after sunset in summer. Between May and August feed intermittently throughout night, sometimes resting in a 'night roost' close to be foraging area. From late August may remain away all night. Less often out in wet, windy weather. Hibernation from September/October–April, but may be interrupted 2 or 3 times a month, and will feed near cave entrance in mild weather. FOOD: Often hunts from a perch by flycatching – locating insects and flying out to intercept them. Mainly large flying insects, such as cockchafers, dungbeetles, moths, which are eaten on the wing; sometimes lands to take food (eg dor beetles) from cow dung. Large prey taken to regular feeding perches. Drinks during low-level flight or while hovering. Feeding range of colonies in England 8–16km. HOME RANGE: Generally travels within 8km of roost at night, but many records over 16km and longest 64km. Winter–summer roost dispersal up to 64km. Sedentary: distance between summer and winter roost usually 20–30km; longest movement 180km. SOCIAL: Winter: gregarious hibernation (complex mechanism of cluster formation possibly to control temperature). Summer: pregnant females form nursery colonies in company of some immature, and non-breeding individuals of both sexes, adult males live in small scattered groups. COMMUNICATION: Relatively deep chirping or scolding calls. Sometimes loud twittering from summer colonies if disturbed. Echolocation: *c.* 80–83 khz.

Breeding. Normally mates in late September–October, but some matings in hibernation over winter. Sperm is stored in the female's oviduct and uterus until ovulation and fertilization in March/April. Births in mid-June until early August depending on area. Maternity roosts occupied from May, although most breeding females do not arrive until early June. Maternity roosts normally 50–500 females, but up to many thousands. Initially males may be present, but usually depart when young born (mid-June and throughout July). Females hang with young either individually or in clusters. Sometimes shares roosts with Mediterranean Horseshoe Bats and Geoffroy's Bats. First young is usually produced in 3–5th summer (average 5.7 yrs) in Britain, third summer in south of the continent, but sometimes not until 9 years old. A few produce in 2 years. Females may not breed every year. Males mature at end of their second year at the earliest. Litter size 1. Litters per year 1. Mammae 2 (+ 2 false nipples). YOUNG: Open eyes at about 4 days. Fly short distances insde roost at *c.* 3.5 weeks, before being weaned; can catch insects at 3 weeks; independent at 7–8 weeks. Juveniles more ash-grey on dorsal side than adults. Weaning age *c.* 5–7 weeks, most juveniles independent by late August. Parental care by female only.

Lifespan. Max. 30 years: the oldest recorded for any European bat species.

Measurements. Head-body length: 57–71mm. Tail-length: 35–43mm. Hind-foot length: 11–13mm. Forearm length: 49–61mm. Wingspan: 290–350mm. Ear: 20–26mm. Condylo-basal length: 20–22mm. Weight: 14–34g (April: male 19g, female 18g. October: male 29g, female 24g). Dental formula: 1/2, 1/1, 2/3, 3/3= 32.

General. Serious decline in numbers (90% decrease in UK between 1950s and 1980s). Decline in numbers due to changes in agricultural practice which has diminished permanent pasture (and thus food supply). Roosts therefore much smaller, and thus less able to maintain body temperature from shared warmth. Therefore, more likely to roost in warm lofts, where more likely to be killed by timber preservatives – a major cause of colony loss. Feeds in winter if conditions suitable. Will wake from hibernation to seek suitable temperature conditions – needs much warmer sites (7–11°C) than other bats.

MEHELY'S HORSESHOE BAT *Rhinolophus mehelyi* Pl. 9

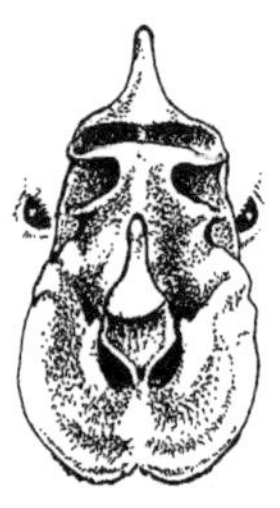

Recognition. Medium sized. Fur relatively thick, base of hairs grey-white; grey-brown on back, almost white below (relatively sharp dorso-ventral border). Dark grey-brown 'spectacles' around the eyes. Horseshoe and lips pale, flesh-coloured, ears and flight membrane grey-brown. Noseleaf: upper connecting process relatively blunt in profile, only slightly longer than the lower connecting process which is wide and rounded when viewed from the front; lancet narrows sharply in its upper half and tapers off to a thin apex. Broad wings. 2nd phalanx

< *Mehely's Horseshoe Bat.* Largest pocket of occurrence in Spain, others throughout S Europe and on Corsica, Sardinia and Sicily.

> *Whiskered Bat.* Europe, except Scotland and parts of Iberian peninsula.

of 4th finger more than double length of 1st phalanx. Body not completely enveloped by wing membranes when at rest. Similar species include Lesser and Greater Horseshoe Bats. Hunts low over ground on warm hillsides, also among bushes and trees. Flight slow, very skilful, with short glides.

Habitat. Appears to prefer limestone areas with close access to water. Cave dweller, both for summer and winter roosts. Sometimes found together with other horseshoes, Lesser Mouse-eared and Schreiber's Bats. Hangs from cave roof.

Habits. Emerges at dusk. No precise information on hibernation available. FOOD: Moths and other insects. May also feed on ground: can take off easily from ground. HOME RANGE: Probably sedentary. SOCIAL: Up to 500 animals in nursery roosts in past (eg Romania). COMMUNICATION: Relatively deep, loud, short chirping or squeaking.

Breeding. Litter size 1. Mammae 2 (+ 2 false).

Measurements. 1st phalanx, 4th finger: 7.7mm; 2nd phalanx, 4th finger: 19mm. Head-body length: 55–64mm. Tail-length: 24–29mm. Hind-foot length: 9–12.5mm. Forearm length: 50–55mm. Wingspan: 330–340mm. Ear: 18–23mm. Condylo-basal length: 16.6–17.5mm. Weight: 10–28g. Dental formula: 1/2, 1/1, 2/3, 3/3 = 32.

General. Distribution only partially known, but probably endangered.

WHISKERED BAT *Myotis mystacinus* Pl. 13

Name. *Mys* = mouse (Gk.); *ous, otos* = ear (Gk.); *mystax, mystakos* = moustache (Gk.).

Recognition. Smallest European *Myotis* species. Easily confused with Brandt's Bat. Long, somewhat shaggy fur, dark grey at base. Dorsal colouring variable, from dark brown or grey-brown to (more rarely) light brown. Underside dark to light grey. Nose, ear and wing membranes black-brown. Outer-ear edge with clear notching, below the tip of a long tragus. 4–5 transverse folds. Tragus straight or concave on outer i.e. posterior edge. Wing membranes relatively narrow, lateral membrane starting on base of toes. Small feet. Calcar shorter than half tail membrane, and usually has narrow keel. Thin penis, unlike Brandt's, not club-shaped at end. Third upper premolar tooth sometimes lacks anterior cusp which is diagnostic of Brandt's bat, cusps on third upper premolar lower than on second upper premolar, first lower premolar obviously larger than second lower premolar. Face much darker than Daubenton's Bat. Similar species are Brandt's Bat, Common Pipistrelle. Flight agile, rapid, weaving.

Habitat. Open meadows and wooded landscape, often near water, but more a house-dwelling than woodland bat, so frequently in parks, gardens, villages. In summer up to 1920m (Alps); in winter up to 1800m (Poland) above sea level. Summer roost: sometimes trees (behind bark) but most regularly in buildings near water. In all types of houses, but particularly older ones – often without insulation, within walls or along ridge in loft space. Colonies of Whiskered or Brandt's may use separate parts of the same roof and may occur with pipistrelles or Long-eared Bats. Winter roost: most underground sites; caves, mines. Prefers cooler entrance parts of caves, 2–8°C. Mainly crevice dwellers. In small numbers (usually less than 100) – main hibernating sites may be elsewhere, but unknown. Often solitary and often hanging exposed on walls or ceiling but also wedged in cracks; sometimes found in small clusters (less than 10), sometimes alongside Brandt's Bat.

Habits. Nocturnal. Generally emerges at sunset; occasionally seen hunting by day (especially in spring and autumn). Peak of activity after sunset and before sunrise. FOOD: Flying insects (midges, mayflies, beetles, moths), but also known to catch spiders etc. from foliage. Often flies along a regular 'beat' over or alongside a hedgerow, meadow or woodland edge, or 1.5–6m above ground in parks, gardens or over flowing water. HOME RANGE: Mainly sedentary, but possibly an occasional migrant: furthest recorded distance 240km. SOCIAL: Summer: females in nursery colonies of 20–50 often in buildings, males apparently solitary. Winter: solitary or small groups. Sex ratio in caves during hibernation male biased (eg 62%), females presumed to be elsewhere. COMMUNICATION: Long, high-pitched scolding or twittering if disturbed.

Breeding. Mating in autumn to spring, thus proportion inseminated increases as hibernation progresses. Females segregate from males in May to form maternity colonies of 20–70 females-April/May–July/August, sometimes until October. Births in June or early July. Nurseries disperse at end of August. Sexual maturity: females generally at 15 months, but some at 3 months (in first year). Litter size 1. Litters per year 1. Mammae 2. YOUNG: Foraging within 6 weeks. Weaning age 6 weeks. Parental care by female only.

Lifespan. Max. recorded 23 years. Average 4 years.

Measurements. Head-body length: 35–48mm. Tail-length: 30–43mm. Hind-foot length: 7–8mm. Forearm length: 32–36mm. Wingspan: 190–225mm. Ear: 14–15mm. Condylo-basal length: 12.3–13.3mm. Weight: 4–8g. Dental formula: 2/3, 1/1, 3/3, 3/3 = 38.

General. Distribution uncertain due to confusion with Brandt's Bat: two species only distinguished in 1970. Two similar species are referred to as sibling species. Study showed Brandt's Bat to be more associated with woodland than Whiskered Bat. Another suggests that Whiskered Bats prefer mountains in heavily wooded areas, while Brandt's nurseries tend to occur at lower altitudes near water. Generally thought to be uncommon over much of Europe, but common in parts of the Netherlands and Belgium.

BRANDT'S BAT *Myotis brandtii* Pl. 13

Recognition. Small. Relatively long fur, dark grey-brown at base. Dorsal side relatively lighter than Whiskered Bat's, being light brown sometimes with a golden sheen. Underside light grey with a yellowish tinge. Nose, ear and flight membranes medium to dark brown, base of tragus and inner edge of ear obviously lighter (not found in Whiskered Bat). Ear shorter than Whiskered Bat's (barely reaches to tip of nose when folded forwards), outer edge with clear indentation, which is below tip of long pointed tragus, 4–5 transverse folds. Tragus has convex posterior margin. Relatively narrow wings with lateral flight membrane starting at base of toes. Calcar shorter than half length of tail membrane and usually narrow 'keel' of skin on outer edge. Club shaped penis in males. Third upper premolar tooth has anterior inner cusp higher than, or of equal size to, second upper premolar; second lower premolar equal in size to first lower premolar. Only reliable way to distinguish from Whiskered Bat is dentition and shape of penis. Similar species in addition to the Whiskered Bat: Daubenton's Bat, Geoffroy's Bat. Hunts at low/medium height in fairly open woodland, or over water. Rapid, skilful flight.

Habitat. Woodland, especially near water, less often near human settlements compared to Whiskered Bat. Up to 1730m above sea level in winter, 1270m in summer (Switzerland). Summer roost: buildings. Winter roost: caves, tunnels, mines, cellars at (on average) 7–8°C. Often roost with Whiskered Bat, mostly hanging free on wall or ceiling, rarely in cracks. Also in clusters with Daubenton's.

Habits. Emerges early, occasionally seen by day. FOOD: Probably flies and other small flying insects. HOME RANGE: Occasional migrant; furthest recorded distance 230km. Two records of winter movements of 2.5km (S. England). SOCIAL: Summer: nursing colonies of females. COMMUNICATION: high-pitched chirping and scolding when disturbed. Young make isolation calls (high twittering) in nursery.

Breeding. 20–60 females per nursery roost (mixed roosts with Nathusius' Pipistrelle in bat boxes recorded). Sexual maturity in females probably 2nd year. Gestation as for Whiskered Bat. Litter size 1. Litters per year 1. Mammae 2. YOUNG: Juveniles closely resemble those of Whiskered Bat: dorsal side dark black-brown to grey-brown, nose and ears black-brown. Parental care by female only.

Lifespan. Max. 19 years, 8 months.

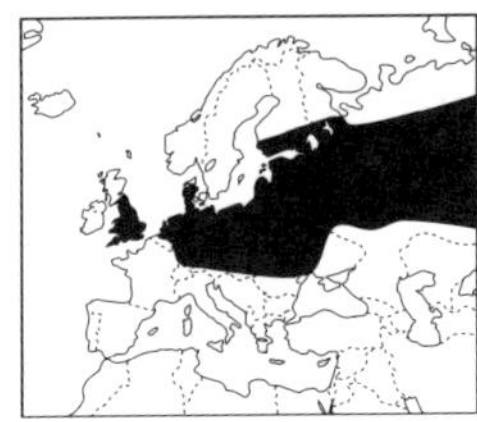

<*Brandt's Bat.* Widespread in England, Wales and C Europe.

>*Daubenton's Bat.* Widespread throughout UK, Ireland, and continental Europe N to about 60° latitude.

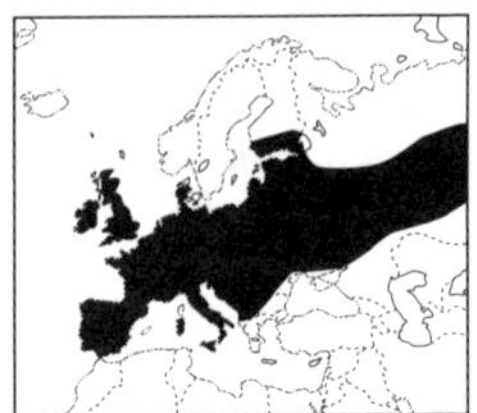

Measurements. Head-body length: 37–51mm. Tail-length: 32–44mm. Forearm length: 31–39mm. Wingspan: 210–255mm. Ear: 11–15.5mm. Condylo-basal length: 13.1–14.4mm. Weight: 4.3–9.5g. Dental formula: 2/3, 1/1, 3/3, 3/3 = 38.
General. Distribution and biology only partially known since previously not distinguished from Whiskered Bat (only separated as a distinct species in 1970). Thought to be widespread from central to northern Europe, absent from S. and W. Widespread over England and Wales, one recorded in Scotland.

DAUBENTON'S BAT *Myotis daubentonii* Pl. 14

Recognition. Small to medium-sized. Fluffy fur with dark grey-brown base. Upper side brown-grey to dark bronze (often with shiny tips), underside silvery grey, often with a brownish tinge, generally a sharp border between the two. Pinkish face, red-brown nose, ears and wing membranes dark grey-brown. All juveniles have a black chin spot; most lose the spot after one year: any Daubenton's Bat without a chin spot is an adult. Shortest ears of this genus. Outer-ear edge slightly notched in lower half, 4–5 transverse folds. Tragus: posterior edge strongly convex (like Brandt's Bat), front edge straight, half height of conch. Lateral membrane starts at base of toes, large feet (more than half length of tibia) with long hairs. Calcar reaches approximately two thirds of the length of the tail membrane, but at three quarters of the length an obvious flap in the tail membrane edge appears like end of calcar. Similar species: Pond Bat, Long-Fingered Bat, Natterer's Bat, Geoffroy's Bat and Brandt's bat. Droppings typically long (1cm), thin, often moist, smooth, slimy and black. Characteristically skims low (within a few centimetres) along the surface of still water for long periods with slow flickering wingbeat (generally said to be reminiscent of Common Sandpiper). Easily seen with spotlight if sweep water surface an hour after sunset (also Pond Bat and Long-Fingered Bat).

Habitat. Predominantly flat countryside in open woodland and riparian landscapes, often hunting low over water. Summer: up to 750m above sea level, recorded up to 1400m in winter. Summer roosts: nursery roosts, usually in tree holes or rock crevices, sometimes with entrance less than 1m above ground, or attics. May share roost with Natterer's, Brown

Long-eared, noctules and pipistrelles. Individuals and small groups of males also found in cracks, under bridges, in walls, but rarely in bat boxes. Winter roosts: underground sites, where they may squeeze into crevices or hang free from wall in large, tight clusters of up to 100 animals; caves and mines (often wriggle into soft scree), cellars, old wells. Prefer high humidity and temperatures of 0–6°C. Several thousand in large hibernacula. Hibernate late September/mid-October (females in hibernaculum before males) to late March/April.

Habits. Emerges *c.* 30 mins after sunset, with activity peaks between dusk and dawn. FOOD: Largely caddisflies, mayflies, various small flies and other insects taken from or just above water surface. Flying insects eaten on the wing. Takes food from surface of water said to use its large feet as gaffs or tail membrane as a scoop. Usually feed within 6km of roost. This species is thought to need to drink more than other species. Closely associated with water, but often travels across land and occasionally feeds away from water. HOME RANGE: Regularly tracked over 2km from roost immediately on leaving at dusk – up to 10km if following linear water courses (eg canals). Longest record: 240km (Czechoslovakia), but mostly a short distance migrant (under 100 km), moving to hibernacula from all directions. SOCIAL: Summer: females in colonies of up to several hundred, often in buildings. Variable composition of individuals at a roost suggest large range with numerous roosts in it – the whole making up the colony. Males found in some roosts in summer but in separate clusters. Winter: singly or small groups in tight crevices (bats often horizontal in caves). COMMUNICATION: In flight no calls other than echolocation ever heard. Chirrups in roosts at times.

Breeding. Mating in September–March, often in hibernacula. Nurseries occupied from May onwards, usually numbering between 20–50 adult females, although can exceed 100. Males may be found individually in nurseries, or elsewhere in groups of up to 20. Births in late June – early July. Colonies often remain until October, though usually a nursery will disperse in August. Sexual maturity: males 15 months, some females 1 year but usually 2 years old. Gestation depends on weather (as with other bats). Litter size 1. Litters per year 1 (but do not always produce every year). Mammae 2. YOUNG: Dorsal side, ears and wing membranes grey-brown at birth, ventral side pink. Born with very thin, short hair on dorsal side, sensory hairs on tail. Birth weight about 2.3g; 5.5g at 21 days. Eyes open from 3rd–4th, complete hair cover from 4th day, hair growth finished at 31–55 days, adult dentition at 37 days. Can fly at third week. Juveniles greyer and darker than adults. Weaning age 4–6 weeks. Parental care by female only.

Lifespan. Max. recorded 40 years.

Measurements. Head-body length: 45–55mm. Tail-length: 31–44.5mm. Hind-foot length: 7.5–11mm. Forearm length: 33–41.7mm. Wingspan: 230–275mm. Ear: 8–14.2mm. Condylo-basal length: 13.2–14.6mm. Weight: 6–15g. Dental formula: 2/3, 1/1, 3/3, 3/3 = 38.

General. Moderately common throughout UK up to N. Scotland, common in western Germany. Population increases reported in some NW European areas. Occasionally hooked by anglers (through mouth or wing) while hawking for insects.

LONG-FINGERED BAT *Myotis capaccinii* **Pl. 14**

Name. *capaccinius* from *capax* = able to hold a lot (Lat.).
Status. Vulnerable.
Recognition. Medium-sized. Light smoke-grey with yellowish tinge above, underside light grey, border between the two blurred. Base of hair dark grey. Nose rufous brown, ears and flight membranes grey-brown. Long, medium-length ears, outer edge to tip of ear slightly notched, 5 horizontal folds, pointed tragus reaches half ear length, inner edge of tragus convex, outer edge serrated slightly. Both sides of wing membrane covered in thick brown downy hair. Feet noticeably large with conspicuous long bristles. Calcar straight and about 1/3 of tail membrane with 'false end' 2/3–3/4 of the way down its length. Hairs in calcar region extend out over tail membrane. Lateral membrane wide, starts 3–5mm above heel. Nostrils more clearly prominent than other European *Myotis* species. Similar species are Daubenton's, Natterer's, Pond, and Geoffroy's: none has hair on tail membrane. Resembles Daubenton's Bat in flight.
Habitat. Limestone areas, frequently near water in wooded landscape. Roosts in caves (both summer and winter), often in crevices.
Habits. Nocturnal. Emerges late dusk. FOOD: Feeds particularly on flying insects caught over water using its tail membrane and/or feet (in same way as Daubenton's Bat). SOCIAL: Colonies of several hundred, occasionally associated with mouse-eared bats. COMMUNICATION: Shrill scolding similar to Daubenton's.
Breeding. Nursery roosts in caves with up to 500 females clustered on cave roof. Births in mid- to late June. Litter size 1. Litters per year 1. Mammae 2. YOUNG: Can fly by mid-July. Parental care by female only.
Lifespan. Max. unknown.
Measurements. Head-body length: 47–53mm. Tail-length: 35–42mm. Hind-foot length: 10–13mm. Forearm length: 38–44mm. Wingspan: 230–

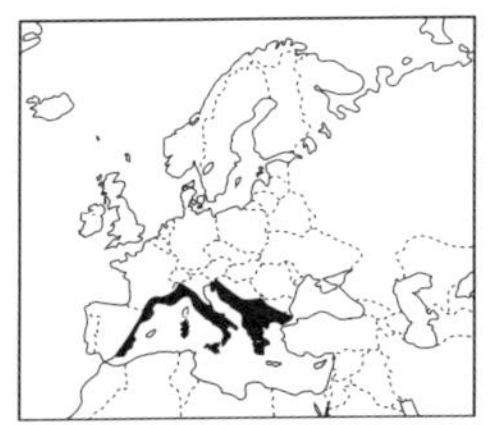

< *Long-fingered Bat.* Mediterranean coast from Seville to Black Sea.

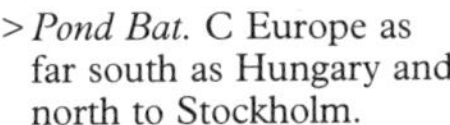

> *Pond Bat.* C Europe as far south as Hungary and north to Stockholm.

260mm. Ear: 14–16mm. Condylo-basal length: 13.9–14.8mm. Weight: 6–15g. Dental formula: 2/3, 1/1, 3/3, 3/3 = 38.
General. Little known, but assumed to be similar to Daubenton's Bat.

POND BAT *Myotis dasycneme* Pl. 14

Name. *dasys* = shaggy; *cnemos* = legged (Gk.).
Recognition. Medium-sized. Thick long fur, brownish or pale grey-brown with silky sheen on back, white-grey to yellow-grey underside (fairly sharp border between). Base of hairs black-brown. Short, red-brown nose; ears and flight membranes grey-brown. Outer ear edge without obvious indentation, 5 horizontal folds, tragus shorter than half ear length (very short for *Myotis* species). Long, broad wings, lateral membrane starts at heel, very large feet with long bristles. Calcar straight and reaches ⅓ of tail membrane, ¾ of way along length is 'false end'. Fine white hairs extend along lower part of leg on underside of tail membrane, and reaches over end of tail membrane at calcar. Similar species are Daubenton's (smaller) and Long-fingered (hair on upper side of tail membrane). Skillful, rapid flight often only 5–10cm above water surface.

Habitat. Wooded areas in lowland regions with access to water and meadows. In winter also found in the foothills of mountains (up to 1000m above sea level, although roosts generally no higher than 300m). Summer roosts: nurseries in roof spaces and attics (females may change site from year to year), individuals also in hollow trees. Winter roosts: caves, mines, cellars: wedged into crevices or hanging from walls and ceilings. Temperatures 0.5–7.5°C, from October–mid March/April.
Habits. Nocturnal. Emerges late, with two foraging periods, one in evening, other in early morning. FOOD: Flying insects such as gnats, mosquitos and moths. Insects sometimes picked from water's surface. HOME RANGE: Undertakes short migrations, usually over 100km from more northern summer roosts south to winter roosts; furthest distance recorded being 330km. SOCIAL: Summer: 40–400 females in nurseries, males usually elsewhere singly or in small groups distributed over a wide area. Winter: usually hangs separately in hibernacula, although small clusters may occur in sites with several hundred animals. COMMUNICATION: alarm – loud scolding similar to blackbird.
Breeding. Mating from late Aug, also occurring in hibernaculum. Nursery roost occupied in May, births from mid-June, dispersal from roost in August. Sexual maturity of females probably in 2nd year. Litter size 1. Litters per year 1. Mammae 2. YOUNG: Juveniles are darker overall than adults, independent from mid-July. Parental care by female only.
Lifespan. Max. recorded 19 years.
Measurements. Head-body length: 55–67mm. Tail-length: 46–51mm. Hind-foot length: 11–12mm. Forearm length: 43–49mm. Wingspan: 200–

320mm. Ear: 16–18mm. Condylo-basal length: 15.7–17.4mm. Weight: 14–20g. Dental formula: 2/3, 1/1, 3/3, 3/3 = 38.
General. Insufficiently known throughout range, but large decline in numbers.

NATTERER'S BAT *Myotis nattereri* Pl. 15

Recognition. Medum sized. Fur short/medium length fluffy, base of hairs dark grey. Dorsally light buff brown, very white by contrast below. Rather bare red-pink face, limbs and wing membranes (gave rise to old name 'red armed bat'). Long, narrow ears extending beyond nose tip when folded forward, five transverse folds and distinct notch exceeded in length by long lancet-shaped tragus which is longer than half length of ear. Broad wing membranes start at base of toes. Calcar reaches half length of tail membrane, S-shaped bend in calcar, conspicuous 'fringe' of stiff, downwardly pointing bristles along edge of interfemoral membrane between calcar tip and tail. Similar species include Long-fingered, Daubenton's, Pond, Geoffroy's and Bechstein's Bats, none of which has S-shaped calcar.
Habitat. Open woodland and farmland, especially with areas of open water or marsh, and occasionally in towns (less strongly associated with woodland than Whiskered or Geoffroy's Bats). Summer nurseries in tree holes, behind bark, cracks on or in buildings (in roof spaces), bat boxes in woodland, cracks under bridges (individual bats also found under bridges). In winter, hibernate in caves, mine tunnels, often very damp sites as need high humidity. May be found squeezed into narrow cracks (sometimes lying on back), in ground scree, or hanging free from wall or roof, sometimes in small clusters (one group of 160 reported in England), often mixed with Daubenton's Bat.
Habits. Nocturnal. Emergence time generally peaks 1 hour after sunset, but varies with light and temperature (prefers warm calm nights). Summer: several bouts of activity each night. Hibernates October-April. FOOD: Small insects, including mainly flies of wing-length 4–10mm, beetles and caddisflies, taken and eaten either on the wing or picked off foliage, sometimes using tail to flick insect into bat's mouth. HOME RANGE: Longest recorded movement: 90km, but generally a very sedentary species. SOCIAL: in summer roosts in large nursery colonies of 20–80 or more females in buildings, sometimes with individual males; otherwise small groups or solitary. COMMUNICATION: chirping and squeaking (deeper than Daubenton's). Deep humming noise if alarmed. High, shrill calls audible in flight.
Breeding. Mating between autumn and spring. Nurseries occupied in April-May prior to births in June-July. Maternity colonies of adult females usually number *c.* 40, but can consist of 100–200. Established May/June, sometimes until September/October. Change nursery roost site frequently. Litter size 1. Litters per year 1. Mammae 2. YOUNG: Juveniles light

< *Natterer's Bat.* Throughout W Europe, including S Scandinavia and British Isles.

> *Geoffroy's Bat.* Most of S Europe, excluding Mediterranean France and Portugal.

greyish-brown in first year. Weaning age 6 weeks. Parental care by female only. Begins foraging for itself *c.* 3 weeks.

Lifespan. Max. recorded lifespan 20 years.

Measurements. Head-body length: 42–50mm. Tail-length: 38–47mm. Hind-foot length: 7.5–9mm. Forearm length: 36–43mm. Wingspan: 245–300mm. Ear: 14–18mm. Condylo-basal length: 14–15.6mm. Weight: 5–13g. Dental formula: 2/3, 1/1, 3/3, 3/3 = 38.

General. Widespread throughout UK and most of Europe. When flying slowly in calm conditions, tail is directed downwards and not in line with the body. Broad wings and tail membrane give great manoeuvrability at slow speeds. Normally flies at slow-medium speed at heights of less then 5m, but occasionally reaches 16m. Can hover for short periods. The bat most frequently found in any cave-like site. Usually turns somersault when flying up to resting place.

GEOFFROY'S BAT *Myotis emarginatus* **Pl. 15**

Name. Notch-eared Bat. *Emarginatus* – referring to angular notch of ear – literally means deprived of its edge (Lat.).

Recognition. Medium-sized. Long, woolly fur. Tricolour dorsal fur: base grey, middle straw-yellow, tips a striking rich rufous brown. Ventral side yellowish-grey. Nose red-brown, ears and wing membranes darker grey-brown. Ears with 6–7 transverse folds, and outer edge with distinct angular notch about two thirds from base. Lancet-shaped tragus with fine notches on outer edge, just over half height of ears (almost reaches notches on outer ear edge). Broad wings, lateral membranes start at base of toes, small feet. Calcar straight and reaches about half length of tail membrane. Only extreme tip of tail protrudes from tail membrane, free edge of latter having sparse, short, straight, soft hairs. Similar species are Natterer's, Brandt's, Whiskered and Daubenton's Bats.

Habitat. Woodland (often forages over water), in limestone areas or populated areas with parks, gardens and water. Prefers warmth: nursery roosts in northern regions in warm attics (36–40°C); in south usually in caves and mine tunnels. Winter roosts in caves, mine tunnels, cellars, rarely below 6–9°C. Mostly hang singly from roof/wall, rarely in small clusters or crevices.

Habits. Nocturnal; emerges late. FOOD: Hunts 1–5m above ground, or over water. Preys mainly on spiders which, together with other prey such as brown lacewings, diptera (eg gnats), hymenopterans, moths and caterpillars, are taken off branches, foliage or ground (flying insects also caught on the wing). HOME RANGE: Predominantly sedentary, movements usually under 40km (longest 106km). SOCIAL: Colonial throughout year: nursery roosts from 20–200 females (Czechoslovakia) to 500–1000 females (France, Balkan states). COMMUNICATION: loud shrill scolding in nurseries.

Breeding. Mating starts in autumn. Nurseries formed in May (often shared with horseshoe bats). Births in mid/late June to early July. Nurseries disperse in September. Sexual maturity: females can mate in first year but birth at end of first year unproven. Litter size 1. Litters per year 1. Mammae 2. YOUNG: Able to fly at 4 weeks. Juveniles much darker than adults, with fur smokey-grey to brown-grey without rufous tinge. Parental care by female only.

Lifespan. Max. recorded 16 years.

Measurements. Head-body length: 41–53mm. Tail-length: 38–46mm. Forearm length: 36–42mm. Wingspan: 220–245mm. Ear: 14–17mm. Condylo-basal length: 14–17mm. Weight: 7–15g. Dental formula: 2/3, 1/1, 3/3, 3/3 = 38.

General. Not recorded in Britain but found throughout most of southern Europe, extending north to the extreme south of the Netherlands, where it is now virtually extinct. Uncommon everywhere and almost extinct in northern areas, including western Germany and Austria.

BECHSTEIN'S BAT *Myotis bechsteinii* **Pl. 15**

Recognition. Medium sized. Fur relatively long, base of hairs darker than tips. Back light to reddish-brown, underside greyish-white. Bare, pink face, nose reddish-brown, ears and wing membranes light grey-brown. Ears strikingly long, fairly broad, extend beyond nose when folded forward (projecting by up to 8mm), outer ear edge with nine transverse folds. Held straight even in hibernation. Tragus lancet-shaped and long (half height of ear). Short, broad wings, lateral membranes start at base of toes, small feet. Calcar straight, reaching ⅓–½ of tail membrane, last tail vertebra free. Similar species are: long eared bats (but longer ears joined at base at front); all other *Myotis* species of similar size have much shorter ears. Fluttering flight, agile even in very confined spaces. Hunts low (1–10m above ground).

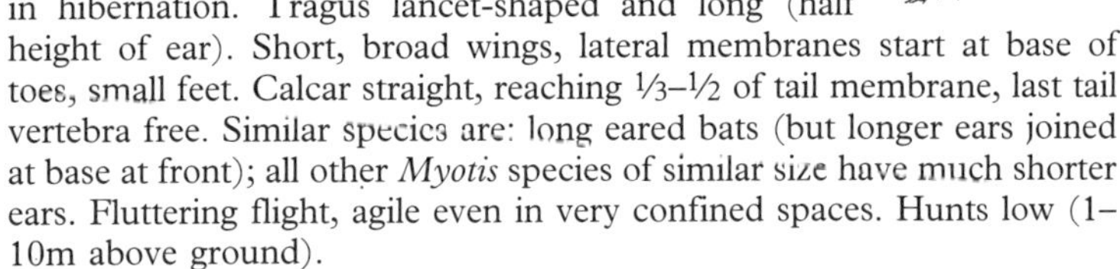

Habitat. Woodland – mainly damp, mixed, but also in pine woods, parks, gardens. Roosts in tree holes throughout year, especially used for nursery roosts (but also bat boxes and less commonly buildings). Frequently changes roost site and roosting associations. In winter also uses buildings, caves, mines. Often hangs free from roof or wall, or more rarely squeezed

<*Bechstein's Bat.* C&W Europe, pockets in S England, S Spain, and on Corsica, Sicily and Balearic Islands.

>*Nathalina Bat.* Recorded from scattered distributions throughout C&W Europe.

into crevices where it hangs head down. Generally found as individuals not clusters. Favours temperatures of 3–7°C, with high humidity.

Habits. Emerges *c.* 20 mins after sunset, preferring calm, warm weather. Hibernates October–March/April. FOOD: Mostly moths, but also mosquitos and beetles, caught in flight or from foliage (gleaning). Feeds in cold and wet, but not windy weather. HOME RANGE: Apparently sedentary, longest recorded movement 35km. SOCIAL: Summer: colonies of 10–30 females in nursery roosts; males solitary. Winter: sex ratio in some hibernating groups strongly male biased (75%). COMMUNICATION: no audible flight calls, but hollow humming or chirping when threatened.

Breeding. Mating between autumn and spring. Occupy nursery roosts from late April-May (changing sites frequently). Births June-early July. Dispersal end of August. Litter size 1. Litters per year 1. Mammae 2. YOUNG: Able to fly beginning to mid-August. Juvenile colouring light to ash grey. Parental care by female only.

Lifespan. Max. recorded lifespan 21 years.

Measurements. Head-body length: 43–55mm. Tail-length: 34–44mm. Hind-foot length: 8.5–10.5mm. Forearm length: 38–45mm. Wingspan: 250–300mm. Ear: 20–26mm. Condylo-basal length: 16–17mm. Weight: 7–13 g. Dental formula: 2/3, 1/1, 3/3, 3/3 = 38.

General. Recorded from much of Europe (in UK found mainly in Dorset, Wiltshire, Hampshire) but very uncommon throughout range. Fairly slow flight, wings held rather stiffly. Flies up to 15m above ground but generally low flight. Normally recorded only once or twice a year in Britain and probably a genuinely rare and localised species. Loss of old deciduous woodland probably responsible for decline; bat skeletons from 3000 year old deposits indicate Bechstein's Bat was then much more abundant and widespread than it is today.

NATHALINA BAT *Myotis nathalinae*

Name. Lesser Daubenton's Bat.

Recognition. Possibly slightly smaller than Daubenton's Bat, but positively identified by microscopic examination of teeth and genitals.

Habitat. Probably similar to Daubenton's Bat, living in riparian habitats and lake regions, in lowlands, roosting in buildings, tunnels, caves and trees.

Habits. Nocturnal. FOOD: Probably feeds over water.
Measurements. As for Daubenton's Bat.
General. Unknown to science until 1977, previously classified with Daubenton's Bat. Differentiation of the two species remains problematic and it is widely believed that this is not after all a separate species.

GREATER MOUSE-EARED BAT *Myotis myotis* Pl. 16

Recognition. Large (largest British bat). Short dense fur, base of hair brown. Light grey-brown above, sometimes with rusty tinge. Underside white-grey. Short, broad nose; nose, ears and wing membranes pinky-brown, face almost bare and pinkish-brown. Thick long broad ears, outer edge with 7–8 transverse folds, front edge distinctly bent backwards, ear tip broad. Tragus broad at base, becoming narrow and pointed nearly halfway up ear. Wings broad, lateral membrane starts at base of toes. Calcar reaches half length of tail membrane, with narrow keel and no post-calcarial lobe. Last tail vertebra protrudes beyond membrane. Similar to Lesser Mouse-eared Bat. Slow, direct flight with 'rowing' wing beat. Forages in parks, fields and meadows, also near housing, 0–10m above ground.
Habitat. Open, lightly wooded country, parks and built up areas (especially in the north). Prefers warmth: house-dwelling in the north, cave-dwelling in the south. Summer: nurseries in north in warm attics, in church towers (temperatures up to 4–5°C), very occasionally in warm rooms below ground. Individuals also in nest boxes or tree holes. Winter: caves, mines, cellars. Temperatures 7–12°C. Usually hangs free, but often in protected space in wide gaps or cavities – rarely in narrow crevices. Often clusters. In past, many thousands of animals in some hibernacula, now rarely more than 100. Females appear in hibernaculum before males. Preference for roosting deep in caves in early winter, moving to sites near cave entrance in spring.
Habits. Nocturnal; emerges late. FOOD: Majority of prey carabid beetles, also cockchafers, dung beetles, moths, caught in flight; large, non-flying beetles, spiders, grasshoppers, crickets, taken from ground. Often actually forages on ground. HOME RANGE: Feed mainly in forested habitats (98% of time away from roost) and returns to particular sites which may be shared by other individuals. Occasional migrant: distance between summer and winter roosts in the north around 50km and movements of more

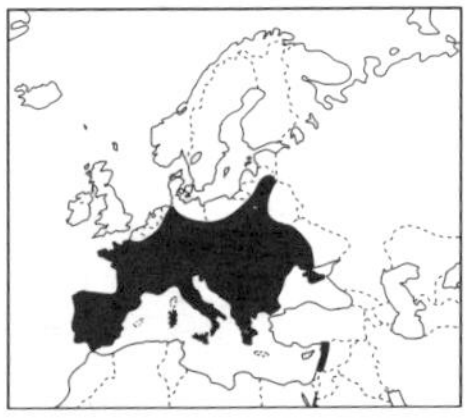

< *Greater Mouse-eared Bat.* Throughout continental Europe, except Scandinavia and British Isles.

> *Lesser Mouse-eared Bat.* Southern Europe with pockets in northern and western France.

than 100km not uncommon (largest migration 390km in Spain). Radio-tracking studies indicate that pregnant females spend more time foraging and possibly forage less efficiently. SOCIAL: Summer: nursery colonies up to several thousand females, males solitary. Winter: generally solitary or small groups. COMMUNICATION: strident shriek when disturbed during hibernation, deep bumble bee-like humming when threatened. In nurseries, loud shrill scolding and shrieking.

Breeding. Females visit solitary males for mating in autumn, each male having a harem of up to 5 females. Females form large maternity colonies in buildings or caves from March onwards (up to a maximum of 2000 females). Occasionally individual males in nurseries but generally live alone during this period. Births from June onwards. Sexual maturity: female 3 months, a small percentage reproducing in first year; male 15 months. Gestation 46–70 days. Mammae 2. YOUNG: Born pink with fine, almost colourless hair dorsally, weigh *c.* 6g. Eyes open 4–7th day. Weight 18–19g on 22nd day when hair cover is complete. Adult dentition complete 30–35th day. Can fly at 20–24 days, fully independent after 40 days (mid-July). Parental care by female only. Mother leaves baby in roost to go hunting on same day as birth (young generally born early morning); babies deposited in groups with some females remaining behind with them.

Lifespan. Max. 20 years (Britain); older individuals recorded elsewhere in Europe (28 years). Average 5 years.

Measurements. Head-body length: 65–89mm. Tail-length: 45–60mm. Forearm length 54–68mm. Wingspan: 350–450mm. Ear: 20–31mm. Condylo-basal length: 22–24mm. Weight: 28–40g. Dental formula: 2/3, 1/1, 3/3, 3/3 = 38.

General. Numbers declining in NW Europe, especially the Netherlands, Germany and Austria, but remnant populations have stabilised in many regions recently. Decline may be due to reduced habitat and food supply. Effectively extinct in Britain with only a single known individual (male) left in 1990. Never common. Only discovered in Britain in 1958.

LESSER MOUSE-EARED BAT *Myotis blythi (oxygnathus)* Pl. 16

Name. *Oxys* = sharp, gnathos = *jaw* (Gk.).

Status. Endangered throughout Europe.Verging on extinction in Austria.

Recognition. Smaller than, but very similar to, the Greater Mouse-eared Bat. Fur short with dark grey base to hair. Upper side light brownish-grey, underside grey-white. Nose seems longer than that of Greater Mouse-eared Bat due to being narrower and more pointed. Ears also narrower and a bit shorter with front edge less sharply bent back, tapering off to more of a point, outer edge with 5–6 transverse folds. Tragus base narrower, lancet-shaped; reaches nearly half length of ear. Wing membrane starts at base of toes, calcar reaches about half of tail membrane length and has a small keel. Slow, even flight – more agile in confined spaces than Greater Mouse-eared Bat.

Habitat. Prefers warm areas with tree and scrub cover (not too dense), limestone areas, parks, urban areas. Nurseries mainly in warm caves (often alongside Schreiber's and horseshoe bats), also attics, hanging free from roof rafters, individuals occasionally in tree hollows. Winter roosts in caves and mine tunnels, temperatures 6–12°C; mostly hanging free.

Habits. Nocturnal. Emerges late in the dusk or after dark. FOOD: Moths and beetles – probably takes prey from ground as well as in air. HOME RANGE: Occasionally migratory, maximum recorded distance 600km (Spain). SOCIAL: Males can have harems of females during mating season. Large nursery roosts of up to 5000 animals. COMMUNICATION: Similar to Greater Mouse-eared Bat: shrill scolding or deep humming if threatened.

Breeding. Mating in autumn and, probably, through until spring. Mixed nurseries with Greater Mouse-eared Bats have been encountered in attics and caves. Litter size 1. Litters per year 1. Mammae 2.

Lifespan. Max. recorded 13 years.

Measurements. Head-body length: 59–74mm. Tail-length: 46–60mm. Hind-foot length: 12–14mm. Forearm length: 48–61mm. Wingspan: 300–400mm. Ear: 19.8–23.5mm (width 13.6mm). Condylo-basal length: 17.2–18.5mm. Weight: 15–28g. Dental formula: 2/3, 1/1, 3/3, 3/3 = 38.

General. Occurs together with Greater Mouse-eared Bat, but details of ecological separation unknown. In evolutionary history, probably separated from Greater Mouse-eared Bats in the Pleistocene, i.e. during the last 2.5 million years.

COMMON PIPISTRELLE *Pipistrellus pipistrellus* Pl. 10

Name. *Pipio* = I squeak (Lat.); *pipistrellus* = little squeaking beast; *vespertilio* = bat (Lat., vesper = evening).

Recognition. Smallest European bat and one of the commonest. Nose, ears and wing membranes black-brown, base of hair dark to black-brown, dorsal side orange to rufous-brown, chestnut brown, or dark brown. Underside is yellow-brown to grey-brown. Ears short and triangular with a rounded tip, outer ear edge with 4–5 transverse folds. Tragus slightly curved inwards, blunt, longer than it is wide (c. 50% height of ear). Narrow wings, lateral membrane starting at base of toes. Calcar approximately one third tail membrane length. Distinct post-calcarial lobe with transverse T-shaped cartilage. Lower leg and tail membrane hairless. First upper premolar tooth is small and partly concealed by canine when viewed from side. Pipistrelle may feign death if disturbed during daytime torpor or hibernation (see vignette p.69).

Similar species: Nathusius' Pipistrelle (relatively larger 5th finger, and has hair on tail membrane); Kuhl's and Savi's Pipistrelle (usually larger, differences in colouring and tooth characteristics, Savi's also with different tragus shape); Whiskered Bat (ear and tragus totally different). Generally hunts 2–10m above ground. Flight fast and jerky as it dodges about in pursuit of insects, often along fixed flight paths. Some people can detect parts of their high pitched, normally inaudible calls which are very different from echolocation cries. These are social calls and although heard throughout the year they are most obvious in autumn when males perform some flights for display purposes. Young people, being more sensitive to higher frequencies, are more likely to hear these calls. Sometimes plenty of droppings below access to roost at height of summer activity, with characteristic smell at close quarters. Oily smears on surfaces close to roost entrances, sometimes abrasion of landing area.

Habitat. Predominantly a house-dwelling species, but otherwise all habitats excluding very exposed areas. Most common urban species. Frequently feeds over water or water meadows, along hedgerows or edges of woods, also common around gardens and streetlights. Has been recorded up to a maximum of 2000m above sea-level. Summer nurseries in very confined spaces around the outsides of buildings in wall crevices and behind noticeboards, eg behind board cladding, panelling, shutters, bat boxes and wall cavities. Roost site for nursery colonies often in housing estates on the periphery of cities or villages and usually below 600m above sea level. Each colony usually has several roosts and uses them according to ambient temperature and probably availability of food supplies. Winter roosts in buildings and trees (regularly flies until late November/December, i.e. later than most species and may fly (and feed) throughout winter). Only rarely found roosting in caves/tunnels in winter, especially in UK, although large numbers may be found in single caves on Continental Europe (eg 100,000 in a cave in Romania). Extreme cold weather forces them to change roost and at such times often found in houses or walls. Found in roof cavities.

Habits. Nocturnal. Leave roost *c.* 20 mins (2–35 mins) after sunset, although this may be dependent on the weather. In May-June they return between midnight and dawn, earlier on cold nights. June-August first activity peak after dark, second before dawn but normally intermittent activity throughout night. Active at feeding sites for up to 8 hours, but dependent on the weather. Females with young may make several feeding trips, particularly around dawn/dusk, and often return to the roost *c.* 60 mins after first emergence. FOOD: Small insects caught and eaten in flight; main food is midges and caddis flies, larger insects, lacewings and occasionally small moths. As many as 3000 insects taken in a single night. HOME RANGE: For colonies average 16 km^2 with colonies averaging 160 bats each. Females move an average of 1.8km from roost to feeding site during pregnancy, 1.3km during lactation. Most populations sedentary, with distances between summer and winter roosts about 10–20km.

Largest recorded movements from two females ringed in Germany: one was recovered 770km SW, the other 540km SE. In UK largest recorded movement is 69km. SOCIAL: In summer females found in nursing colonies from April/May (sometimes shared with Nathusius' Pipistrelle). Immature males sometimes also found in these nursery colonies. Irregular movements at each roost (especially in newer housing), as Pipistrelles often move between several sites within a small area (even with babies). Average membership of nursing colony is *c.* 80, but can exceed 1000. In contrast, males solitary or in small groups. Even during the nursery season, males occupy individually established mating territories and defend these against other males during mating season (end August-end November). Smell strongly of musk during this period. Females visit these mating roosts temporarily – one male will have up to 10 females. In winter all ages and sexes found singly or in small groups of 10–20 which change in composition, or in colonies of several hundred to 100,000. May hunt in groups of 10–20. Daily variation in numbers per roost in some instances. Breeding females have high roost fidelity – only 1% recaptured not in roost of origin. COMMUNICATION: Very noisy chattering and snuffling pre- emergence. Shrill scolding when disturbed. Social calls: undulating loud 'shout' between 35–18 kHz, emitted by males during mating flights.

Breeding. In Britain mating is late August–late November at well established 'mating' roosts. Thereafter females store sperm in the uterus until ovulation and fertilization in April, Births in June–mid-July (occasionally late May to August). Sexual maturity: females when 2–3 months old; males during the summer following their birth, together with majority of females. Gestation 44–80 days. Strong synchrony of births over 2 weeks within the nursery colony. Under adverse climatic conditions with lack of food, bats become torpid and hence extend period of gestation. Litter size 1 (occasionally 2). On average one young per two years per female. YOUNG: Dorsal side pinkish. Weight: 1–1.4g (new-born), 4–4.3g (day 30). Eyes open 3–5 days. Baby is fed solely on mother's milk (see vignette on p.66 for suckling position) – makes first flight within 3 weeks, can forage for itself by 6 weeks. Individuals of less than 12 months usually darker colour and greyish. Remains in nursery roost for 5 weeks. Weanlings may emerge before adults (due to greater need for food). Weaning age 3–6 weeks. Parental care by female only; mother and offspring remain in contact for extended period.

Lifespan. Max. recorded 16 years 7 months (Czechoslovakia), average 4 years. Estimated annual mortality of 64% for one population of adult females. Annual survivorship of territorial males in Sweden is 44%, 54% for females.

Measurements. Head-body length: av. 36–51mm (extremes 33–52mm). Tail-length: av. 23–36mm. Hind-foot length: 4–7mm. Forearm length: 28–35mm. Wingspan: 180–240mm. Ear: 9–13.5mm. Condylo-basal length: 11–12mm. Fifth finger: 36–42mm. Weight: 3.5–8.5g (5g at end of hibernation; August-October: male av. 6.3g, female av. 6.8g). Dental formula: 2/3, 1/1, 2/2, 3/3 = 34.

< *Common Pipistrelle.* Throughout British Isles, Europe and most Mediterranean islands.
> *Nathusius' Pipistrelle.* Widespread in C&E Europe, scattered populations in W Europe, with a few records from SE England.

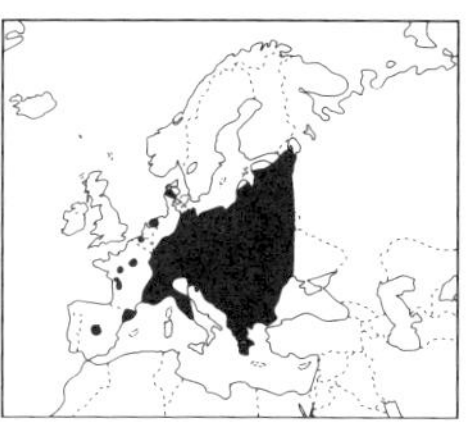

General. One of the most common bat species in northern and central Europe (commonest bat in Britain, despite 60% decline during 1980s). Some evidence of larger individuals in cooler regions (and smaller on Is. Rhum). Organochloride pesticides probably substantial mortality factor. As with all bat species, will fly at any time during the winter if warm enough.

NATHUSIUS' PIPISTRELLE *Pipistrellus nathusii* Pl. 10

Recognition. Small, though slightly larger than Common Pipistrelle. Upperside red to chestnut brown in summer, darker brown (often with obvious grey tips) after moult in July/August. Underside light brown to yellow-brown. Base of hair dark brown. Tail, ears and flight membranes brown-black. Ears short, triangular, rounded tip. Outer ear edge with 4–5 transverse folds. Tragus short, slightly bent inwards, blunt. Wings relatively broader than Common Pipistrelle, and longer. Tail membrane hairy on upper side up to half length and on underside along lower leg. Calcar reaches approximately one third of tail membrane, post-calcarial lobe smaller than that of Common Pipistrelle, with obvious T-shaped cartilage. Teeth conspicuously longer than Common Pipistrelle; first upper premolar is well developed, visible from outside. Feigns death like Common Pipistrelle (see vignette). Rapid flight, often deep wingbeats when flying in a straight line: more manoeuvrable than Common Pipistrelle in confined spaces.

Habitat. A woodland bat (both in damp deciduous woods and dry pine forests); frequently in riparian habitats. Often seen on farmland, parkland and along woodland edges, but less common near human settlement. Prefers lowlands. Summer roost: nursery roosts in hollow trees, tree cracks, crevices on hunting towers (where the erection of bat boxes or boarding has been very successful); occasionally in buildings. Sometimes shares roost with Common Pipistrelle or Brandt's Bat. Winter: cliff crevices, cracks in walls, hollow trees, rock fissures/caves.

Habits. Nocturnal. Emerges in early dusk. FOOD: Small to medium-large flying insects. Forages 4–15m above ground, along rides, paths and woodland edges. Also over water. HOME RANGE: Highly migratory; several records of more than 1000km. (max. 1600km from Russia). From mid-August–September the whole population in the north of eastern Germany migrates to France, Switzerland and western Germany, returning in April–May. The females migrate earlier in autumn than do the males. SOCIAL: Males have mating territories, with transient harems of 1–10 females – behaviour in mating territories similar to Common Pipistrelle. Female pipistrelles (and Noctules), like horseshoe bats, form a vaginal plug after mating. Common Pipistrelle males arrive at the mating ground about 2 months before onset of mating, whereas Nathusius' Pipistrelle males establish territories just before the start of the mating period. They perform song flights, and also advertise with calls while sitting in the roost. Females often return to nursery roosts of their birth, and have strong loyalty to their breeding area, although several changes of nursery site per summer are possible. COMMUNICATION: 'Sharp' whispering in roost (often heard from bat boxes). High-pitched scolding when disturbed (similar to Common Pipistrelle).

Breeding. Mating from early September–November. Males have noticeable increase in testes and epididymis size, and glandular swellings on the nose. Nursery roosts occupied in April–May (50–200 females). Births in second half of June–July. Mothers leave nursery roosts from late July onwards to seek mating roosts (as much as 15km away). Sexual maturity: females in first year, males in second. Litter size 1–2. Litters per year 1. Mammae 2. YOUNG: Newborns are pink, weighing 1.6–1.8g, 10th day = 3.2–4.5g, 20th day = 5.2g. Eyes open on 3rd day, can fly at 3 weeks. Juveniles dark grey-brown. Parental care by female only.

Measurements. Head-body length: 44–58mm, average 46–55mm. Tail-length: 30–44mm, average 32–40mm. Hind-foot length: 6.5–8mm. Forearm length: 31–37mm. Wingspan: 220–250mm. Ear: 10–14mm. Condylo-basal length: *c.* 14mm. Fifth finger: 43–48mm. Weight: 6–15.5g. Dental formula: 2/3, 1/1, 2/2, 3/3 = 34.

General. About 10 records of this bat in Britain (a vagrant) but several from oil rigs in the North Sea and one from Channel Isles. Recent research (studying bat boxes) in the Netherlands has greatly increased the number of records of this species. Occurs eastward to the Urals.

KUHL'S PIPISTRELLE *Pipistrellus kuhlii* Pl. 10

Recognition. Small. Fur colouring very variable. Dorsal: medium brown to light cinnamon brown, also yellow-brown. Underside light grey to grey-white. Base of hair dark brown. Ears, flight membranes and nose dark to black-brown. Short, roughly triangular ears, rounded at the tip and with 5 transverse folds on outer edge. Tragus not enlarged at tip, rounded, slightly bent inwards. Relatively narrow wings, lateral membranes start at base of toes. Calcar with post-calcarial lobe divided by T-shaped cartilage. First upper incisor tooth with 1 point, second upper incisor very small, first upper premolar set inside tooth row and not visible from outside. Conspicuous glands in mouth corners. Similar species: Savi's Pipistrelle (tail-tip free of membrane by 3–5mm, tragus shorter, and wider in upper part, second upper incisor has two points). Rapid, agile flight.

Habitat. Quite closely associated with human settlements, but also in areas of limestone. Lowlands. Summer (nursery roosts) predominantly in crevices on and in buildings. Single animals also in cliff cracks. Winter: crevices in cliffs, and cellars.

Habits. Nocturnal. Emerges in late dusk or darkness. FOOD: small flying insects. Forages at low/medium height above ground around street lamps, over water surfaces in gardens. SOCIAL: Small nursery colonies (c. 20 females).

Breeding. Sexual maturity: females in first year. Litter size 1–2. Litters per year 1. Mammae 2. YOUNG: Parental care by female only.

Lifespan. Max. lifespan 8+ years.

Measurements. Head-body length: 40–47mm. Tail-length: 30–34mm. Hind-foot length: 5.5–7mm. Forearm length: 31–36mm. Wingspan: 200–240mm. Ear: 12–13mm (width 10mm). Condylo-basal length: 12–13.2mm. Weight: 5–10g. Dental formula: 2/3, 1/1, 2/2, 3/3 = 34.

General. Known to be spreading northwards. Habits thought to be similar to Common Pipistrelle. Kuhl's Bat makes a 'feeding buzz' i.e. an increase in call repetition rate when prey are detected. A study of hunting around street lights showed that over 98% of flights were further than 1.5m from the light, although the greatest density of insects was within that distance. This is presumably because it is difficult to single out a single prey in the heart of the swarm. Although insects occur at every street light, 4–5 bats tend to group at one light. Bats in foraging groups of 4–5 made more feeding buzzes than bats in smaller groups. This advantage of social hunting seems to arise because the bats' feeding calls cause the moths to take evasive action, causing them to break away from the dense aggregation. It seems that the bats swoop on the light in parallel or linear formation, so that an insect swerving to avoid one bat falls prey to the next. However, there is no evidence that these hunting parties are socially stable; they may simply be formed by any passing individuals for short-term expediency.

<*Kuhl's Pipistrelle.* S Europe, including Adriatic coast south to Greece and including Crete.

>*Savi's Pipistrelle.* N coast of Mediterranean Sea, with an isolated pocket in Bulgaria.

SAVI'S PIPISTRELLE *Pipistrellus savii* Pl. 10

Status. Status largely unknown but possibly extinct in Austria and vulnerable in southern Germany.

Recognition. Small. Relatively long fur. Pelage varying from pale yellow-brown to dark brown with glossy gold tips above. Underside: pale white-yellowish to grey-white. Clear demarcation between dorsal/ventral fur. Base of fur black-brown. Ears and nose black-brown or black, flight membranes dark brown. Ears wider and rounder than other European pipistrelles, outer edge having four transverse folds. Tragus short, widening slightly towards upper part. Length of tragus edge almost equals greatest width. Blunt tip of tragus directed inwards at base of outer edge of ear, where there are two serrations, one above the other. Lateral membrane starts at base of toes, calcar with narrow post-calcarial lobe, 3–5mm of calcar tip free of membrane. First upper premolar tooth displaced (or occasionally absent) and not visible from outside; first upper incisor has two cusps. Similar to Common and Nathusius' Pipistrelles, and Northern Bat (similar colouring but larger). Flight straight, not very quick, sometimes above houses and tree tops.

Habitat. Valleys of precipitous, rocky mountains (up to 2600m above sea level), fringes of alpine meadows and woods, also towns and villages, limestone areas. Roost: summer, nursery roosts in tree holes, rock crevices, houses and outbuildings; winter, low-lying valleys, in rock crevices, tree hollows, caves, buildings.

Habits. Nocturnal. Emerges shortly after sunset and forages almost all night. FOOD: Small flying insects. HOME RANGE: Probably migratory, furthest record over 250km. SOCIAL: Summer: nursery roosts with 20–70 females. Winter: hibernates alone.

Breeding. Mating in late August–September. Births in mid-June to early July. Litter size 1–2. Litters per year 1. Mammae 2. YOUNG: Parental care by female only.

Measurements. Head-body length: 40–54mm. Tail- length: 31–43mm. Hind-foot length: 6–7mm. Forearm length: 30–38mm. Wingspan: 220–250mm. Ear: 12–15mm. Condylo-basal length: 11.9–13.6mm. Weight: 5–10g. Dental formula: 2/3, 1/1, 2/2, 3/3 = 34.

LEISLER'S BAT *Nyctalus leisleri* Pl. 12

Name. *Nyx, nyctos* = night (Gk.); *ala* = wing (Lat.)

Recognition. Medium-sized bat, smaller than Noctule. Long fur, hair bi-coloured: black-brown at base, dorsal side rufous brown, underside yellow-brown. Face, ears and wing membranes black-brown. Ears broad, triangular, rounded tip, outer edge with 4–5 transverse folds, much broader at base. Tragus mushroom-shaped, short. Ears extend more than halfway to nose tip when folded, nose slightly more pointed

<*Leisler's Bat.* S Europe, with pockets in Iberian peninsula, England and Ireland.

>*Noctule.* C&W Europe, England, Wales, S Scandinavia, with pockets in Spain.

than Noctule's. Wings long, narrow, with dense hair covering membranes along body and behind forearm on underside of wing. Calcar reaches half length of tail membrane; wide postcalcarial lobe with visible T-shaped piece of cartilage. Relatively small lower incisors, the first upper premolar is larger than that of the Noctule and visible from side view. Similar species are Noctule (larger, uniform hair colour) and Nathusius' Pipistrelle (smaller, tragus not mushroom-shaped). Fast, high flight over trees and woodlands, often with dives.

Habitat. Woodlands, similar to Noctule. Up to 1920m (in Alps) in summer. Summer roosts: holes and crevices in trees and buildings; found mostly in attics, on or close to the wall plate; also bat boxes (sometimes with Noctules). Winter: tree holes and cavities on and in buildings.

Habits. Nocturnal. Emerges shortly after sunset, occasionally flies by day (predominantly diurnal on Azores, in absence of raptors, where it is the only species of bat). Two nightly activity periods of *c.* 1 hour each, second ending *c.* 30 mins before dawn. Hibernates from end September–beginning March. FOOD: Medium and large insects eaten on the wing. As with other bats, will feed throughout winter if conditions are suitable. HOME RANGE: Migratory, longest recorded movement being 810km. Holds record for long range movement in Britain (one was found 200km from where it had been ringed four years previously). Movements probably from NE to SW. SOCIAL: Summer colonies of usually 20–40 bats, but up to several hundred breeding and immature females separate from males. Adult males very territorial in autumn, and establish mating roosts which females visit. Harem of up to 9 females found at males' mating roost. COMMUNICATION: Short, shrill calls, similar to Noctule.

Breeding. Similar to Noctule. Mating in late August–September. Maternity colonies formed in April: 20–50 females in tree holes, up to 500 females in buildings; change roost site frequently. Births in June–July. Litter size usually one in the west, although twins regularly recorded. Litters per year 1. Mammae 2. YOUNG: Weaning age *c.* 6 weeks, when can forage for themselves. Parental care by female only. Mother leaves young in a creche while she feeds, but will carry young if changes roost site during lactation.

Lifespan. Max. recorded 9 years, although likely to live longer.

Measurements. Head-body length: 48–68mm. Tail- length: 35–50mm. Hind-foot length: 7–10mm. Forearm length: 38–47mm. Wingspan: 300–

340mm. Ear: 12–16mm. Condylo-basal length: 14.7–16mm. Weight: 11–20g. Dental formula: 2/3, 1/1, 2/2, 3/3 = 34.
General. Found nearly all over Europe, but with only sporadic records except for Ireland, where it is common and widespread. Possibly under-recorded in England, as several new roosts have been found there over last few years.

NOCTULE *Nyctalus noctula* **Pl. 12**

Name. *Nox, noctis* = night (Lat.).
Recognition. One of Europe's largest bats. Fur short, smooth and uniform colour: rufous brown above (and glossy in summer), duller, lighter brown below. Ears, nose and wing membranes dark brown. Ears as for Leisler's Bat but perhaps somewhat broader. Heavily used tree holes sometimes marked by streak of bat urine and faeces. Flight is fast (up to 50km per hr), high (10–40m and up to 70m or more above ground) and generally straight with fast turns and dives. Flight silhouette has long, slender, often distinctly angled wings and wedge-shaped tail membrane, the latter with a smooth rear edge if spread out.

Habitat. Various, but especially in woodland and larger parks: characteristic bat of urban parks and gardens. Predominantly in lowlands but up to 1920m (Alps) during migration. Summer roost: tree holes made by woodpeckers or where trunk has rotted or split, at 1–20m above ground. Numbers per hole may change. Occasionally buildings, rock crevices, street lights (concrete), bridges, bat boxes. Winter roosts: well insulated tree hollows, rock crevices, buildings (inc. wall cracks) and bridges. Up to 1000 individuals may roost together. Can tolerate 0°C for short periods.
Habits. Nocturnal and crepuscular. Often emerges before sunset, with intermittent activity throughout the night with up to 3 foraging flights per night: forages mainly at dusk for up to an hour, and at dawn for half an hour or so, somtimes well into morning daylight. Hibernates from beginning of October–mid-November to mid-March–beginning of April. FOOD: Large insects, caught and eaten in flight, although occasionally prey taken from the ground; also, small flies such as midges; may forage over rubbish dumps for crickets. Forages above meadows, lakes, refuse tips, tree tops. HOME RANGE: 1.5–2.5km radius. Each individual seems to have a favoured feeding area; one radio-tracked male travelled 10km to a feeding site each night. Roosts used by one individual may be spread out within an area of about 185ha. Western Europe: many individuals hibernate close to summer roosts (but some move up to 900km). Eastern Europe: migrations of 500–1600km (20–40km per day), to spend summer in Russia before migrating south in winter. May travel during day, and has been observed hawking for insects together with swallows. SOCIAL: Summer (from April onwards): males solitary if holding mating

roost, or small groups if not breeding. Females begin to form maternity groups in mid-May, changing roost site frequently. Usual size of colony, 20–40 bats, but can be 70 or more. Autumn: males in territorial mating roosts visited by 4–5 females (occasionally up to 20), each female remaing for a couple of days. Winter: hibernate in mixed sex roosts of up to 1000, forming clusters where they often sit on top of each other, like roof tiles. COMMUNICATION: loud metallic sounding calls ('zick' or 'bick') in flight can be heard up to 50m away and are almost painful if the listener is close. Male display is made from trees. Near roosts, loud scolding or twittering heard, often before emergence, and even in winter.

Breeding. Mating in August–October (sperm storage by both sexes for up to 7 months). Births in June–July. Sexual maturity: some females in first autumn (3 months old) but majority, and all males, in second autumn (15 months old). Gestation 70–73 days. Litter size usually 1 in UK, twins regularly recorded elsewhere in Europe, rarely 3. Number of twin births lowest in SW, increasing in N. Litters per year 1. Mammae 2. YOUNG: Born naked, pink, weight 7.5g (36th day = 15g). Eyes open at 3rd–6th day, at 14 days completely covered with grey to silvery-grey hair, turning brown by 36 days. Can fly from 4th week, independent when adult teeth developed from 5–7th week. First year juveniles much darker, duller coloration than adults. Weaning age about six weeks. Parental care by female only. Mother leaves young in a creche when she goes to feed. Carries young if changes roost site during lactation period.

Lifespan. Max. recorded lifespan 12 years, although likely to live longer. Some mortality due to low winter temperatures with evidence of up to 50% of animals freezing to death in unsuitable hibernacula.

Measurements. Head-body length: 60–82mm. Tail- length: 41–60mm. Hind-foot length: 10–12mm. Forearm length: 47–58mm. Wingspan: 320–450mm. Ear: 15–21mm (width: 14–17mm). Condylo-basal length: 17.4–19.9mm. Weight: 19–40g. Dental formula: 2/3, 1/1, 2/2, 3/3 = 34.

General. Declining everywhere, but still common in woodland and pasture areas. Can survive –4°C in hibernation. Starlings may expel (and kill) Noctules from roost holes. Feed throughout winter if conditions are suitable.

GREATER NOCTULE *Nyctalus lasiopterus* Pl. 12

Name. *Lasios* = hairy (Gk.); *pteron* = wing (Gk.).

Recognition. Only distinguished from Noctule by larger size.

Habitat. Woodland, often deciduous, up to 1920m (Alps). Summer (nursery) roosts and hibernacula in tree holes and especially rock crevices. Summer roosts may be shared with Noctule, Nathusius' Pipistrelle or other pipistrelle bats.

Habits. Nocturnal. HOME RANGE: Migratory, moving

< *Greater Noctule.* SE Europe, north of the Black Sea, pockets on Adriatic coast and N Italy, W France, Spain and Portugal.
> *Serotine Bat.* C&W Europe north to S Scandinavia and S England.

SE in autumn in at least part of its more northern range. SOCIAL: Small nursery roosts (up to 10 females) often shared with other species.
Breeding. Births recorded from late July. Litter size 1–2. Mammae 2. YOUNG: Newly born weight 5–7g. By about 40 days, fur like adults' and can fly.
Measurements. Head-body length: 84–104mm. Tail-length: 55–66mm. Hind-foot length: 12–14mm. Forearm length: 62–69mm. Wingspan: 410–460mm. Condylo-basal length: 22–23.6mm. Weight: 41–76g. Dental formula: 2/3, 1/1, 2/2, 3/3 = 34.
General. Largest species of bat in Europe, but distribution only partly known from occasional records. Appears to be very rare throughout its range. Colonies known in Spain. Biology largely unknown but presumed to be similar to the Noctule bat. Dental characteristics to identify the species are the larger crown area of the second upper incisor and the extremely small first upper premolar which is invisible from the sideview.

SERIOTINE BAT *Eptesicus serotinus* Pl. 11

Name. *Epten* = flying (Gk.), *Oikos* = home (Gk.); *serotinus* = late (Lat.).

Recognition. Dorsal fur dark brown, hairtips sometimes slightly golden, shiny with oils. Underside yellow-brown (blurred border of dorsal/ventral colours). Fur long, base of hair dark brown. Ears and nose black, wing membranes dark brown-black. Broad wings with lateral membrane starting at base of toes, 5–8mm of tail tip free; calcar reaches *c.* ⅓–½ of tail membrane length, narrow post-calcarial lobe without visible T-piece. Ears relatively short and triangular; outer edge narrow with 5 transverse folds, ending shortly before corner of mouth. Tragus up to ⅓ length of ear, blunt, concave front edge, convex rear edge. Not easily confused with other species, but cf Northern and Parti-coloured Bats (smaller, different dorsal colouring and Parti-coloured has mushroom-shaped tragus like Noctule): Noctule (rufous-brown, tragus mushroom-shaped), Greater and Lesser Mouse-eared Bats (dorsal side pale grey, ear and tragus longer and of different shape). One of the largest British species. Access to roost in building often at/near the gable apex or the lower eaves, and droppings are usually present in large numbers at the gable ends or around chimney base. Often much squeaking before emerge at night. Droppings large and relatively short and broad. Slow, erratic, fluttering flight with steep, fast dives, at 0–10m above ground. Flies in large loops in gardens, along woodland edges, above rubbish tips, around street lights. Broad wings hardly angled and tail appears rounded, not wedge-shaped.

Habitat. Lightly wooded parklands and grass (including urban). Feeds over pasture, parkland, along woodland edges (sometimes inside wood) and tall hedges, and along roads. A house-dwelling bat. Summer (nursery) roosts mainly in buildings. Often in ridge timbers or hollow walls, usually behind boarding or beams, hang free in roost. Rarely in trees. Individuals (usually males) also in gaps between roof beams or behind facia boards and weather boarding, rarely in nest or bat boxes. Very few serotines found in winter, but likely that most hibernate in buildings. Large-scale winter roosts unknown: usually solitary males found, either wedged in cracks or hanging free on roofs or walls, also in gravel on the ground. Very rarely found in the coldest parts of caves, since prefer hibernation temperatures of 2–4°C with humidity relatively low. Sometimes seen hunting on warm nights in winter.

Habits. Nocturnal. Emerges 15–20 mins after sunset; most of colony emerges within 10 mins – total emergence time rarely exceeds 40 minutes. Early in season, return after about 30 minutes, circling roost before entering. More time spent away from roost as season progresses and there may be a secondary peak of activity before dawn. Sometimes uses a night roost for rest while foraging. Hibernates November–end March/April. FOOD: Earlier in year, feeds on a variety of flies and moths, later takes larger beetles (including chafers) and moths. Mostly caught and eaten in flight, but sometimes lands on foliage with wings outstretched. Most food caught within 2km of roost. HOME RANGE: Generally within 5km of roost. Up to 300km recorded between winter and summer roosts. SOCIAL: Summer (May onwards): nursery roosts of females usually contain between 10–50 bats, but some of up to 100 and one roost of 200 recorded in S. England; adult males usually solitary or in small groups, but occasionally found with females in spring or autumn. Maternity colony usually disperses early September, until early October (in some roosts). COMMUNICATION: Frequently active 'squeaking' in roost. Loud, high chirping or 'tsicking' calls when alarmed.

Breeding. Mating in September–October, nursery roosts occupied from April/May. Mating may last for several hours. Births normally from the second half of June, but occasionally through to mid-August. Sexual maturity of both sexes in first summer. Litter size 1. Litters per year 1. Mammae 2. YOUNG: New born weigh 5.2–6.2g. Eyes open at 3–4 days. Make first flight within 3 weeks, and can forage for themselves within 6 weeks. Juveniles up to 12 months very dark, almost black. Nursery roosts disperse at the end of August. Weaning age 6–7 weeks. Parental care by female only.

Lifespan. Max. recorded 19 yrs 6 months.

Measurements. Head-body length: 58–80mm. Tail- length: 46–57mm.

Hind-foot length: 9–12mm. Forearm length: 48–55mm. Wingspan: 320–380mm. Ear: 15–20mm. Condylo-basal length: 19–22mm. Weight: 15–35 g. Dental formula: 2/3, 1/1, 1/2, 3/3 = 32.

General. Found throughout most of Europe and may be one of few species actually expanding its range, although declining in abundance. One of Britain's less common species but locally abundant in S.E. England. Roost building sometimes shared with pipistrelle and/or Long-eared Bats. Also known to roost with Natterer's, Whiskered and Noctule Bats. Particularly threatened by chemical woodworm treatments, disturbance and loss of nursery roost sites.

NORTHERN BAT *Eptesicus nilssoni* Pl. 11

Recognition. Medium-sized. Dorsal fur: long and rather shaggy, the hairs having a dark brown base irregularly tipped with light, glossy yellow-buff or golden-buff. Underside is yellowish-brown, only sharply demarcated with the darker dorsal fur on the neck. Nose, ears and wing membranes black-brown. Ears relatively short, outer edge with 5 transverse folds, getting broader towards base and outer edge nearly reaches corner of mouth. Tragus short, broad, slightly bent towards inside, rounded tip. Calcar reaches about ½ length of tail membrane: narrow post-calcarial lobe. 3–4mm of tail vertebrae free. Lateral membrane starts at base of toes. Similar to Savi's Pipistrelle and Parti-Coloured Bat. Rapid, agile flight with fast turns.

Habitat. Predominantly foothills of mountains or medium-high mountainous regions (up to 2290m in the Alps in summer, in winter 2200m in caves), in areas of fairly open scrub, woodland, farming villages. Summer (nursery) roosts: usually in or around buildings, especially houses covered with slate or sheet metal, behind chimney cladding, shutters and in roof timbering cracks. Single animals also in tree holes and woodpiles. Winter: caves, mine tunnels, cellars. Preferred temperature 1–5.5°C. Hides among boulders on the ground, hangs free from walls or roofs, or wedges itself into a crevice. Does not form clusters.

Habits. Emerges early, two hunting flights per night at dusk and dawn. Frequently flies in daylight in spring and summer. Hibernates from October to March or April, with regional differences. FOOD: Flying insects. Small flies, also beetles, moths, crane-flies and caddis flies. Prey in size range 3–30mm. Prefers to forage over lakes rather than woods or farmland, and often forages around street lamps. May defend a feeding territory. Usually hunts in open areas, above water, or at tree-top height. Hangs from branches to rest. HOME RANGE: Not especially migratory. SOCIAL: 20–60 females in nursery roosts. COMMUNICATION: Sharp, loud pulses at 10 kHz, not for echolocation.

Breeding. Nurseries occupied from late April, births from second half

<*Northern bat.* Throughout Scandinavia and C Europe.

>*Parti-coloured Bat.* C Europe and S Scandinavia.

of June. Litter size 1–2. Litters per year 1. Mammae 2. YOUNG: Able to fly mid-to end July. Juveniles darker than adults, dorsal side without golden sheen, hair tips more silvery, belly grey. Parental care by female only.
Lifespan. Max. recorded 14.5 years.
Measurements. Head-body length: 48–70mm. Forearm length: 38–47mm. Wingspan: 270mm. Condylo- basal length: 14–15.5mm. Weight: 8–14g. Dental formula: 2/2, 1/1, 1/2, 3/3 = 30.
General. Widely distributed from eastern France to Central Europe and much of Scandinavia, including the Arctic Circle. Most northerly European bat – extending beyond Arctic circle. Vagrant to Britain – only one record (male, Surrey, 1987).

PARTI-COLOURED BAT *Vespertilio murinus* Pl. 11

Name. *Murinus* = of a mouse (Lat. *mus* = mouse).
Recognition. Medium sized. Long dense fur has distinctive 'frosted' appearance on the dorsal surface: base of hairs black-brown with silvery white tips. Ventral side white-grey, throat white – both contrast sharply with dorsal colouring. Ears, nose and wing membranes black-brown. Ears short, broad, slightly rounded, outer edge with four transverse folds. Ear extends with wide fold to below the corner of the mouth and them comes back up to it again. Tragus short and somewhat mushroom-shaped. Wings narrow, lateral membrane starting at base of toes, calcar longer than ½ length of tail membrane, distinct post-calcarial lobe 3.5–5mm (last two vertebrae) of tail free. Only European bat to possess two pairs of mammary glands, 4–5mm apart from each other. Similar species are Noctule and Leisler's Bats. Fast, high flight (10–20m above ground) in a direct line.

Habitat. Originally probably cliffs – now found in large cities on tower blocks, as well as the more usual wooded hilly areas and steppe regions (eg Alps) up to 1920m. Summer: mainly cracks in/around buildings, including large tower blocks in cities, fissures in rock. Winter: deeper caves, cellars, occasionally high in buildings, maybe hollow trees.
Habits. Nocturnal. Emerges in late dusk; hunts throughout night. Hibernates October–March. FOOD: Flies most important, but also moths and lacewings. HOME RANGE: Highly migratory (three long distance records: 360, 800, 850km), with movements from nurseries (mostly in northern

and eastern Europe) to hibernacula (mostly in south and west). SOCIAL: Hibernate in clusters. Nursery roost of 30–50 females. Males form large summer colonies of up to 250 animals or more. COMMUNICATION: rapid, loud, shrill chirp (4.3 per sec) used as a mating call in autumn.
Breeding. Mating from August, when males' testes are greatly enlarged. Births in late June to early July. Litter size generally 2, occasionally 3. Litters per year 1. Mammae 4 (2 pairs). YOUNG: Juveniles darker and more grey-black than adults, hair tips dirty grey-white, belly yellowish-white.
Measurements. Head-body length: 48–66mm. Tail-length: 37–44.5mm. Hind-foot length: 8–10mm. Forearm length: 39–49mm. Wingspan: 260–330mm. Ear: 12–16.5mm. Condylo-basal length: 13.9–15.7mm. Weight: 11–24g. Dental formula: 2/3, 1/1, 1/2, 3/3 = 32.
General. Thought to be rare throughout range except in Scandinavia. Vagrant to Britain (2 records this century in 1980s for mainland), also occasional sighting in North Sea (oil rigs and ships) and in the Shetlands. Highly migratory from Baltic to S. Europe. Normal distribution from S. Scandinavia, Poland and central Europe, to Switzerland and the eastern edge of France, and eastwards to the Pacific. Sites of known roosts are protected, but otherwise protective measures are difficult as little is known about the species.

HOARY BAT *Lasiurus cinereus* Pl. 9

Name. *Lasios oura* = hairy tail (Gk.); *cinereus* = ashy (Lat. *cinis* = ash).
Status. Vagrant – only few records in Europe.
Recognition. Mottled, frosted pelage. Dorsal surface of tail membrane covered in dense fur.
Habitat. Open or mixed country, well covered with foliage above but open 3–5m above ground. Edges of clearings. Mostly below 1200m but recorded at 4000m. Very rare in houses or caves.
Habits. Nocturnal. FOOD: Two specimens examined had eaten mosquitos, bugs, grass and a snake skin, but these were most likely starving, since Hoary Bats are thought to feed normally on flying insects, particularly moths. HOME RANGE: Migrant in USA. SOCIAL: In summer sexes are segregated; in autumn they roost together. COMMUNICATION: chattering in flight.
Breeding. Births in mid May-early July. Litter size 2. YOUNG: Weaning age at least 48 days. Young cling to mother by day; by night left clinging to foliage.
Measurements. Forearm length: 46–58mm. Wing-span: 380–410mm. Condylo-basal length: *c.* 17mm.
General. Vagrant from USA, specimens found in Iceland and Orkneys.

BARBASTELLE BAT *Barbastella barbastellus* **Pl. 17**

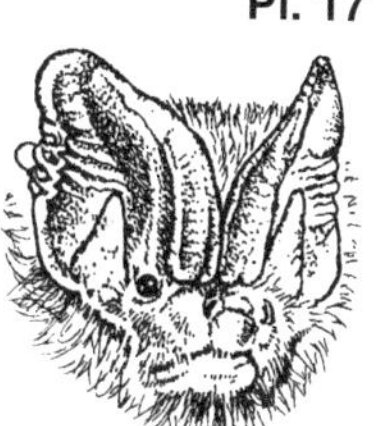

Name. *Barba* = beard (Lat.). *Barbastelle* = little beard.

Status. Vulnerable. Endangered or threatened with extinction over much of Europe. Very rare throughout Britain – no nursery roosts known.

Recognition. Medium-sized. Fur long and silky, base of hairs black. Dorsal side appears frosted due to black-brown fur having whitish or yellow-white tips. Underside dark grey. Naked parts of face and ears black, wing membranes grey-brown to black-brown. Ear conche wide, opening faces forwards, inner edges joined together across forehead; can fold forwards to 4mm beyond nose tip. Outer ear edge with 5–6 transverse folds, and roughly in the middle a distinct knob-like skin flap appears to be patched on, but can also be absent. Tragus triangular with long rounded tip and slightly more than half height of conch. Eyes small, gape of mouth very narrow, teeth small. Short, snub nose, nostrils opening upwards. Wings broad with pointed tips, short; lateral membrane starts at base of toes; calcar reaches roughly half length of tail membrane, the latter exceptionally large in area; very narrow post-calcarial lobe. Slow skilful flight. Hunts low over water, or at tree top height along woodland edge, in gardens and avenues. Rarely heard in flight as it is a 'whispering' bat producing very quiet echolocation sounds through nostrils. Behaves like Long-eared Bat.

Habitat. Wooded river valleys, foothills and mountainous regions, parks and areas of human settlement. Nurseries in roof spaces, cracks in buildings (often behind shutters); individuals also in tree holes, nest boxes or in caves near the entrance. Winter in caves, mines, trees. Cold-resistant, tolerating temperatures of 0–5°C; often found in open or near entrance to roost even in severe weather. Cryptic colouration gives the bat an advantage when roosting in exposed sites. Narrow cracks as well as hanging free from wall or roof, sometimes in large clusters.

Habits. Sometimes emerges before sunset, intermittent activity throughout night (probably two peaks of activity). Hibernates October/November to March/April. FOOD: Observed feeding on flies over water, also takes other small, delicate insects (moths, small beetles) often gleaned from foliage: cannot manage larger insects with hard chitinous shells because of small gape and weak teeth. HOME RANGE: Movements of up to

<*Barbastelle Bat.* W Europe, excluding parts of Denmark, Belgium, Netherlands, Scotland and Ireland.

290km recorded, but usually not this far: an occasional migrant. SOCIAL: Summer: sexes segregate, females in small nursing colonies of 10–20, rarely up to 100; males in small groups apart from nurseries. Winter: in large hibernating colonies up to 1000 individuals (eg W. Poland). Sex ratio of hibernating groups is male biassed (eg 68% males). COMMUNICATION: High-pitched chirping and sometimes also dull humming if disturbed.

Breeding. Mating in autumn, and sometimes in hibernaculum. Births from mid-June. Sexual maturity of females in 2nd year. Litter size 1, rarely 2. Litters per year 1. Mammae 2. YOUNG: Juveniles that can fly are somewhat darker than adults, and develop whitish tips to dorsal fur in their 1st year. Parental care by female only.

Lifespan. Max. 23 years.

Measurements. Head-body length: 45–58mm. Tail-length: 38–54mm. Hind-foot length: 6–7.5mm. Forearm length: 36–44mm (eg male: av. 38.8mm, female: av. 39.9mm). Wingspan: 245–290mm. Ear: 12–18mm. Condylo-basal length: 12.0–14.7mm. Weight: 6–13.5g. Dental formula: 2/3, 1/1, 2/2, 3/3 = 34.

General. Partial albinos (patches or flecks of white) quite common (2.1% of Czechoslovakian sample – especially affecting males). Usually alights head upwards but sometimes somersaults to hang by feet. Few (eg 3) records from Britain every year.

GREY LONG-EARED BAT *Plecotus austriacus* Pl. 17

Name. *Pleko* = 1 fold (Gk.) *ous, otos* = ear (Gk.) reference to obvious folds in ear; *austriacus* = auster (Lat.) = south wind.

Recognition. Medium-sized. Long fur with dark grey or black base to hair. Dorsal side grey with occasional slight brownish tint, ventral side white with black bases. Nose and upper lip dark chocolate-brown/black, grey 'mask' noticeable especially around eyes. Ears and wing membranes blackish. Ears like Brown Long-eared Bat, about 22–24 transverse folds, longer and more rounded. Tragus grey, not translucent and at least 5.5–6mm at widest point. Wings broad, lateral membrane starting at base of toe, calcar nearly half length of tail membrane. Thumbs, thumb claws and feet small: thumb less than 6mm. Penis club-shaped at end. Similar in general appearance to Brown Long-eared Bat and, but less so, Bechstein's Bat. Flight similar to Brown Long-eared Bat. Often hunts in open spaces, also around street lights. Uses feeding perches.

Habitat. Lowland, especially cultivated landscape or in warm valleys in highland, usually below 400m. Usually associated with human settlements in the north. Avoids larger woodlands. Nursery roosts in buildings, hidden in cracks and cavities on beams or visible on ridge timbers. Often share with Greater Mouse-eared and Lesser Horseshoe Bats. Winter roosts in

< *Grey Long-eared Bat* C&S Europe (including S England), eastwards to Himalayas. In the SW it overlaps with the > *Brown Long-eared Bat* which is widespread throughout C&W Europe, and British Isles.

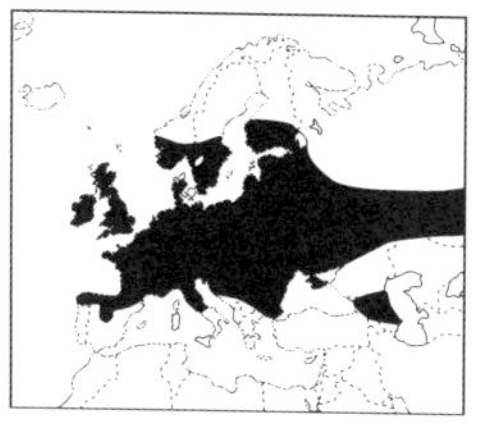

underground sites such as caves, mine tunnels (may share with Brown Long-eared Bat). Temperatures 2–9°C. Hangs free on walls more frequently than Brown Long-eared Bat but also found in crevices; usually solitary, occasionally 2–3 animals together.

Habits. Nocturnal. Emerges late, active intermittently throughout night. Hibernates September/October to March/April. FOOD: Moths, flies, beetles. HOME RANGE: Sedentary, distance between summer and winter roosts less than 20km. Furthest recorded movement 62 km. SOCIAL: Summer: segregation of sexes from June-September, 10–30 females in small nursery roosts, hanging separately or in small groups. Male territorial, especially in autumn mating season. Winter: sex ratio equal at hibernation sites. COMMUNICATION: chirping or humming calls when disturbed.

Breeding. Mating in autumn, births in mid-to late June. Sexual maturity: male 1 year; female 2–3 years. Litter size 1.

Lifespan. Max. recorded 14.5 years. Estimated annual mortality of 45% for males (predicted lifespan 5 years), 62% for females (predicted life span 9 years).

Measurements. Head-body length: 40–58mm. Tail-length: 37–55mm. Hind-foot length: 6–8mm. Forearm length: 37–45mm (male: 39.9mm, female: 41.4mm). Wingspan: 255–300mm. Ear: 31–41mm. Condylobasal length: 15–17mm. Weight: 7–14g (e.g. male: October 11.3g, January 10g, April 8g; female *c.* 1g heavier than male) Dental formula: 2/3, 1/1, 2/3, 3/3 = 36.

General. Widespread and abundant in southern Europe, but much rarer than Brown Long-eared Bat in central Europe and rare in north west (very rare in Britain). Recognised as separate species in Europe in 1960.

BROWN LONG-EARED BAT *Plecotus auritus* Pl. 17

Name. Common Long-eared Bat. *Auritus* = with ears (Lat.).

Recognition. Medium-sized. Long fluffy fur. Base of hairs dark brown; fur light brown dorsally and on border, ventral side on neck often lighter, yellowish-brown spot; ventrally light brown, sometimes with yellow tinge. Lips light flesh colour, nose and eye area light brown or pink, ears and wing membranes brown. Strikingly long ears (up to ¾ size of head-and-body length), ear conch thin with 22–24 transverse folds, inner ear-edge broadened with a fringe of fine hairs, knob-like projection near ear base. Sleeps with ears folded back, often tucked under wing, especially when hibernating, or curled back like ram's horns, erected only shortly before flying off and in flight. The long lancet-shaped tragus (half height of ear conch) projects forwards even when the ear is folded; tragus is pale pink (appears translucent) with light grey pigmentation at tip; tragus width less than 5.5mm. Eyes relatively large; muzzle bulbous at sides. Wings broad, lateral membrane starts at base of toes, calcar reaches about half length of tail membrane. Feet large, thumb and claws long; thumb length greater than 6mm. Similar in general appearance to Grey Long-eared Bat and Bechstein's Bat, but ears much longer than the latter. Feeding 'perches' – frequently inside porches or barns and indicated by the piles of insect remains, particularly moth wings, beneath them. Small accumulations of droppings beneath favoured roost sites, and sparse scattering of droppings generally over floor of roof space. Flight slow, fluttering and low; can hover and very agile in confined spaces.

Habitat. Sheltered, fairly open deciduous and coniferous woodland and parkland, well wooded gardens. Summer roosts (nurseries) in tree holes, bird and bat boxes and attics (more frequently visible in roof space than most other bats). Individuals also in caves, behind shutters, cracks in buildings. Winter roosts in buildings, underground sites, very occasionally in well insulated tree hollows. Often occupy same roost throughout year. Temperatures 0–5°C, resistant to cold, hence often nearer cave entrances than *Myotis* species. Wedges into cracks, down narrow pipes, occasionally hangs free from wall (wing membranes partially cover belly and chest). May form mixed associations with other species.

Habits. Nocturnal. Generally emerges in dark, but sometimes as early as 20 mins after sunset (especially in north). First flight lasts *c.* 1 hour, followed by intermittent flights throughout night, returning before dawn. After awakening may spend up to 75 mins grooming, having emerged from crevices in preparation for flight. Like Natterer's, often flys around in roost for some time before emerging to feed. Hibernate October/November to end of March/April. FOOD: Insects, mainly moths, but also beetles, large caddisflies, flies, bugs, taken in flight or more often gleaned from from foliage; also feed on earwigs, spiders, etc. Sometimes land on ground for catch. Small insects eaten on the wing, larger taken to a perch. HOME RANGE: Not known. Records of dispersal over 42km, but sedentary species with usually only a few kilometres between summer and winter roost. SOCIAL: Summer: females and some immatures segregate from adult males in nursery roosts in June, 10–50 or sometimes over 100 females. Young rejoin males after weaning in late August. Usually solitary in hibernation, very occasionally in small clusters of 2–3 animals. COMMUNICATION: Spring and autumn call sequences in flight similar to 'tsick-tsick'; deep chirping and humming in defence. Observations of captive animals have suggested visual perception of prey as well as listening. May

often hunt by passive listening for prey wingbeats (rather than echolocation) and its hearing is especially sensitive to low frequency sound.
Breeding. Mating normally in autumn, although can occur in winter or even spring. Nursery roosts occupied May/April. Births in June and July. Some females breed alternate years for first five years. Record of a female bearing young for 11 consecutive years. Breeding roosts may be used until October or even for the entire year. Significant numbers of males in maternity groups. Autumn invasions of up to 10 animals in living space of houses are juveniles looking for roosts. Sexual maturity: 75% of females at 2 years, 100% at 3 years; males at 1 year. Litter size 1, rarely 2. Litters per year 1. Mammae 2. YOUNG: Eyes open on 4th day, ears erected 11th day, able to fly mid- to end July. Juveniles have pale grey fur without brown tones, dark face. Fully weaned by six weeks. Mother feeds young solely on milk for first 3 weeks, leaving them in a creche when she goes out to feed.
Lifespan. Max. recorded 22 years, although average 4.5 years. Estimated annual survival rates of 75% for females and 54% for males.
Measurements. Head-body length: 37–53mm. Tail-length: 34–55mm. Hind-foot length: 6.5–10.5mm. Forearm length: 34–42mm (eg females: 38.3, males: 37.5 in south UK). Wingspan: 240–285mm. Ear: 29–41mm. Condylo-basal length: 13–15.6mm. Weight: 5–12g. Dental formula: 2/3, 1/1, 2/3, 3/3 = 36.
General. Common and widespread throughout Central and Northern Europe and in Britain (where it is the second commonest bat). Increases in size to north. Habit of landing on the ground makes them vulnerable to predators. Echolocation calls emitted through nose with mouth closed. Low amplitude echolocation pulses, hence term 'whispering bat' – probably a response to prey being able to detect louder calls. Most prey caught by listening to wing beats, hence big ears.

SCHREIBER'S BAT *Miniopterus schreibersii* Pl. 13

Name. Bent-winged Bat. *Minium* = bright red (Lat.); *pteron* = wing (Gk.).
Recognition. Medium sized with very short nose and domed forehead. Fur on head short, dense and erect. Dorsal fur short, grey-brown to ash-grey, sometimes slight lilac tinge. Underside lighter grey. Nose, ear, wing membranes grey-brown. Ears short, triangular, do not project above top of head, very far apart, 4–5 transverse folds. Tragus yellowish-white to faint grey, short, bent inwards and rounded at tip. Long, narrow wings. 2nd phalanx of 3rd finger about three times length of first phalanx: at rest 3rd and 4th fingers are bent inwards at the joint between 1st and 2nd phalanx. Lateral membrane starts at heel. Feet and tail relatively long, calcar reaches about 1/3–1/2 tail membrane length, no post-calcarial lobe. Easily distinguished

< *Schreiber's Bat.* S Europe from Iberian peninsula to Black Sea.

> *European Free-tailed Bat.* S Iberian peninsula around Mediterranean coast and W Italy, including Sardinia, Sicily and Crete.

from other European species. Flight very fast (50–55km per hr), reminiscent of that of swallows.

Habitat. Open, rocky landscapes (lowlands as well as up to 1000m), limestone areas. A cave dweller. Summer (nursery roosts): caves, but also buildings (especially large attics) in the north. Hibernates in caves from October–March: preferred temperatures 7–12°C, hanging free from roof or wall, sometimes in clusters. Hibernacula may be changed even in winter.

Habits. Nocturnal, emerging shortly after dusk. FOOD: Moths, gnats, beetles. Hunts at 5–20m above ground in open areas, often far from roost. HOME RANGE: Seasonal migration between winter and summer roosts, but generally not long distance (although max. recorded 350km). SOCIAL: Highly social. Summer: several hundred young in breeding caves, (females roost in clusters apart from the young). Summer roosts often number over 1000 females: up to 40,000 individuals in some instances (including young and males which are also using the nursery roost). COMMUNICATION: short, shrill scolding noises when disturbed. Within a resting group make deep 'whispering' noises (similar to calls of ducklings).

Breeding. Mating in autumn. Delayed implantation, unlike all other European bats. Sexual maturity of females in 2nd year (probably males similar). Gestation 8–9 months because of delayed implantation. Litter size 1. Litters per year 1. Mammae 2. YOUNG: Parental care by female only.

Lifespan. Max. recorded 16 years.

Measurcments. Head-body length: 50–62mm. Tail-length: 56–64mm. Hind-foot length: 9.5–11mm. Forearm length: 42–48mm. Wingspan: 305–342mm. Ear: 10–13.5mm. Condylo-basal length: 14.1–15.5mm. Weight: 9–16g. Dental formula: 2/3, 1/1, 2(3)/2, 3/3 = 36 (38).

General. Locally common in small areas of France and Switzerland. Vagrant in Germany, has become threatened with extinction in Austria. Often roost in association with *Myotis* species.

EUROPEAN FREE-TAILED BAT *Tadarida teniotis* Pl. 16

Name. *Tenuis* = thin (Lat.); *ous, otos* = ear (Gk.).

Status. Vulnerable.

Recognition. Very large. Fur almost mole-like: short, soft, fine. Dorsal side black-grey to smokey-grey with brownish sheen. Underside somewhat lighter grey. Ears, nose and wing membranes black-grey. Ears long

and broad, projecting forwards beyond eyes and face, their bases touching at the front; outer ear edge wider at approximately eye-height with an obvious, almost rectangular skin flap (anti-tragus). Muzzle long, upper lip with five folds, nostril front facing, eyes large. Tail membrane short, ⅓–½ of the tail protruding beyond the membrane, calcar without post-calcarial lobe. Wings very narrow and long, lateral membrane starting at heel. Smells strongly, described as reminiscent of a mixture of musk and lavender. Short, strong legs: can scurry about and climb in cracks. Unmistakable in Europe. High, fast, straight flight; needs wide open air space, and may hunt circling over water.

Habitat. Mountains with steep cliffs and gorges, also in areas of human settlement. Recorded up to 1920m (Alps). Summer roosts in rock crevices at foot of limestone cliffs and caves (crevices and on roof), and crevices on buildings. Hibernation sites unknown.

Habits. Nocturnal. Emergence usually late in dusk. Unknown whether or not this species hibernates for long periods of time. FOOD: Flying insects. Seems to ascend to passes to feed on migratory insects. Hunts intermittently in winter. HOME RANGE: Probably migratory or occasional migrant. COMMUNICATION: loud, sharp 'tsick' in flight, also whistles.

Breeding. Sexual maturity: females probably in 1st year. Litter size 1. Mammae 2. YOUNG: Young independent after 6–7 weeks. Juveniles greyer in colouration than adults.

Lifespan. Max. 10 years.

Measurements. Head-body length: 81–92mm. Tail-length: 44–57mm. Hind-foot length: 10.5–12mm. Forearm length: 57–64mm. Wingspan: *c.* 410mm. Ear: 27–31mm. Condylo-basal length: 20.9–24mm. Weight: 25–50g. Dental formula: 1/3, 1/1, 2/2, 3/3 = 32

Order Primates – Monkeys, Apes and Man

Origins: The primates make their first appearance in the fossil record in the Cretaceous in the form of *Purgatorius*, a rat-sized mammal found in the Rocky Mountains. The early primates resembled today's tree shrews. In Europe, a squirrel-sized, squirrel-like fossil prosimian, *Plesiadapis*, was found in the Palaeocene (*c.* 60 mya) deposits of France. Ancestral primates were arboreal, omnivores weighing around 150g, moving around on all fours and obtaining their food from the ground and in the lower levels of tropical forests. Their ancestry can be traced directly from the Insectivores.

Main features of the order: Primates are distinguished by their big brains, and tree-dwelling adaptations: 5 flexible digits on each foot, usually with nails not claws, sensitive pads on undersides of fingers and toes – designed for grasping objects with power and precision. They have mobile limbs, foreshortened muzzles and flattened faces, traits which indicate a decline in the importance of smell and increased reliance on vision. The eyes face forward giving stereoscopic vision, and they have colour vision. Primates breed slowly, and have a long period of parental care, delayed sexual maturity and an extended lifespan. They also have complex societies. They generally eat a high proportion of fruit and foliage and little animal matter. They range in size from 100g (Dwarf Bushbaby) to more than 100kg (Gorilla). Generally, primates are limited to lower latitudes due to their reliance on fruits, shoots and insects, all of which are scarce in a temperate winter. However, the macaques of Gibraltar are provisioned in winter.

Primates in Western Europe: Two families of primate occur in Europe: the Cercopithecidae (Barbary Ape) and the Hominidae (Man). These families are quite closely related; they are both Old World primates, or catarrhines, i.e. they have paired, downwardly directed nostrils, close together, usually only 2 premolars in each jaw, the anterior upper molars always with 4 major cusps (platyrrhines, the New World monkeys, have flat, well-separated, outward facing nostrils, 3 premolars per jaw and 3–4 major cusps in the anterior upper molars). Cercopithecidae first appeared in the Miocene. They all have tails, except for the Barbary Ape, which serve as a balancing organ. The face is generally bare, the palms and soles are naked, and the underparts sparsely haired. The buttock pads are often brightly coloured. The muzzle is elongated, rounded, and usually larger in males. Some genera (including the Barbary Ape) have cheek pouches. The forelimbs are shorter than the hindlimbs, the hands and feet grasping with flattened nails. Cercopithecine monkeys are mostly diurnal, with good sight, hearing and sense of smell and they use many facial expressions in communication. The dental formula is always 2/2, 1/1, 2/2, 3/3 = 32. Hominidae are quite similar to Cercopithecidae, though they are more closely related to the great apes (chimps, orangs and gorillas) i.e. the Pongidae. Mankind is the only surviving hominid. There are 52 modern genera of primates, 79 extinct genera and 181 modern species.

BARBARY APE *Macaca sylvanus* **Pl. 64**

Name. Barbary Macaque. *Silva* = wood (Lat.).
Status. Vulnerable

Recognition. No other non-human primate in Europe. Moults in summer, shedding thick winter coat.
Habitat. Rocky cliffs and scrub on Gibraltar. Caves and rock crevices.
Habits. Diurnal. FOOD: Omnivore: primarily fruit-eaters but their diet may include seeds, flowers, buds, leaves, bark, gum roots, insects, lizards, birds and mammals. Most food is caught and or gathered with thc hands. Within the troop foraging information is passed from adults to juveniles. SOCIAL: In the Barbary Ape, babies are the focus of a rich repertoire of behaviour among troop members of both sexes and all ages. This social interaction between babies and adults is notable because males of this species interact more with unweaned youngsters than do males of other Old World Monkeys. Males not only protect babies against predators but also undertake 'maternal' chores such as grooming and carrying infants. Males regularly emigrate from their natal groups at puberty. The group is centred around females who are the permanent core of the social unit. Kinship and rank clearly influence the nature of social interactions in the group. Barbary Ape groups usually have higher numbers of adult males than other macaques, up to 10 adult males.
Breeding. Most births in summer, but oestrous cycle one month long. Conspicuous swelling of sexual skin at oestrus (see vignette). Sexual maturity 3–4 years. Gestation *c.* 6 months. Litter size 1. Litters per year 1. Mammae 2. YOUNG: Weaning age *c.* 6 months. Independent at 1 year.
Lifespan. 25 years (in captivity).
Measurements. Head-body length: 60–70cm. Tail length: 45cm. Hindfoot length: 14.5–17cm. Shoulder-Height: 45cm. Weight: 11–15kg. Dental formula: 2/2, 1/1, 2/2, 3/3 = 32.
General. Artificially maintained on Rock of Gibraltar, population of *c.* 30–40. Fossils indicate widespread through Europe during Pleistocene 2.5 mya. Claims that some may have survived in Southern Spain until 1890s but unlikely to be correct.

<*Barbary Macaque.* Naturalised population on Rock of Gibraltar, originating from North African stock.

Order Carnivora

Origins: The Carnivora arose about 58 million years ago, and are distinguished by the possession of carnassial teeth (i.e. teeth with a specialised shearing edge to slice meat as would scissors). Other groups have evolved carnassial teeth but they all went extinct. Only the true Carnivora evolved these scissors out of their most rearward upper premolar and most forward lower molar. The Carnivores' success may owe much to adapting these two teeth, rather than others of their cheek teeth, which were left free to evolve a number of other specialisations which facilitated an omnivorous diet. Not all members of the Carnivora are carnivorous; many are omnivores and others, like the pandas and the Kinkajou, are vegetarian. The Carnivora are closely related to the Pinnepedia, which descended from them about 25 million years ago. Some authorities acknowledge this link by classifying the Pinnipedia as a suborder of the Carnivora; this is probably correct in evolutionary terms, but it makes for such a large and diverse assemblage that it is often convenient to treat them separately.
Main features of the Carnivora: Carnivores are very diverse in form and function, ranging in size from the 35g female Least Weasel to the 650kg Polar Bear. In addition to their carnassial teeth, carnivores tend to have strong jaws. Two bones in their feet are fused to form the scapholunar, and they have reduced clavicles (collar bones). They may walk on the flat of their feet (plantigrade bears) or on their toes (digitigrade foxes). A penis bone (baculum) is found in all species except for hyaenas.
Carnivores in Western Europe: Carnivores include at least 223 extinct genera, 93 modern genera, and 236 modern species. There are 27 European species in 15 genera, and of the eight families of Carnivore seven are represented in W. Europe (the hyaenas are missing). The Carnivora are divided into two super families, depending on the architecture of the bones of the inner ear (the auditory bulla). These are the Canoidea (also called the Arctoidea) which include the dogs, bears, raccoons and weasels, and the Feloidea (also called the Aeluroidea) which include the cats, hyaenas, civets and mongooses.

The Canidae or dog family originated in the Eocene. The first canids were long-bodied and short-limbed, much like their arboreal ancestors. They adapted for fast pursuit of prey in open grasslands, so they generally have lithe builds, long bushy tails, long legs and digitigrade 4-toed feet, with non-retractile claws. Other adaptations to running are the fusion of wrist bones (scaphoid and lunar) and locking of the front leg bones (radius and ulna) to prevent rotation. Males have a well-developed baculum and after mating animals can remain fixed together in a 'copulatory tie'. Gestation lasts about 8–9 weeks in most species, and they produce one litter per year. Parental care is highly developed, both parents bringing food to cubs and, in some species, it is regurgitated. Canids tend to be opportunistic and adaptable, and they include both solitary and social species. Senses of smell, sight and hearing are all very good. They communicate vocally, with facial expressions and with scent. Canids are generally active throughout the winter, except for the Raccoon Dog which undergoes a period of winter lethargy.

The Ursidae or bears evolved more than 30 mya. Early bears were small, but recent species are massive, with more rounded teeth, and include the world's largest terrestrial carnivores – the Grizzly and Polar Bears. Bears, within and between species, are larger at lower latitudes and males are much bigger than females in polygynous species. Modern bears have large, heavily built bodies, thick, short, powerful limbs and short tails (rarely more than 12cm). They have a plantigrade gait on broad, flat feet with 5 long, curved non-retractile claws, used while foraging or for climbing. They can run fast over a short distance. Bears are largely herbivorous, apart from the Polar Bear which is strictly carnivorous. The sense of smell is much their best sense, and they use few vocalizations and facial expressions in communication. Most species are promiscuous, male home ranges overlapping those of several females. They exhibit delayed implantation and 1–3 young are born nearly naked (except in the Polar Bear), helpless, very small (200–700g – adult bears average 270–650kg). Bears in cold regions are dormant in winter.

The Procyonidae or raccoon family have been introduced into Europe from the Americas. They are descended from the same lineage as canids and are small, long-bodied animals with long tails and often have distinctive markings. They have 5-toed feet and a plantigrade gait. Procyonids usually live around seven years in the wild, have a gestation of 2–4 months, producing litters of 3–4 poorly developed young weighing 150g. Only females provide parental care. Procyonids are generally nocturnal (except coatis) and generally solitary.

The Mustelidae includes the weasels, polecats, mink, martens, otters, skunks and badgers. This is a large, widely distributed, diverse family, whose representatives occupy most habitats, including fresh and salt water, on all continents except Australasia (though introduced to New Zealand) and Antarctica. They include the smallest carnivores – many are under 1kg – but Giant Otters and Wolverines may both be 1000 times heavier than the Least Weasel. Mustelids have long bodies, short legs, and are often skilful climbers. They have 5 toes per foot, and sharp, non-retractile claws. Males are generally larger than females (in weasels and polecats male body weights can be up to 120 % greater). Anal glands are important in communication and have developed into a form of defence in skunks and some Old World polecats. Most species are solitary. Induced ovulation appears to be widespread in mustelids, and at least 16 of the 67 species also show delayed implantation.

The Felidae or cats are the most carnivorous Carnivores. They have excellent night vision and hearing, and good senses of smell. Whiskers indicate an important sense of touch and are especially useful when hunting at night. Over three quarters of the felids are forest dwellers and agile climbers. They are mostly solitary, secretive, and live in inaccessible, remote places. A major threat to their survival is the fur trade.

Viverridae and Herpestidae are, respectively, the civets and the mongooses. They closely resemble the earliest of the Carnivora, the miacids, but they have some modern specialisations (highly developed inner ear). Most civets and genets look like spotted, long-nosed cats: they have long slender bodies, pointed ears and short legs, and tails equal to or longer than their body length: mongoose tails average half to three-quarters of their body length. Males are usually slightly larger than females. These families have excellent vision and hearing and are generally omnivorous. Civets and genets are generally spotted or striped, have pockets on the lateral margins of the ear flaps and a scent gland behind the genitals. Mongooses differ in many ways from civets. They vary less – all have long bodies, short legs, small rounded ears, 4–5 toes/foot, non-retractile claws, reduced or absent webbing between their toes, and rounded ears placed on the side of the head, which rarely protrude above the head's profile.

WOLF *Canis lupus* Pl. 19

Name. *Canis* = dog; *lupus* = wolf (Lat.).

Status. Vulnerable.

Recognition. Reminiscent of big Alsatian (i.e. German Shepherd dog), but shallower chest, broader head, thicker neck. Pelage: grey to greyish fawn with brown-reddish colour on the back and the head/ears (especially in Italy). Hairy ruff of hair on cheeks. Northern individuals may have reddish tinge on head, ears, shoulders and legs, but not on back. Moult in early summer. Footprint very difficult to distinguish from equal-sized dog, but normally longer (dog prints tend to be rounded). Pack travels single file in deep snow. Urine smell, scratch marks (10–20cm × 2.5–3.5cm) on ground from scraping with hind feet after urination.

Habitat. Adaptable to all terrestrial habitats. Generally open country with cover. Open woodland, tundra, dense forest and mountains (especially as refuge when persecuted). Den in burrow or cave, sited in dense cover, between tree roots, under rock, in cave. Sometimes digs own den, or extends burrow of other species (Red Fox or Eurasian Badger).

Habits. Largely nocturnal in Europe (due to persecution by Man). FOOD: Opportunistic, taking ungulates, lagomorphs, rodents, birds, carrion. Larger mammal prey includes Beaver, Elk, Reindeer, Roedeer, Wild Boar, Red Deer and domestic stock. Human garbage can be important element (for example in Italy). HOME RANGE: Pack may travel 100–1000km^2, depending largely on food supply. Generally territorial. Population density of 1 per 80km^2 in Italy, 1 per 50–60km^2 maximum. Adolescents of 1 to 2 years old tend to leave parents' territory especially if dominated by siblings and other pack members. SOCIAL: Variable, depending on food supply. Larger packs where large prey predominate (up to 30 in Alaska,

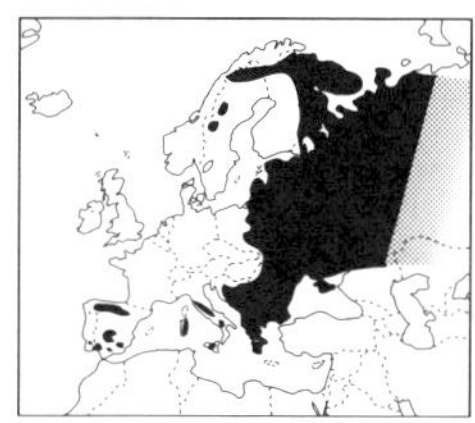

< *Wolf.* E Europe, with relict populations in Italy, Spain, Portugal and Sardinia.

> *Golden Jackal.* Spreading out of Middle East and Turkey into Balkans, Yugoslavia and N Italy.

where hunting Moose), small packs (less than 10), where feed on small deer; forage alone in territory of family group where food items very small (rodents, garbage). Pack dominated by 'alpha' pair; other members largely their adult offspring. COMMUNICATION: Vocalisations include howling for long distance communication and 'get-together'. Posture and expressions complex and similar to Domestic Dog. Urine marking with raised leg signifies dominance within pack and indicates territorial ownership. Droppings often deposited at trail junctions and similar vantage points. Smell excellent, hearing good, sight good at close range, only moving objects detected at long range.

Breeding. Mating February–April (later in north), births March–May. Sexual maturity at 2 years (but rarely breeds until older due to social constraints). Gestation 63 days. Litter size 3–7. One litter per year. Mammae 8. YOUNG: Weaned at 8 weeks. Both parents (and other pack-members) feed and tend pups. Male brings food to nursing mother. Young remain with parents for at least one year.

Lifespan. Max. 15–20 years in captivity, less in wild. Mortality high in pups. Causes of death include shooting, poison, rabies, parvovirus.

Measurements. Head-body length: 90–150cm. Tail- length: 30–50cm. Hind-foot length: 20–26.5cm. Shoulder height: 65–80cm. Condylobasal length: *c.* 22cm. Weight: male 20–60kg; female 18–50kg. Dental formula 3/3, 1/1, 4/4, 2/3 = 42.

General. Exterminated by Man in most of former European range (eg England and Wales *c.* 1500). Now relict populations imperilled by hybridisation with dogs. Highly cooperative species, hunting in packs and sharing care of pups. Widespread accounts of damage to stock and threat to humans grossly exaggerated. Ancestor of domestic dogs, with which produces fertile hybrids. Previously the carnivore with the largest natural distribution. Formerly common in whole of Europe including Britain and Ireland.

GOLDEN JACKAL *Canis aureus* — Pl. 19

Name. *Aureus* = golden (Lat.).

Recognition. Similar to Wolf, but smaller, and more tawny. Also, more slender than Wolf, shorter legs, smaller paws and bigger ears.

Habitat. Semi-arid country with cover (eg steppes, reedbeds, marshes), open country including grassland and farmland. Den is characteristically a burrow in dense cover.

Habits. Largely nocturnal, spending day in cover or den. FOOD: Omnivore, eating vertebrates (eg rodents, ungulate fawns, birds), insects, fruit (eg grapes), carrion and garbage. HOME RANGE: 0.5–2.5km^2 territories. Records (in East Africa) of individuals emigrating from family territory only to return several months later (presumably having failed to find territory of their own). SOCIAL: Basic social unit is mated pair, sometimes developing social group of adult offspring of both sexes. Maximum group size *c.* 30, but families of 5 more common. Cooperative hunting and care of young. Social hierarchy determines reproductive behaviour within group, with breeding generally confined to dominant pair aided by non-breeding helpers. COMMUNICATION: Vocalisations include howling, often at dusk, and associated yelps and barks. Postures and expression similar to Domestic Dog and Wolf. Mated pair urine mark in tandem; faeces left on visually conspicuous beacons (eg shrubs or boulders) or, in some populations, in piles near territory border.
Breeding. Mating early spring, births summer. Sexual maturity at 11 months (although reproduction may be postponed due to social constraints). Gestation 63 days. Litter size 1–9 (usually 2–4). One litter per year. YOUNG: pups remain in den for 3 weeks, during which mother spends most of her time with them. Weaned at 8 weeks.
Lifespan. Max. 16 years (captivity).
Measurements. Head-body length: 65–105cm. Tail- length: 20–24cm. Hind-foot length: 15–16.5cm. Shoulder-height: *c.* 50cm. Condylobasal length: *c.* 16cm. Weight: 7–15kg. Dental formula: 3/3, 1/1, 4/4, 2/3 = 42.
General. Recent sightings outside previous range in NE Italy.

ARCTIC FOX *Alopex lagopus* — Pl. 18

Name. Polar Fox. *Lagos* = hare, *pous* = foot (Gk.); *Alopex* = fox (Gk.), hence *alopecia* = baldness from fox-mange.
Status. Threatened with extinction in Fennoscandia, where it has total protection.
Recognition. Two phases exist, one is greyish-brown and turns white in winter, the other is the 'Blue Fox', which is brown throughout the year but more bluish in winter. Overall, Blue Fox is rarer (generally under 5% of the population) but predominates in coastal, less snowy, areas. Muzzle shorter and ears shorter and rounder than Red Fox, coat very thick; hairy undersides of paws. Considerably smaller (up to 25%) than Red Fox. Two moults: May–June and September. Moult from back down, males and non- breeding vixens often having remains of winter fur on tail and sides. Difficult to see other than at dens, but tolerant to Man's presence in areas with none to low hunting pressure. Fieldsigns include prints, faeces, uneaten prey remains. Warning calls from Ravens and other birds may call attention to the presence of Arctic Foxes. Fieldsigns similar to Red Fox.
Habitat. Arctic and alpine, mostly above the tree line. Also open tundra woodland in winter. Higher densities on coast. Several dens may be found

in a favourable area; most however not utilized at any one time. Occupied dens are generally more than 1km apart, usually distributed so one family every 32–70 km^2. Sometimes a very extensive burrow system with dozens of exits, which may have been used for centuries. Because of accumulated droppings and food remains, den area becomes covered with luxuriant vegetation and distinct from surroundings. Usually 4–12 entrances and a network of tunnels covering *c.* 30m^2. Site: rock or fissure or hole among boulder screes, or moraines and eskers for excavations. If suitable habitat missing, may breed in open (eg during years of peak population in Siberia).
Habits. Active by night and day (in daylight during the arctic summer). FOOD: Voles, lemmings, birds and eggs in summer. Carrion in winter; shellfish at coast in winter. Also small rodents, seabirds, Ptarmigan; may scavenge on Polar Bear kills; shore invertebrates, fruit and berries. HOME RANGE: 8–19 km^2. May migrate 100s km southwards or to the coast in winter, depending on food availability. May travel singly or, reputedly, in groups when dispersing. In coastal habitats of Iceland, mean distance of dispersal is approximately 25km. SOCIAL: Lives in pairs or small groups, generally composed of one adult male and several vixens with offspring, but many other combinations found including 2 males and 3 females (of which two bore litters and one was a helper). Female group-members probably relatives. COMMUNICATION: Vocalisations include hoarse barking, shrieks and yelps especially during mating season. Probably more vocal than Red Fox all year. Ear and tail important in postures: in submissive greeting, tail lashed from side to side in submission, ears folded back and lips retracted in 'grin'. Urine and faeces left by both sexes on conspicuous mounds and tussocks as scent marks throughout territory.
Breeding. Mating March–April, births May–June; may mate for life. Sexual maturity at 10 months. Gestation 53–54 days (in fur farms, gestation of less than 53 days or more than 54 days is regarded as pathological with high cub mortality). Litter size in Iceland is normally 5–6, but up to 9. In Fennoscandia, Soviet arctic tundra, litters of 10–12 are normal, but up to 18. One litter per year. Mammae 12–16 (usually not symmetrical). YOUNG: All born dark brown, eyes open at 14–16 days. White and blue phases can be distinguished at age 2 weeks, because of grey hairs inside ears and around snout of white foxes. Play and fight intensively – during food shortage may kill and eat siblings. Weaned at 4–10 weeks (see vignette of 6 wk old cub). Food brought by both parents from 3 weeks of age. Non-breeding helpers may also bring food. Most young disperse in autumn, but some (generally females) remain as helpers for next litter. Males defend den against attacking Red Foxes.
Lifespan. Max. recorded 11 years in wild (but up to 15 in captivity). Mortality approximately 50% among adults in Iceland, varies enormously elsewhere depending on rodent cycles. High juvenile mortality except during peak lemming years.

<*Arctic Fox.* N Scandinavia, Iceland and Greenland.

>*Red Fox.* Widespread throughout Europe; absent from Iceland.

Measurements. Wide geographical variation is size. Head-body length: male 55–75cm; female 50–65cm. Tail-length: 25–40cm. Hind-foot length: 11.4–12.5cm. Shoulder-height: *c.* 30cm. Condylo-basal length: male 119–144mm; female 112–131mm. Weight: 2.4–6kg, average 3.6kg (Birth 57g). Dental formula: 3/3, 1/1, 4/4, 2/3 = 42.

General. Buries food in carefully hidden caches. Rarely trots, usually gallops, a hindfoot often placed close to a front footprint in each group of bounding jumps; other times hindfoot set wide apart, similar to hare print. Remarkable cold-tolerance: starts shivering at -40°C. The protozoan *Encephalitozoon cuniculi* (nosematosis) may contribute to population regulation in Iceland. Harvested for fur – both captive and wild. Vector of rabies eg in Soviet Arctic. Susceptible to sarcoptic mange. In Iceland, distemper twice affecting population in 19th century. Reduced population in Fennoscandia due to overhunting and poisoning for Wolves, but population has not increased to former size despite protection, perhaps due to competition with Red Foxes.

RED FOX *Vulpes vulpes* Pl. 18

Name. *Vulpes* = fox (Lat.).

Recognition. Usually rich reddish- brown but variable colour (sandy beige to russet red). Ears erect and pointed with black backs. Tail long, thick and bushy; some individuals of both sexes have conspicuous white tip to tail and many have at least some white hairs at tail tip. Slender muzzle is white on upper lip and many have a black or brown 'teardrop' on the cheeks which is variably conspicuous. Many have white bib on throat, but some individuals dark steely grey verging on black under throat and belly (such individuals often have a white star on the chest). In breeding season underside fur of vixens has pink tinge. Black 'socks' on lower legs (rarely splashes of white on the lower legs). Ears longer than those of Arctic Fox. Wolf and Golden Jackal are larger and higher on the legs with a relatively shorter tail. Various colour morphs known in the wild; although the vaste majority of European foxes are variations on the red theme some 'silver' (i.e. black) foxes occur at high latitudes and are occasionally reported further south as are white Red Foxes (some are albino) which are superficially similar to the white morph of the Arctic Fox. Very occasional occurrence of specimens lacking guard hairs, thus appearing woolly, called Samson foxes. Eyeshine in torchlight generally blue or white, but reddish if viewed at an angle. Moult in spring is conspicuous, sometimes giving piebald appearance. New hairs grow first on lower legs then spread upwards, up to flanks by early July, back and tail by late August. In early winter the summer coat continues to thicken. In winter, dens ('earths') may be cleaned out so that fresh soil is in evidence, associated with the smell of fox (urine) and faeces on the spoil heap. Outside breeding dens uneaten prey remains (eg bird wings, rabbit feet, whole shrews and moles) may accumulate. When cubs are in residence the smell of rotting food within, and attendant flies, are additional clues (also occasionally high pitched, rhythmic murmuring whine of cubs). Spoil heap at den, plus muddy furrows at field borders, paths under fences and across ditches are all good places to spot footprints. These are more oval than those of most dogs (but similar to some sheepdogs), *c.* 5cm long, 3–4cm wide (hindfoot slightly smaller than forepaw, walking prints spaced *c.* 30cm apart). All four toe prints on each foot associated with pin-prick claw marks (which tend to register more markedly than those of hares, and less than the five toes on the broad feet of Badgers). In contrast to the more rounded tracks of most Domestic Dogs, the two central claws of each fox track often register close together, and fur between the pads leaves a stippled pattern on soft mud. In snow, the track of a trotting fox appears as a beaded straight line (in deep snow may use the same foot holes repeatedly on successive journeys to save energy). Barbed wire, brambles etc may snag hairs beside passageways (underfur has crinkly appearance, guard hairs banded black, reddish and white). Paths in mud or through tall grass tend to be narrower than those (otherwise very similar) of Badgers. Faeces often deposited on prominent sites, such as molehills, stones, tussocks of grass and at the junctions of trails. They are sometimes similar to Domestic Dog faeces, but generally darker, have a quite different smell and have more tightly-packed consistency, composed of visible wadges of undigested fur or feather, and, in season, traces of beetle elytra and fruit pips. Fur in fox faeces sometimes forms curly ends. May also contain lots of soil (from gut contents of earthworms, as may those of Badgers, but generally much less liquid than Badger faeces, and not associated with dung pits). Sometimes cache prey carelessly, leaving feathers etc protruding from soil. Characteristic treatment of prey, eg quills of feathers chewed through (cf. plucked by birds of prey), heads of poultry sometimes chewed off and cached separate from body, pelt sometimes skinned back over discarded limbs (eg Rabbit's leg or haunch with skin partially rolled back like glove off hand), or hedgehog eaten out to leave casing of spines (similar for Badger). Frequently blamed for killing lambs, often on inadequate evidence: typical injury on dead lamb includes tooth punctures over shoulders and crushed cervical vertebrae, wounds from upper canines spaced at *c.* 3cm, lower canines at *c.* 2.6cm (bruising indicates whether prey alive when bitten versus scavenged).

Habitat. Almost limitless adaptability to European habitats, including especially mosaic patchwork of scrub, woodland and farmland. Also abundant on moorland, mountains (above tree line), sand dunes, suburbs and cities. In UK cities most abundant in affluent suburbs characterised by detached houses with ample gardens. Large conifer plantations constitute good habitat while ground vegetation remains, but once mature are probably principally used largely for shelter. Den in burrow, either self-dug or in enlarged Rabbit hole or Badger sett (sometimes cohabiting with both these species). Den dug into bank, or in rocky crevice, drain or under garden sheds. Den often has several (eg 2–4) entrances. Dens used only sometimes, except by breeding vixens (and even they may occasionally give birth above ground in secluded spots). Signs of digging at a den does not guarantee a fox is within. Commonly spend day in cover (in rocky scree, woodpiles, under tree roots, in ditch culverts etc) above ground outside breeding season. Breeding den has no bedding.

Habits. Nocturnal and crepuscular, but more diurnal where undisturbed. In urban areas may postpone major activity until traffic density low late at night. FOOD: Opportunistic: rodents, lagomorphs, birds, insects (especially beetles), eggs, earthworms (caught above ground on warm, moist nights). Tends to reject shrews and moles, but eats hedgehogs. Scavenges in rural and urban areas, eg on sheep afterbirths, bird-tables, compost heaps and refuse. Fruit (windfall apples, plums etc) and berries (especially blackberries) in summer and autumn. Sheep carrion may be vital in Scottish highlands, as are deliberate feeding sites by some householders in urban areas. Food requirement: approximately 500g (120 kcal) daily. Caches surplus food (each item scattered into a separate cache). Probably originated as specialist rodent hunter, but highly adaptable. HOME RANGE: Great variation with extremes of 20–40ha in cities such as Oxford and Bristol (UK) and 4000ha in Scottish Highlands. Examples from agricultural land commonly 200–600 ha, with smaller ranges in more diverse, mixed farming landscapes. Upland ranges larger, eg English uplands *c.* 2000ha, mid-Wales up to 1500ha. Juveniles (6–12 months old) disperse October January inclusive. Dispersal distances may be huge (eg 250km), but commonly 5–10 home range diameters with a greater proportion of males than females dispersing and dispersing further, eg rural mid-Wales: males 14km, females 6.5km; urban Bristol: males 2.3km, females 0.8km. Proportion of each sex dispersing varies between habitats and perhaps depends on extent of mortality (eg rabies and Man); all or most males invariably disperse, but variations between habitats in proportion (30–80%) of females dispersing. Population densities similarly variable, with an average of one family per km^2 of farmland as an approximate rule. However, extremes vary from 5 families (perhaps 20 adults) per km^2 in some suburban settings to one family (perhaps two adults) per 40 km^2 in barren uplands. SOCIAL: Groups of one adult and several vixens (max. proven adult group is 6). Female members probably relatives. Non-breeding females may act as 'helpers' at the den, guarding, playing with and provisioning the cubs. Hierarchical system between vixens, with reproduction sometimes being restricted to most dominant females within a group. In some populations *c.* 20% of vixens undergo late-term abortion rather than failing to conceive – perhaps this is a mechanism for subordinate vixens to keep their reproductive options open until the last minute. Where more than one vixen breeds within a group they may pool their litters and suckle them communally. COMMUNICATION: Vocalisations have been described in detail, at least 28 different categories of sound. These include characteristic 'wow-wow-wow' bark, sometimes with several animals calling back and forth, also the eerie shriek commonly referred to as the vixen's scream (but occasionally heard from males too). This shriek is commonest in the mating season (eg January). Different voices clearly recognisable. Close contact ratchet-like noise called gekkering characteristic of aggression, a wailing, warbling whine indicative of submission. A gruff, monosyllabic bark is used by adults to warn cubs of approaching danger. Postures include movements of ears, mouth, tail and whole body. Ears folded flat back against head in extreme submission, rolled sideways in aggression. Mouth opens and lips retracted but not wrinkled in submission, gaped wide in aggression. Tail lashed sinuously from side to side in submissive greeting. Back arched and rump swung into oncoming protagonist in 'sideways barge' during mild scuffles, but intense aggression results in both parties rearing on hind legs and pushing against each other's shoulders with their forepaws, with mouths gaping and gekkering. Scent-marking involves urine and faeces (both left on visually and olfactorally conspicuous landmarks throughout territory, but most in well-used areas and especially along paths). Both males and females 'cock' legs, but urine marking sometimes confined to dominant females within social groups. Subordinate females and juveniles of both sexes generally squat to urinate. On snow, signs of urine of male squirted forward, that of female backward, with respect to rear paw prints. Sometimes sprinkle urine on other group members (especially male on female). Secretions from anal sacs (paired openings visible on either side of anus) evacuated when alarmed and sometimes coat droppings. Violet or supracaudal gland highlighted by conspicuous ellipse of dark fur (and yellowish bristles and greasy skin below) on top of tail, 7-10cm, from base. In males violet gland (reputed, incorrectly, to smell like violets) more active during breeding season, and scent may be wafted around when tail lashed in greeting. Other glands around lips and angle of jaw, occasionally rubbed, along with saliva, on vegetation. Scent glands between pads of feet.

Breeding. Males seasonally fecund, mating December–February when six-fold increase in testis size (females in oestrus for 3 weeks when vulva swollen, pink and moist, fertilization possible during 3 days). Post-copulatory tie of up to 90 mins. Births in March–May. Sexual maturity 10 months. Gestation 52–53 days. Litter size 4–5. Maximum of 12 foetuses found in one vixen. Food availability governs whether they breed and litter size. Litters per year 1. Mammae 8 (but sometimes 7, 9 or 10). Sex ratio at birth 1:1, but some evidence of more males (1:0.76) at birth in high density populations. YOUNG: At birth 100g. Dark brown velvety fur at birth (with white tail tag in those that will have it as adults); reddish tinge to face at 4 weeks when snub nose elongates, and ears grow fast (and first emerge from den), milk teeth complete at 7–8 weeks and appearance as miniature, somewhat fluffy adult with black facial teardrop mark. Almost indistinguishable from adults in field after 6 months. Blind and deaf at birth and need mother's body heat for first 2–3 weeks. Eyes open 11–14 days and blue before one month old, then brown to amber with vertical pupil. Weaning age *c.* 6 weeks (up to 12 weeks), and eat solid food from *c.* 4 weeks. Subsequently, breeding vixens tend to have larger nipples than those that were barren. Cubs grow 50g per day from 4–10 weeks. Both parents diligent in care of young. Male carries food to den entrance for female while she is confined for 2–5 days after birth. After weaning male and female carry food to young, including some stored near den in anticipation. Female spends progressively more time lying above ground to escape attentions of cubs in den. Male also plays with cubs, and grooms them. Where groups develop commonly only one, sometimes two of up to five females present will breed. If two females have cubs they may share a den. Non-breeding females may feed, groom and tend cubs, and adopt them if orphaned. Non- breeding determined by low social status. Some evidence that social suppression of reproduction lessened in disturbed populations such as urban populations subject to heavy road traffic mortality. Stay with mother until autumn.

Lifespan. Max. 9 years (wild). Mortality varies, but may be 80% in first year. 50,000–100,000 foxes killed per year in Britain. In some populations, where mortality is high, almost half the foxes may be in their first year and few survive three years, elsewhere, where mortality is low, only 15% may be in their first year, and 60% may be 5 years old or more. During rabies epidemics of 1970-80s hundreds of thousands reported rabid annually in Europe (and those reported probably less than 10% of those dying from rabies). Individuals that disperse, and especially those that disperse further, have lower life expectancy than those that remain in their natal area (small male cubs from larger litters, and female cubs from larger litters are more likely to disperse).

Measurements. Head-body length: 58–90cm; UK average: male 67cm, female 63cm. Tail-length: 32–48cm; UK average: male 41cm, female 38cm. Hind-foot length: 12.5–17.8cm (paw print 50–60mm). Shoulder-Height: 35–40mm. Ear: 8.5–10.2cm. Condylo-basal length: male 135–165mm, female 127–155mm (most adult males have larger sagittal crest on skull than do females); UK averages: male 14.5cm, female 13.7cm. Weight: 6–10kg; UK averages: males 6.7 (5.5–9.3)kg, females 5.4 (3.5–7.8)kg (exceptionally more than 15kg). Dental formula: 3/3, 1/1, 3/4, 3/3 = 42.

General. Most widespread, and abundant, wild carnivore in the world (throughout northern hemisphere, from Arctic to subtropical latitudes, and introduced to Australia). Major vector of wildlife rabies on continental Europe; traditional attempts to contain rabies by killing foxes did not noticeably slow its spread across the continent. Recent attempts to control rabies by feeding foxes an oral vaccine (delivered in bait such as chicken heads) has drastically reduced the disease (eg Switzerland, Germany, France). Sarcoptic mange can kill foxes, which lose fur and develop skin lesions due to scratching at the irritant mite and may therefore die of cold. Susceptible to pesticide accumulations: 1959–61 large-scale mortalities from this in E. England due to foxes eating birds that had fed on seeds dressed with chemicals. Generally trots at 6–13km per hour, but clocked at 60km per hr in brief sprints. Notorious habit of 'surplus killing' when get into poultry house, pheasant rearing pen or other situations where prey cannot escape (eg ground-roosting colonies of Black-headed Gulls on dark, windy nights). This behaviour often inappropriately judged according to human morals (eg wicked, vindictive), probably due to 'pathological' situation of prey that fail to run away (circumstances not often met in nature and therefore fox not evolved to cope with it). Where only a few surplus prey are killed the fox caches them and has extraordinary memory for finding the larder again later, thus an opportunistic adaptation to uncertain food supplies. Recent attempts to quantify pest status of foxes suggest that unlikely to affect the economics of farming (especially lambing) significantly, and inconsequential nuisance in towns (where warmly welcomed by many people, and where risk to domestic cats is minimal), but may be significant competitor of people breeding and releasing game birds for sport shooting.

RACCOON DOG *Nyctereutes procyonoides* Pl. 19

Name. *Nyx, nyctos* = night, *ereutes* = seeker; *pro* = before; *kyon* = dog; *oides* = looking (Gk.).

Status. Introduced. Populations scattered, probably expanding.

Recognition. Distinctive black facial mask somewhat reminiscent of Raccoon, but differs because of larger size, short ears, uniformly coloured short tail, mask not continuous between eyes although colouring on bridge of muzzle sufficiently dark brown to make mask look continuous from distance. Feet like foxes'. Dense pelage (guard hairs up to 120mm long in winter). In winter, fat deposits combined with thicker coat give very rotund appearance. Thick underfur grown during September–November, and replaced by thinner summer underfur during May/June. Whelps moult at age 4–5 weeks to new summer pelage. Burrow distinguished from Red Foxes' by narrower entrance (20cm wide; Red Foxes': 25–30cm wide). Reputedly creates more fan-shaped, rather than rounded (like Red Fox), heap of soil at entrance. Each foot has 4 digits – forefoot 5 × 4.5cm, hindfoot 4.5 × 3.5. Droppings twisted, 5–8cm long, 1cm wide.

Habitat. Deciduous woodland, especially damp areas close to rivers and lakes. Favours woodland with dense understorey. Often takes over den dug by another species (eg Red Fox), or uses space under tree trunks, in dense bush or rock crevice.

Habits. Nocturnal. Inactive during winter and said to hibernate, although body temperature does not decrease. FOOD: Omnivore. Vertebrates (eg rodents, amphibians, fish), fruit, bulbs, nuts, and insects. Scavenges carrion on seashore or river banks, waste and garbage. HOME RANGE: 100–200ha overlapping and, apparently, not territorial. Dispersal of year's young starts September, but some pups stay with parents during winter, and long range dispersal possible after disturbance. SOCIAL: No firm evidence of social ties between animals sharing home ranges, but in captivity there is indication of social ties (grooming care of pups, common dung piles, attacks against intruders); feed together amicably around clumped food. Social unit probably monogamous pair and associated previous young. COMMUNICATION: Growls, whines, reported not to bark. Faeces in communal middens (several per home range, not all used by each individual). Size of latrine ranges from 25–150cm in width. Unusual posture with tail held in inverted U position during some encounters. Keen hearing and sense of smell, poor vision.

Breeding. Mating in February–March, births in April–May. Sexual maturity 9–11 months. Gestation 59–64 days. Litter size 2–19 (usually 5–8). Litters per year 1. Mammae 6–8. YOUNG: Capable of independent existence at 4–5 months. Weaning age 8 weeks. Both parents feed young (which take solid food from 25–30 days old).

Lifespan. Max. 11 years (captivity). Causes of death include hunting, trapping, rabies and enteritis, bacterial diseases and parasites. Also starvation, flood, road traffic, mechanised harvesting of maize (Poland). Predators – Wolf in Russia, feral dogs in Poland.

Measurements. Head-body length: 55–80cm. Tail- length: 15–26cm. Hind-foot length: 9–12cm. Shoulder-height: *c.* 20cm. Condylo-basal length: *c.* 11cm. Weight: 5–10kg (birth 60–90g). Dental formula: 3/3, 1/1, 4/4, 2/3 = 42.

General. Unique among canids in hibernating (and consequent ability to incubate rabies virus over winter – important vector in Central Europe). Vast amount of subcutaneous fat stored in the autumn can dra-

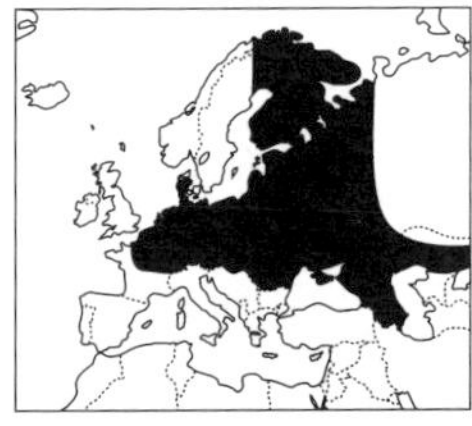

< *Raccoon Dog.* Spreading throughout C&E Europe and north into Scandinavia.

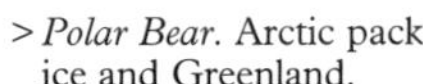

> *Polar Bear.* Arctic pack ice and Greenland.

matically alter body shape. Introduced through fur-farm escapees, originally from East Asia (Japan and China). Voluntarily introduced in Russia during 1950s, *c.* 5000 released in the Western Republics. Considered pest in France.

POLAR BEAR *Ursus maritimus* **Pl. 20**

Name. *Ursus* = bear; *maritimus* = maritime (*mare* = sea) (Lat.).
Status. Vulnerable.
Recognition. Very large, long neck, creamy fur – white yellow/grey or even almost brown, according to season and light conditions, as well as oxidation in the summer sun, and occasionally from staining by seal oil. Male's nose more Roman in profile than that of female, often with considerable scarring. Nose, lips (and skin) are black. Fieldsigns include remains of seal kills. Scats on ice are oily, black liquid splatters, but on land if feeding on vegetation they are similar to those of Brown Bears.
Habitat. Distribution is circumpolar, occurring only rarely in northern Iceland. Nearly always in association with sea ice. In general, most numerous within 300km of the coast over continental shelves, where wind and currents keep ice in motion and there is greater biological productivity, (i.e. thinner annual ice and open water permit more light penetration which stimulates greater productivity, so more seals). Maternity dens are generally located in drifted snow along coastlines. Hole in hard-packed snow: tunnel slanting uphill, behind which is lair. Den usually single-chambered, but occasionally several chambers. In the southern Beaufort sea populations, 80% of denning is in the semi-permanent ice 200–300km offshore. The extent of this behaviour in other populations that border on the polar basin is not known.
Habits. Active in daylight or darkness, according to time of year. Diurnal rhythm dependent on availability of seals, ice conditions and weather. FOOD: Carnivorous: seals, mainly Ringed (especially during the breeding season), but Bearded Seals are also important. Taken in breeding lairs, at breathing holes or resting on ice, striking them with lethal blow with front paw and biting. The whitish coat camouflages the hunting bear, which has very sharp claws for gripping slippery seals. May also take Harp and Hooded Seals, and scavenge whale, Walrus and seal carcasses; may also kill Walruses, White Whales and Narwhales. Fish, sea-birds and eggs, Reindeer, Musk-Ox, Arctic Hares also recorded rarely in diet. During ice-free periods they move inland and some eat small amounts of berries and grasses. During winter and spring also stalk seals by crawling on ice or swimming in the 'leads'. It is rumoured, implausibly, that Polar Bears lying in ambush cover their black noses with their paws. HOME RANGE: Polar Bears can travel great distances but apparently have weakly defined home ranges which can have a diameter of 300–400km. Males

and females have similar sized home ranges. May travel up to 8–20km in a day: an Alaskan bear moved 1119km in a year. They are facultative migrants: around the Beaufort sea, for example, they move south in winter, north in summer with the extension of the drift ice border. They may migrate inland in summer in areas where the ice has disappeared and bears are stranded on the land. SOCIAL: Solitary; occasionally congregate, showing tolerance, at good food source, eg 60 bears observed at a whale carcass. Also congregate and tolerant in absence of food, such as where waiting together on shore for freeze-up which enables them to hunt again. Males aggressive to one another in breeding season and also occasionally kill cubs and other bears. COMMUNICATION: Not noisy – sometimes snorts, can roar and scream: vocalisations in adults primarily used in aggressive interactions. Sense of smell particularly well developed.
Breeding. Mating in March–June, with peak in April (delayed implantation: implants presumably October–November). Births in December–January. Sexual maturity shows geographical variation: male 6 years (reach adult weight at 8–10 years); female 5–6 years (reach adult weight at this time), and may experience reproductive senility at about 20 years of age. Gestation 200–250 days. Litter size usually 2 (variation 1–4). Observed average litter size after emergence from den about 1.6–1.8. Average breeding interval about 3.5 years. Mammae 4 (functional mammae on thorax). YOUNG: Cubs weigh roughly 0.6kg at birth, are blind and covered with hair (about 5mm long). Weigh 9–11kg (though this can range from 5–20kg) when they emerge from the dens in March or April, aged about 3 months old. Weaning age 2–3 years of age. Parental care by female only; the milk is highly concentrated in solids (*c.* 47.6%), with a similar calorific content to milk of cetaceans and seals, although the fat content (*c.* 33%) is higher than that of cetaceans (and that of other bear species) but lower than that of seals. Frequently nursed outside den (see vignette). Young remain with mother for up to 28 months after birth, during which time they learn hunting behaviour. Females with young avoid adult males due to the possibility of predation upon the cubs.
Lifespan. Max. recorded 40 years in captivity, 32 years in wild. Mortality high in first years of life, normally from starvation. Causes of death among adults little known.
Measurements. Head-body length: male 190–240cm; female 170–200cm (Birth 25cm). Tail-length: 8–10cm. Hind-foot length: female length 30cm, width 20cm; male length 35cm, width 23cm. Shoulder-Height: 120–150cm. Condylo-basal length: male 335–344mm; female 380–407mm. Weight: male 300–600kg; female 150–300kg; males can reach 800kg and females up to 500kg, but this is unusual (Birth *c.* 0.6kg). Dental formula: 3/3, 1/1, 3/2, 2/3 = 38–42. The dental formula can vary with premolar count = 2–4/2–4 although 3/2 most common.
General. World population not depleted: estimated at perhaps as many as 40,000 in 1988. Occurs in separate subpopulations. Managed according to national legislations and international rules in the 1973 Oslo Agree-

ment on Conservation of Polar Bears. Only pregnant females hibernate. Females with small young and other bears may use temporary dens during inclement weather and doze through very bad weather, digging in or getting covered with snow. Known to be able to dive underwater for at least 72 seconds. Using large oar-like paws, can reach speed 2–3km per hr at surface when swimming. Pelage consists of thick layer of underfur (5cm long) and tufted intermediate guard hairs (15cm long). Guard hairs slippery, and the pelt well adapted to shedding water and absorbing solar radiation. Use both fat and pelt for insulation. Thickness of subcutaneous fat on adults varies from 5–15cm. Adult females particularly fat just prior to denning. Heat dissipation via foot pads, face, ears, shoulders (conduction) and panting (evaporative cooling). Polar Bears thought to have originated from a segment of the Siberian population of Brown Bears (*U. arctos*), isolated during glacial advances of the mid-Pleistocene. Their former distribution was circumpolar, on the edge of the pack ice. With few exceptions, the geographic ranges of Polar Bears and Brown Bears do not overlap.

BROWN BEAR *Ursus arctos* Pl. 20

Name. Grizzly Bear (USA). *Arctos* = bear (Gk.).
Status. Threatened.

Recognition. Very large size, heavy build, short rounded ears, absence of visible tail. Pale fawn to dark brown. Mostly brown but some light greyish-yellow to black. Moults once a year in summer (*c.* April–June), new coat thickens in autumn to give winter coat. Distinctive footprints of adult hindfoot about same size as Man, but broader and with prominent claw marks. Size of footprint from hindfoot loosely correlates with age: cubs aged 0.75–1.5 years having lengths of 13–14.5cm, 16cm at 2.5, 18–19.5cm at 3.5–4.5 years, up to 22cm at less than 5 years. Prints of large bears (eg old males) measure 27–30cm. Faeces contain many undigested traces of diet, especially vegetable matter. Signs of excavations when feeding on ant hills, wasp nests; scratches and chunks torn out of rotten wood; stones overturned. Scrapes on tree bark. When using a lair, will flatten plant cover or press down soil or snow in a 1–2m radius.
Habitat. Mixed woods, usually spruce forests in mountains nowadays, due to persecution. Extends to tundra in north and open pasture above tree-line in mountains. In open areas needs some dense cover and shelters there by day, mostly in remote inaccessible forests. Hibernates in underground hole or cave lined with twigs, moss and other dry vegetation. May make more than one such den before winter. In summer in warm climates such as southern Europe, dens in natural hole or cave among rocks or under a tree on a sheltered slope (average volume of the bed-chamber: *c.* 2 m^3).
Habits. Mainly nocturnal (more diurnal where undisturbed) and when feeding on berries in autumn, but seasonal variations. FOOD: Berries,

grasses, herbs, carrion, insects. Also roots and, in summer, green plant materials; occasionally Elk, Reindeer, sheep, cows (bears can run on encrusted snow which cannot support elk). In autumn eat large quantities of berries, acorns and beech mast prior to hibernation. HOME RANGE: Scandinavia average 25km^2, although male ranges can vary between 250–4000km^2 and females between 150–1000km^2; Italy 56km^2; USA 80–280km^2 (with extremes of 6km^2 and 2655km^2), male 24.4–1054km^2 with male ranges 3–4 times size of female. Czechoslovakia 13–30km^2 (but as low as 5 km^2). Densities in Europe (per 1000km^2 of forest) also vary enormously, from 1–3 in Norway, 6–11 in parts of Sweden to 135–190 in the Abruzzo National Park (Italy). Elsewhere they are equally varied: 37–44 Yukon territories, Canada; 28–29 Yellowstone National Park, USA; 100–300 Caukasus Nature Reserve, Russia; 3–17 isolated population in the Russia. Annual increment to populations varies from 5% to 14–26%. Bears normally travel on average 2–3.5km a day, their speeds varying from a walk (5–6.0km per hr) to a trot (10–12km per hr). Max. speed record: 50–60km per hr or a gallop at 22–51km per hr. Long range movements include 120–150km (Romania) and 300–400km (Kavkaz mountains). Young males may disperse over more than 100km after leaving their mothers, and adult males also travel extensively. SOCIAL: Generally solitary but may gather in large numbers at food sources. Three generations of females have been known in same range. Adult males are solitary but their ranges encompass those of several adult females, and overlap with those of other males. No lasting social bonds, except those between females and young. Males may fight over females during the breeding season. Adopt bipedal stance to intimidate rival as well as to scan environment. Siblings sometimes maintain an association for 1–2 years after leaving their mother. COMMUNICATION: Vocalisations include grunts or howls when angry or frightened. Utters an explosive sound when surprised. Mother calls young with a bleat. Scent marking with tree scrapes/scratches. Sense of smell especially good, also hearing, but sight poor.

Breeding. Mate in May–July. Implantation delayed 4.5–7 months. Births in January–February. Sexual maturity 5 years at earliest. Gestation 210–255 days including delayed implantation: the effective gestation is actually only about 60 days. Litter size 1–3 (often twins), occasionally up to 5. Litters at intervals of 3 years and more. YOUNG: 350g at birth: squirrel-sized, blind and hairless. Young remain in the den for 4 months, then accompany mother for 1.5–3.5 years. Weaned at 1.5 years of age. Immature bears have paler fur on collar of neck.

Lifespan. Max. more than 30 years in the wild; up to 50 in captivity. Mortality especially high for cubs in first year. Cannibalism observed. Annual mortality rates: cubs 30–40%; subadult 15–35%; adult and subadult females 17%; adult and subadult males 23%.

Measurements. Head-body length: 170–280cm. Tail- length: 6–21mm. Hind-foot length: 13–30cm. Shoulder-height: 90–150cm. Condylo-basal

<*Brown Bear*. Main stronghold in N Scandinavia, adjoining Russia. Remnant populations in Iberian peninsula, C Italy and SE Europe.
>*Raccoon*. Spreading from escapes in Germany and eastern Russia.

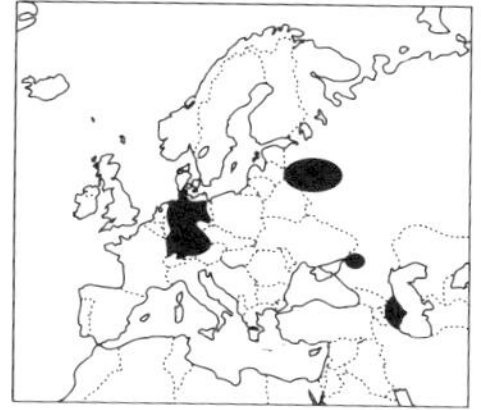

length: 288–337mm. Weight: male 100–315kg (although may be heavier: Kodiak Bear reaches maximum 780–1500kg); female 60–200kg. Dental formula: 3/3, 1/1, 4/4, 2/3 = 42.
General. Listed as threatened in IUCN Red Data Book (1976). Protected in Europe by Bern Convention (1984), also in Poland and Bulgaria. In Sweden, individual bears causing nuisance can be hunted under licence. Formerly everywhere in the Old World, from British Isles and Spain in West to Kamchatka in East.

RACCOON *Procyon lotor* **Pl. 21**

Name. *Pro* = before, *cyon* = dog (Gk.); *lotor* = washer (Lat.).
Status. Introduced.
Recognition. Black highwayman's mask – accentuated by grey bars above and below; grizzled grey-brown coat and bushy tail with rings (usually 5). Moults mid-April–end June, but complete renewal of pelage takes whole summer. Signs include scratch marks outside tree holes, prints on muddy river banks (for example) clearly show long digits of Raccoon's foot. Droppings are the same size and shape as these of a medium-sized (20–35kg) dog but often show vegetable material (acorns, corn, apple seeds) and insect remains. Often defecates at base of den trees or on fallen logs. Groups of 2–4 droppings accumulate (but not big concentrations like Badger latrines).
Habitat. Proximity to aquatic habitats characteristic, and close to woodland eg oak. Many different dens used by one individual, often in a tree hole, sometimes at a considerable height, or among rocks or overturned stump or burrow made by another animal, or in buildings (eg chimneys, sewers).
Habits. Nocturnal, with crepuscular peaks in activity. FOOD: Acorns especially in winter; berries and fruit, oats, plums etc, crayfish, molluscs, fish and frogs. Very fond of maize. Gathers up oats into sheaves, holds under one arm and eats. Also eats insect larvae, eggs and young of ground-nesting birds, small rodents and hares. HOME RANGE: Female ranges probably not exclusive for related individuals, and territorial defence not well developed. Ranges of adult males largely exclusive of each other, but commonly overlap the ranges of 1–3 adult females and up to 4 yearlings (in North Dakota prairies). Probably polygynous. In

Eastern Canada, home ranges 120–380ha. One per 5–43ha in rural areas and 4–8 per km^2 typical population density in urban Canada (with instances of 100 per km^2 in some urban areas) but probably less in Europe. One individual travelled 266km, though general movements not extensive. 50–5000ha year in range with most foraging over *c.* 800ha. 0.2–4946ha St. Catherine's Island, Georgia, USA, annual average 65ha for males, 39ha for females. SOCIAL: Unrelated animals tend to avoid each other, though up to 23 individuals found together in same winter den; more than one adult in one den rare, and most Raccoons denning together are probably relatives (groups of females and young): males den alone usually. COMMUNICATION: various calls, most with little carrying power, except for a chattering, scolding call when squabbling among themselves, which carries over 100m. Fingers have great sensitivity and are dextrous in handling prey.

Breeding. Mating in late January–early February, births in March–May. Also records of late summer birth. If female loses a newborn litter she may ovulate a second time during the same season. Sexual maturity 1 year. Gestation *c.* 63 days. Litter size 3–7, average 4–5. Litters per year 1. Mammae 8. YOUNG: Born in hollow high up in tree. Dispersal in autumn, winter or spring (related to latitude and physiology). Weaning age 1.5–4 months. Young follow mother for first year. In Minnesota young kept in den in hollow tree till 7–9 weeks old, then moved to one of a series of ground beds. Short trips with mother at 10–11 weeks and family begins to move together after another week.

Lifespan. Max. recorded 16 years (wild) but only 1% reach 7 years. Max. in captivity 20 years, 7 months. Main causes of death related to human activities (hunting, trapping, poisoning, road traffic). Otherwise starvation (in winter), parasites and disease (parvovirus and distemper) and predation of young by carnivores and raptors.

Measurements. Head-body length: 48–70cm. Tail-length: 20–26cm. Weight: *c.* 8kg (cf up to 12kg in N. America). Dental formula: 3/3, 1/1, 4/4, 2/2 = 40.

General. May stay in den in ground for 1–4 weeks or in trees for 2–3 days ('roosts') if weather poor, but will come out on tree limbs if sun shines, even if snowy and cold. If temperature is above freezing, will be active, if below, will be inactive. May lose up to 50% of their weight over winter. Agile climber, competent swimmer. Northern populations have individuals with longer and thicker coats, heavier weights, fewer breeding yearlings, larger litters, mutually exclusive territories among males and winter denning. Over 4 million Raccoons harvested annually by hunting and trapping in North America, notably in the east, where sport of 'coon hunting' prevalent September–December each year. Exceptionally adept climber, urban raccoons can climb the vertical walls of homes, clinging to crevices in the brickwork. Relocation of problem urban Raccoons to rural areas carried out in Canada, but research reveals high mortality of translocated animals in first 3 months from starvation, dogs, poisoning

(50%) add up to 75% overwinter as juveniles lose a lot of weight prior to entering winter denning period. Relocated animals show extensive movements until adjusted to new habitat (but no homing tendencies observed), and hence have high potential for transmission of infectious diseases, as has been the case when Raccoons translocated from south Florida to Virginia for hunting purposes during the 1980s leading to an epizootic of Raccoon rabies. Currently most valuable wild furbearer in US; introduced into France, Netherlands, Germany and various parts of Russia because of its commercial value; reached France from Germany in 1970s, now settled in area close to Laons, France. As with Raccoon Dog, considered a pest in France. In Canada, prospers in many urban centres where sometimes viewed as a nuisance: denning in attics and chimneys causes damage, routinely raids garbage bins. Major rabies vector in mid-Atlantic states of USA (i.e. on the Eastern Seaboard), but not yet significant in European rabies. Carries pathogens causing leptospirosis and tularemia. The name *lotor* = washer, applies to food-washing behaviour that may not be washing at all: one idea is that the immersion may enhance the sensitivity of the Raccoon's fingers for kneading and manipulating food, in order to reject inedible parts. The name 'raccoon' derives from an Algonquin Indian world, *aroughcoune*, which means 'scratches with hands', and this may be more apposite than the technical name.

STOAT *Mustela erminea* **Pl. 25**

Name. *Mustela* = weasel (Lat.), *Mus* = a mouse; *telum* = a spear (Lat.) i.e. spear-like mouse.

Recognition. Long sinuous body with end of long tail invariably black. Pelage: in summer, chestnut brown above, yellowish-white below, white rim to ears; in winter, partly or (in northern regions) completely white (except black tip to tail). Pronounced sexual dimorphism in size, the male being larger than the female. Moults in spring and autumn; in mild climates winter underfur thicker but colour unchanged; in northern and alpine populations, the winter (eg November) moult is to white pelage, changing back to red/brown in March/April. Autumn moult starts on the belly, then the flanks and back, then across the body and finally the head. The spring moult starts on the head, then across the body and finally underneath. Thus the belly always loses its old fur last. Faeces *c.* 4–8cm long, *c.* 5mm diameter, twists of fur at ends and full of hair, feathers, pieces of bone etc (sometimes piled in midden at den). Found on walls and stones along paths used by Stoats, especially at points where these cross.

Habitat. Most habitats with some cover, including open moor and marsh near woods, lowland farmland and most habitats from shoreline to mountain. Dens in hollow trees, burrows, rock crevices. Dens may be in nests of prey, and lined with rodent fur.

Habits. Spells of activity alternating with rest throughout day and night. FOOD: Largely carnivorous, mostly killing rodents, lagomorphs and birds.

Seasonal and sexual differences in diet (eg females specialise on voles, males on young rabbits, hares and birds in addition to small rodents) and consumption (female 23% of body weight daily, male). Kill by precision bite to back of neck. HOME RANGE: 2–200ha depending on habitat (food supply) and sex (eg Sweden in autumn: male 8–13 ha; female 2–7 ha). Where hunt from stone walls, female patrols 1.5km, male 5km (in autumn). Larger territories divided into subunits of *c.* 10ha each used for a few days at a time. Distance travelled in one hunt up to 8km. Young disperse July–August of first year (during mating season), particularly males (who may also disperse at other times), females generally settling close to their birthplace. SOCIAL: Outside the breeding season both sexes defend separate territories (each containing 2–10 dens). During the breeding season female territorial, roaming male non-territorial, with overlapping home ranges spanning territories of several females. Older males search far afield for females; younger males try to monopolise one resident female. Female and 5–8 well grown young may hunt together in autumn. COMMUNICATION: Shrill alarm 'kree kree' repeated; female trilling contact call to her young, also used by females at mating. Aggressive chattering. Musky scent from anal sacs (expelled when alarmed) taints faeces (which are left on conspicuous objects, eg boulders, tussocks) and scent-marking by body-rubbing and anal drag. Sight, smell and hearing all good – smell most important when hunting underground, hearing when above ground.

Breeding. Mating in May–June in post-partum oestrus. Induced ovulation caused by mating. Delayed implantation *c.* 280 days. Implantation March. Births in April–May. Male fertile: mid-May to mid-August. Females sexually mature at 5 weeks (pre-weaning, induced ovulation), males sexually mature at 1 year. Gestation 21–28 days (after delayed implantation). Litter size 5–12. Litters per year 1. Mammae 8. YOUNG: Newborn blind, deaf and covered in fluffy pale fur, with temporary darker brown mane which female grabs when moving them. This mane may cover scent-glands which help adult males to identify young females. Full prey-killing behaviour developed by 12 weeks. Weaned at 5 weeks, independent at 12 weeks. Female alone feeds young prey from 4 weeks old.

Lifespan. Max. 10 years, average 1.5 years.

Measurements. Head-body length: 160–310mm, male *c.* 297mm; female *c.* 264mm. Regional variation (eg central and western Alps, head-body length up to 200mm). Tail-length: 95–140mm, male *c.* 117mm; female *c.* 110mm. Hind-foot length: 35–50mm. Condylo-basal length: 38.8–49.6mm. Weight: 90–445g (male 50% heavier than female, eg UK male 200-445g, female 140–280g). Regional variation: smaller in northern than central/eastern Europe. Dental formula: 3/3, 1/1, 3/3, 1/2 = 34.

General. Common. Numbers vary greatly according to availability of food (primarily small rodents and Rabbits). Climbs well. Prey followed by scent, killed by bite at back of neck (larger prey gripped at nape, enfolded by forelegs, scratched with hindlegs). Winter pelts (ermine) formerly valuable to fur trade. Growth of new fur in spring and autumn is

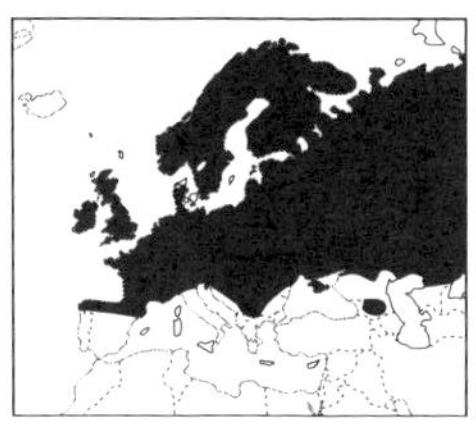

<*Stoat.* Widespread in N Europe and British Isles, absent from Mediterranean countries.
>*Weasel.* Common Weasel E&S of line; Least Weasel W&S. An intermediate zone spans the Mediterranean countries.

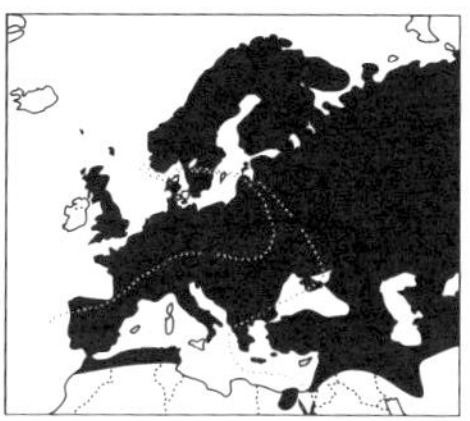

triggered by day length, but the winter whitening of the fur is triggered by a combination of temperature and heredity. Exceptional mating system involves males mating with suckling young while they are still in the nest; these young females therefore are pregnant (delayed implantation) throughout their infancy.

WEASEL *Mustela nivalis* Pl. 25

Name. Common Weasel (Central/Western Europe), Least or Snow Weasel (Northern Scandinavia and Russia). *Nivus* = snow (Lat.); *nivalis* = of the snow.
Recognition. Small, cylindrical body. Weasel from northern Europe is a distinct subspecies (Least, Pygmy or Snow Weasel *M.n. nivalis*) as opposed to the Common Weasel

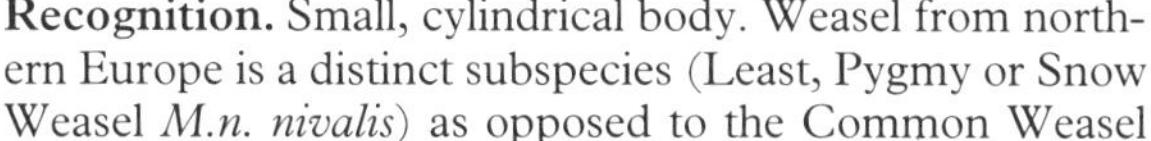

M.n. vulgaris. The differences between size, summer coat and winter whitening between northern and southern subspecies are linked (see table). In general, body size smaller in populations further north, but all measures highly variable. Pelage is chestnut brown above, white below (irregularly and less clearly demarcated than in Stoat) in Common Weasel. The Common Weasel also has a brown spot, the gular spot, behind the angle of the mouth, which is lacking in the Least Weasel. There have been unconfirmed stories of the existence of two species of Common Weasel, one being smaller. The only plausible explanation so far has been that the smaller individuals may be late-born young that were overtaken by the winter before they were full grown. However, there is no proof that growth in weasels depends upon season rather than food supply. In Least Weasel and some Mediterranean weasels, the line between the brown flanks and pale underside is straight. Distinguished from Stoat by the shorter tail, absence of black tip to tail. Frequently stands upright on hindlegs. Northern and Eastern varieties turn white in winter; central, western and southern European varieties stay brown. As well as the difference in appearance between the two subspecies, there are difference in reproductive cycles and therefore in population dynamics, leading to the suggestion that they are in fact separate species. However size and colour pattern in *nivalis* are not reliably linked and are inherited independently and both Common and Least forms share the same chromosomes which are fully inter-fertile. Faeces 3–6mm long, diameter 2–3mm, full of hair, feathers, pieces of bone etc, may accumulate in dens.

Table 5. Characteristics of the two sub-species of Weasel *M. nivalis*

	M. nivalis nivalis	*M. nivalis vulgaris*
Common name	Least, Pygmy or Snow Weasel	Common Weasel
Range	N Scandinavia, Russia, USA	Central & Western Europe and Mediterranean (the latter may be a third sub-species)
Summer coat	Regular demarcation line	Irregular demarcation line with spots between dorsal/ventral colouration (Mediterranean includes some populations with straight line combined with large size)
Winter coat	White	Brown
Size	Very small	Medium to large

Habitat. Very variable – anywhere providing cover and prey, from sand dunes and grassland to woodland and mountains. In UK, weasel less common on moorland/mountains than stoat, but ranges to higher altitudes in Europe and into the Karakul deserts of S. Russia. Burrow taken over from prey, eg in a hole in wall or under tree roots, lined with fur from its prey in cold climates.

Habits. Active both day and night, a few hours activity alternating with a few hours rest. In spring, females less active than males, saving energy for pregnancy by remaining in the nest and feeding from caches. FOOD: Principally rodents (eg 60–80% voles and mice), supplemented with birds (and eggs in season), and occasionally rabbits or water voles. However, diet varies with habitat. Male eats 33% body weight daily (non-breeding female 36%). Must eat every 24 hrs to avoid starvation. Number of voles and mice removed by weasels, however, is only a small proportion of the normal total population turnover. HOME RANGE: 1–25ha, varying with sex (eg UK woodland, female 1–4ha, male 7–15ha) and region. Larger home range of male overlaps several small territories of females. Local extinctions and recolonisations are frequent. Little data on dispersal of young, thought to disperse during first summer/autumn. SOCIAL: Normally solitary – no pair bond between adults, no parental care by male. Families break up at about 8–10 weeks. COMMUNICATION: Vocalisations include alarm hiss, high-pitched trill, short sharp bark if threatened. Scent marking and associated behaviour well developed. Senses acute, range of hearing 61kHz; eyes adapted for both daylight and night vision with good form discrimination.

Breeding. First litter born: April–May, unless food is scarce, when breeding may fail altogether. No delayed implantation (cf Stoat). When prey abundant, adults can produce second litter in July/August, and early-born young females can produce their first litters at this time, but this does not happen every year (but facilitates rapid increase in population when prey are abundant). Sexual maturity *c.* 3–4 months (in first summer if born early and food is abundant). Gestation 34–37 days. Litter size averages 4–6 (live kits), range 1–16 (embryo counts – highest in arctic populations during peak lemming years). Litters per year 1–2, but in boreal populations during lemming peaks go on breeding under the snow all winter. Mammae 3–4 pairs. YOUNG: Spring litters grow fast (at 8 weeks males 70–90g, females 50–55 g); autumn litters (if any) are slower. Deciduous teeth erupt *c.* 2–3 weeks, eyes open 4 weeks. Weaning age 3–4 weeks. Young can kill efficiently at 8 weeks. Parental care by female only and said not to include tuition in killing, which is instinctive, but young reared with the mother are more proficient at earlier age. Family splits up at *c.* 9–12 weeks.

Lifespan. Max. recorded 3 years (wild), 10 years (captivity). Mortality documented as 63% mortality in first year; 96% mortality in 2nd year.

Measurements. Head-body length: means of local population in UK; male 202–314mm, female 173–181mm. Males bigger than females, eg in Sweden: *M.m. nivalis* male 166mm, female 148mm (89% difference); *M.m. vulgaris* male 189mm, female 154mm (81% difference). Regional variation pronounced, eg mean head-body length male 166mm (N. Sweden), 189mm (S. Sweden), 202mm (Sussex), 214mm (Scotland). Tail-length: male 60–125mm, female 30–88mm, depending on country. Hind-foot length: male 30–35mm, female 25–30mm, depending on country. Shoulder- Height: *c.* 40–50mm. Condylo-basal length: 34–42mm – means of males from Scandinavia to Italy. Weight: Difference in size in body weight much more pronouuoced than for head-body length, eg in Sweden: *M.n. nivalis* male mean 54g, female mean 35g (65% difference); *M.n. vulgaris* male mean 73g, female mean 30g (49% difference). Dental formula: 3/3, 1/1, 3/3, 1/2 = 34.

General. Sufficiently small to follow rodents along their burrows. Biology of weasels in south (eg UK) very different from that in the far north, even though they are considered the same species (in terms of capacity for fertile interbreeding). Attempts have been made to use weasels as biological control of various pest species such as voles and Rabbits. However it seldom worked.

EUROPEAN MINK *Mustela lutreola* Pl. 22

Name. *lutreola* = small otter (Lat.).

Status. Extinct in Western Europe, except Spain and France, where it is vulnerable, although two regions of France (Brittany and SW France) harbour significant populations.

Recognition. Glossy brownish-black. White on lower and upper lips and under chin. Distinguished from

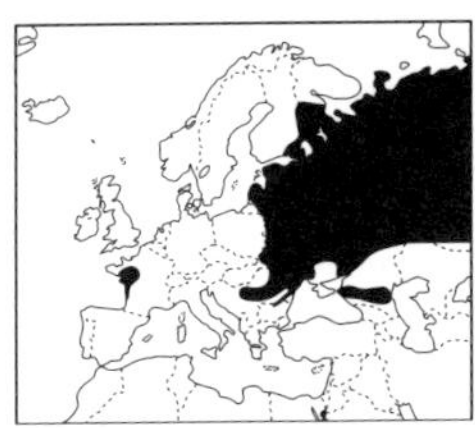

< *European Mink.* E Europe: pockets in W France, and N Spain.

> *American Mink.* Escapee. W Iceland, much of British Isles, Scandinavia and pockets of Spain, France, Belgium and Netherlands.

American Mink which rarely has white on upper lip. Distinguished from Polecat because more uniformly dark colour. Slightly webbed feet. Well-trodden paths from den to water's edge. Prints similar to Polecat's, but European Mink has 5 digits on each foot linked by one pad. Droppings 6–8cm long, 5–8cm wide – shape varies with diet but tend to be more cylindrical than those (very similar) of polecat.

Habitat. Vicinity of slow-moving fresh water, especially in or near woods (eg ponds, sluggish rivers). Den in hollow tree, burrow in bank (access by horizontal tunnel in from above water level, then sloping steeply upwards). Den lined with feathers or other soft material.

Habits. Nocturnal. FOOD: Carnivorous. Vertebrates, principally rodents (eg Water Voles, Musk Rats, rats), water-birds, amphibians, fish, reptiles and invertebrates, principally crayfish (and molluscs and insects). Food caught in or around water. HOME RANGE: Not known. Family disperses late summer (males known to travel 50km or more). SOCIAL: Largely solitary. COMMUNICATION: Vocalisations include an alarm shriek and a chuckling call.

Breeding. Mating in February–March, births in April–June (generally no delayed implantation, cf American Mink). Sexual maturity 2nd year. Gestation 35–42 days. Litter size 4–5. Litters per year 1. YOUNG: Weaning age 10 weeks.

Lifespan. Max. recorded 7–10 years.

Measurements. Head-body length: 30–40cm. Tail-length: 12–19cm. Hind-foot length: 5–6cm. Condylo-basal length: 57–65mm. Weight: 550–800g. Dental formula: 3/3, 1/1, 3/3, 1/2 = 34.

General. Amphibious habits. Formerly found in northern parts of Europe and Asia and NW Europe as far as SW of France.

AMERICAN MINK *Mustela vison* **Pl. 22**

Status. Introduced.

Recognition. Glossy brownish–black, although a small percentage of individuals may show the silver blue or pale brown mutations that are bred for in captivity. White on chin, throat, underside and as many as one in three have small amounts of white on upper lip (cf European Mink). Slightly webbed feet. Semi-aquatic. Adult (i.e. 10 months or older) females may have white hairs on back of neck due to mating bites. Tail slightly bushy, approximately half body length. Sexual dimorphism in size, males are slightly longer than females and almost twice as heavy, otherwise sexes difficult to distinguish. Elongated body and short legs. Two moults per year regulated by day length. Growth of summer pelage begins late March, completed mid-July. Winter coat begins mid-August, prime by mid-November. Prints measure 2.5–4cm long, 2–4cm wide. Splayed 'star-shape', often only 4 out of 5 toes leaving an impression. May form paths leading to water. Occasionally signs of tail dragging in mud. Faeces are similar to other mustelids, with strong smell (due to secretions from anal glands). Cylindrical, tapered ends 5–8cm long, usually 1cm diameter. Faeces frequently found on conspicuous sites, sometimes stations in bowls of waterside trees such as pollarded willows, and around den. Mink scat sites may be confused with otter spraints although mink scats may be more tightly twisted and lack the musky 'sweet' aroma characteristic of otter spraints. Prey remains often stored in hollow trees or beneath rocks by waterside. May find puncture marks from canines in back of prey's neck: 9–11mm apart.

Habitat. Fresh water near cover, eg sluggish rivers, ponds, marshes, canals, lakes and some coasts and estuaries, especially if rocky coasts with abundant cover and rockpools. Young born in nest lined with fur, feathers and dry vegetation in burrow in tree root crevice, among stones or hollow tree. Nearly always in existing cavity – rarely excavates own den.

Habits. Nocturnal or crepuscular (female rather more diurnal when nursing), but may be active at any time. On the coast, activity may be influenced more by the tidal cycle than day or night but this is unproven. FOOD: Carnivorous and opportunistic with broad diet, dictated by availability. Crustacea, fish and birds form the staple mink diet as well as rodents, frogs, lagomorphs, insects and molluscs. The importance of crayfish and crabs increases in summer and autumn, birds such as ducks and coots in autumn and winter whilst fish including Gobiids (bullheads, sculpins etc), wrasse, carp and pike are also more heavily predated in winter. Males tend to hunt larger and more terrestrial prey than females. This is partly due to their larger size and greater dispersal from den areas. Diet in North America: principally Musk Rats. HOME RANGE: Ranges largely linear along rivers and shores (eg 1–6km long) or irregular in marshland (eg 9ha). Along shores both sexes make trips of up to 2km inland, even in absence of streams, to areas of high food (Rabbit) abundance. Males' range larger than females. Territorial, but males show wide ranging movements during the mating season. Juveniles disperse from natal range from mid- August – males further than females. SOCIAL: Largely solitary, although ranges of males and females may overlap. Males very aggressive in mating season and fights over females are common then. COMMUNICATION: Vocalisations include alarm shriek and mating purr. Strong smelling jelly-like secretion from anal gland, deposited as blob

or as a smear by anal dragging. Terrestrial vision superior to that underwater. Possible colour discrimination ability.
Breeding. Mating in February–April, later in higher latitudes (eg May in Canada). Delayed implantation: 13–50 days. Births in April–May. Sexual maturity: 10–11 months. Gestation *c.* 28 days after implantation (total, including delayed implantation, 39–76 days, generally 45–52 days). Litter size 4–7 (up to 17 in captivity). Litters per year 1. Mammae 5–8. YOUNG: Weaning age 5–6 weeks. Parental care by female only: brings food to den and will change den if disturbed. Teaches young to hunt until age 10 weeks, at which point they gain some independence within natal range before dispersing about 13–14 weeks.
Lifespan. Max. recorded 10 years but average is far less in the wild. Most mortality probably results from failure to acquire a territory. Many hunted with hounds or trapped as pests. Also injuries during fights in the breeding season.
Measurements. Head-body length: 30–47cm; male 32–45cm; female 32–37cm. Tail-length: 13–23cm. Hind-foot length: 5–6cm. Condylobasal length: 58–73mm. Weight: male 0.84–1.8kg; female 0.45–0.81kg. Dental formula: 3/3, 1/1, 3/3, 1/2 = 34.
General. Accidently introduced to Europe via escapes from fur farms. Range expanding – concern that it will oust the European Mink. Especially rapid spread in UK thought to be due to absence of European Mink and low numbers of Polecats (i.e. the main potential competitors). Reputed to have damaging effect on native fauna (eg Polecat, Water Vole, Otter, Moorhen) and to damage shooting and fishing stocks – both claims require further investigation, particularly as Otter and Polecat now increasing range in western Britain in areas well populated by American Mink. Adaptations to amphibious habits. Evidence exists of local reduction in water vole in English uplands due to mink predation, but no conclusive evidence of damaging effect to sport fishing stocks.

STEPPE POLECAT *Mustela eversmanni* Pl. 23

Recognition. Very light flanks. Short ears. Much lighter colour than Western Polecat; almost beige, especially on back and sides. Head almost white.
Habitat. Range overlaps with Western Polecat. Fields and rough open country; farmland. Unbranched burrow excavated by itself or taken over from sousliks or hamsters. May occupy burrows for several years and extend then into a complex warren.
Habits. Nocturnal. FOOD: Includes sousliks, hamsters, voles, marmots and other rodents. Food sometimes stored. HOME RANGE: Known to cover up to 18km in one winter night. Local migrations in response to food shortages or extreme snow depth. SOCIAL: Solitary. COMMUNICATION: Vocalisations mainly loud, quickly-repeated hoarse barks; also described as high pitched. Also growls.

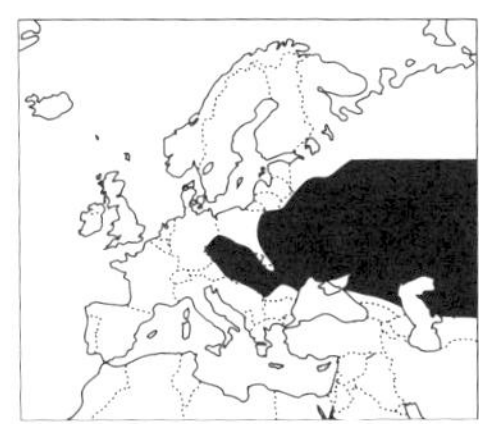

<*Steppe Polecat.* Principally in S Russia, but W end of range is in E Europe.

> *Western Polecat.* Widespread in C&W Europe. Found in Wales, Corsica and Sicily.

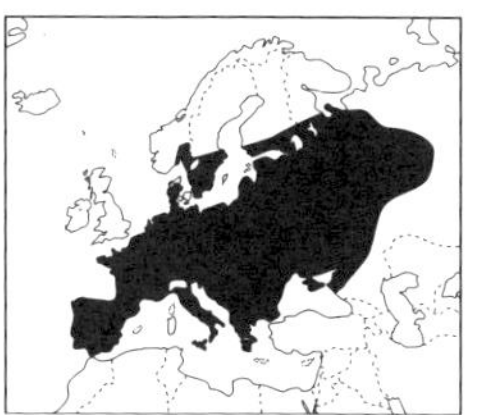

Breeding. Mating in February–March, births in April–May. Sexual maturity 9 months. Gestation 36–42 days. Litter size normally 3–6 (max. 18). YOUNG: Weaned at 1.5 months, and disperse at 3 months.
Measurements. Head-body length: male 32–56cm, female 29–52cm. Tail-length: 10.5–17.5cm (male 8–18cm; female 7–18cm). Weight: male up to 2050g; female up to 1350g (Birth: 4–6g). Dental formula: 3/3, 1/1, 2–4/2–4, 1/1–2 = 28–38.
General. Fur commercially important but not as valuable as *M. putorius*.

WESTERN POLECAT *Mustela putorius* Pl. 23

Name. *Putor* = foul smell (Lat.).

Recognition. Dark brown with white on tip of snout, between eye and ear. In winter pale yellow underfur gives distinctive lustre, showing through darker, sparse hair of outer coat. Lighter coloured flanks, and ear has pale edge. In summer much darker. Juveniles in first winter dense buff, moult May–June. Droppings a useful guide to presence – often found around farm buildings. Black with fur and bone fragments, up to 70.5mm, often twisted and with tapering ends. No fish bone (cf mink). Bound is shorter than Pine Marten: 40–60cm in Western Polecat; up to 50–100cm in the Pine Marten. Grouped in prints of 3 and 4.
Habitat. Mainly lowland wooded country; also river banks and marshes; frequently around farms. Also wooded steppe. For den digs own burrow, or takes over old rabbit warren, or rocky crevice. Den lined with grass and moss.
Habits. Largely nocturnal. FOOD: Rodents, Rabbits, frogs, birds, earthworms, insects, carrion. UK – mammals and frogs most frequently found. Range of prey sizes: hares and hare-sized birds to lizards, insects and earthworms. Uses sense of smell most; said to raid hen-houses. Makes food stores. SOCIAL: Solitary or female with young. COMMUNICATION: Chatters and growls when frightened, also hisses and screams loudly. Anal glands produce copious secretion, used for scent marking but also evacuated as defence when alarmed.
Breeding. Mating in March–June. A few records of young born in May. At mating, male grasps female by neck and drags her about, presumably stimulus to ovulation, which is induced. Copulation can last up to one hour. Sexual maturity in year following birth: male March–May; female

end of March. In NW Russia, sexual maturity at 22–23 months reported. Gestation 42 days. No delayed implantation. Litter size 2–12 (usually 3–7). Litters per year 1. Mammae max. 10. YOUNG: At birth have fine silky white hair; at 3–4 weeks darker coat develops. Some evidence that the size of males exceeds that of females at birth. Weaned at one month.
Lifespan. Max. recorded 4–5 years in wild; 14 in captivity. Causes of death in UK: 90% road traffic and traps; 10% dogs, shooting, snares, other. Life expectancy of females at birth is 8.1 months. High mortality of juveniles in August–October of first year of life.
Measurements. Head-body length: male 30.5–46cm; female 29–35.5cm. Tail-length: male *c.* 14cm; female *c.* 12.5cm. Hind-foot length: male *c.* 5.7cm; female *c.* 5.4cm. Condylo-basal length: male 67mm; female 60.4mm. Weight: male 502–1522g; female 442–800g (birth 9–10g). Dental formula: 3/3, 1/1, 3/3, 1/2 = 34.
General. Some feral Ferrets and recent Ferret/Polecat hybrids probably indistinguishable on external features. Several tens of thousands of pelts still imported annually into Great Britain for fur trade. Reputation for frog slaughter: on banks of rivers or near its den, dozens of frogs and toads found killed or paralysed by a bite on the spinal cord (cf the Otter, which also eats frogs and toads but is said rarely to kill them in surplus).

MARBLED POLECAT *Vormela peregusna* Pl. 23

Status. Subspecies *V.p. peregusna* considered vulnerable. This subspecies' range is central Europe to Greece and Turkey.

Recognition. Body light coloured with dark mottling; hindlegs, throat, chest, forequarters dark. Head dark with light mottling. Uniformly dark below. Short-legged. Ears fairly large. Tail black-tipped. Mottled brownish-black and white or yellowish-white. Has broad white band running above eyes from cheek to cheek and ears edged with white. Underside dark brownish-black. Tail bushy. Moult in March, when the yellowish spots on the back of males may change to orange. Fieldsigns include the foul smell of secretion emitted when alarmed, Footprints are best clue because this is a very elusive species: front print 2.5–3.8cm, hind 3.6–4.8cm. Den entrance: diameter typically about 7cm, usually with footprints at entrance.
Habitat. Shows a preference for dry and open biotopes, but is found from woods to wet and bushy areas to semi-arid areas. Cultivated land also, eg fruit plantations, open fields. Den usually renovated tunnels dug by rodents; enlarges tunnel after 60–100cm to create a sleeping chamber. In winter (breeding season), the chamber is lined with grass and this is where the young are reared. In summer, the chamber is bare.
Habits. Chiefly nocturnal and crepuscular but also active by day. FOOD: Carnivorous: hamsters, sousliks, birds, frogs, insects, rodents, reptiles.

<*Marbled Polecat.* W end of range in Greece and Bulgaria

>*Pine Marten.* Throughout most of W Europe, local pockets in British Isles, and Corsica, Sicily, Sardinia and Balearic Islands.

Hoards food. Also may take seasonal fruit and grass. HOME RANGE: 0.5–0.6km^2. Home ranges of animals of both sexes in same area overlap. At the occasional meetings between individuals they usually behave aggressively to one another. Nomadic within home range, and usually occupy a shelter only once. Young disperse at *c.* 61–68 days old. SOCIAL: Solitary. But in captivity can be kept in groups, forming a well-defined social hierarchy. COMMUNICATION: Shrill alarm cry; grunts, submissive long shriek. Contents of enlarged anal sacs evacuated to give foul alarm smell. Sense of smell excellent, hearing good, sight fairly poor.

Breeding. Mating in April–June, births January–March. Sexual maturity: female 3 months; male 1 year. Gestation 23–45 days, but with long delayed implantation of 6–9 months, total period is 8–11 months. Litter size 1–8. Litters per year 1. Mammae 10. YOUNG: Weaning age *c.* 50–54 days. Parental care by female only.

Lifespan. Max. recorded 8 years and 11 months (captivity). Causes of death include road traffic, particularly in mating season, and secondary poisoning via rodenticides. In captivity very susceptible to flu, rotten food and unknown paralysing diseases. Hunted for fur in past.

Measurements. Head-body length: 27–38cm. Tail- length: 12–22cm. Hind-foot length: 3.0–4.5cm. Shoulder-height: 5–7cm. Condylo-basal length: 4.65–5.47cm. Weight: 265–520g, depending on distribution. Dental formula: 3/2–3, 1/1, 3/3, 1/2 = 32–34.

General. When threatened throws head back, bares teeth, erects tail hairs and curls tail over back (see vignette) – fullest display of contrasting body colours – pattern exposed is a warning associated with foul smelling secretions. Frequently sits up on hind legs, sometimes stands on them. Lives mainly on ground, but can climb. Russian Ministry of Agriculture classifies it as rare. Main cause of the species' decline thought to be the decrease of steppe rodents in association with human activities in many areas of its former range; European continental steppe is a dwindling habitat.

PINE MARTEN *Martes martes* Pl. 24

Name. *Martes* = marten (Lat.).

Recognition. Pelage rich chocolate brown and throat bib creamy yellow. Distinguished from most other mustelidae by longer-legged appearance. Long bushy tail and soles of feet densely furred. Distinguished from Beech Marten by longer and broader ears. Most reliable distinction from

Beech Marten is colour of snout between eyes and nose; also, Stone Martens have light under-fur on the flanks, whereas that of Pine Marten is darker. Sometimes said that hairs between foot pads of Pine Marten longer than those of Beech Marten. Faeces are commonly found at the side of forest roads and paths, prominent boulders in burns are often a good place to look. Conical latrines are found in association with dens used frequently, usually female cubing dens, more often latrines are confined to a small area within 2 metres of the main entrance. The shape and form of the dropping is very much dependent on the diet, eg a small mammal yields 'characteristic' twisted dropping while a meal of tripe from a gralloched deer produces a dropping more similar to that of a goose! Assuming the droppings are fresh, confusion with fox and mink can be avoided by smell, the musky smell is absent from marten droppings. Droppings usually 7–15mm diameter. Droppings very twisted, often folded and tapered at one end and found every 100–200m on forest paths, on soil or protrusions, or sometimes in tree forks. Paired prints at 10–100cm bounds. In snow (due to hair on feet) prints twice as large as in other media (4–5 × 3–4cm). Trails may be similar to those of hare or fox, but if foraging the trail zig-zags to investigate fallen tree roots etc., whereas if travelling the trail moves from one large tree to another, often pausing at the base of each.

Habitat. Mature forest, usually coniferous or mixed, occasionally deciduous. Individuals in forested areas often avoid entering clearings or open hillside, free of canopy cover, during daylight. Also, open rocky ground and cliffs (especially where their niche vacated by polecats). Up to 2000m in Alps and Pyrenees, but not usually found above treeline. Dens in hollow trees (under roots and stumps), old squirrels' dreys, large (eg for owl) nest boxes, also rock crevices; in the colder parts of its range becomes diurnal in winter and dens below the snow at night to avoid temperatures of below -30C. In Scotland martens will use old pigeon nest or similar, lining them with moss, on dry days throughout the year. Infrequently use parts of buildings, eg the attics of holiday cottages have been used to raise young. Dens scattered throughout home range, half a dozen dens used regularly, many others used only once or irregularly. Occasionally use vacated (or occupied) Badger setts.

Habits. Mainly crespuscular and nocturnal, but in summer also active during daylight, unlike Beech Marten. Diurnal movements usually confined to mother and cubs from June—September. In winter very seldom active during daylight. FOOD: Carnivorous. Principally Field Vole (*Microtus agrestis*), between 20–80% of diet depending on time of year, shrews 5%, Bank Voles and Wood Mice in very small numbers. Rabbits can replace *Microtus* if availability permits. Passerines up to 30% of diet from April–September (wrens being most common species), grouse, waterfowl, pigeons and woodpeckers occurring infrequently throughout the year. Frogs represent 10–20% of the diet April–September. Bumble bee nests consumed for honey represent over 30% of diet August–September. Also beetles (scarabids and carabids), mushrooms, and berries in late summer. Carrion representing 30–40% of the winter diet. In Scandinavia where small rodents available, Red Squirrel only 10% diet, but in poor vole years, squirrel may be up to 50% diet. However, in subarctic coniferous forests Red Squirrels are seldom common due to infrequent (once in 10yr) good spruce seed crop and therefore seldom reach 10% of Pine Marten diet. Voles and lemmings may constitute up to 100% of diet when abundant. Daily food intake is *c.* 20% body weight. In some areas, eg parts of Ireland and Switzerland, may be almost exclusively frugivorous for 6–7 months of the year, with insects (ants) and earthworms also forming a major component of the diet. HOME RANGE: 3–82km^2. Population density: 1 per 82ha (mature plantation) to 1 per 10km^2 (young plantation). 0.3–28km travelled per hunting trip. Non-residents travel 16–25km, even over 30km, per night. SOCIAL: Solitary, male and female occupying adjoining territories – not highly territorial. The home ranges very often overlap partially or even totally. Female may mate with several males while on heat. COMMUNICATION: Usually silent, but calls 'tok-tok-tok'. Shrill yowl like cat in mating season. Playful chasing and growling prior to mating. Rub anal gland secretion onto sticks, bushes, branches. Faeces on roots or trunks of fallen trees and similar conspicuous sites, especially near territorial borders. Urine on trails (including those of skis). All these scents left on mounds of snow. Abdominal scent gland increases in size and activity during the breeding season. Pine Martens frequently wipe their abdominal gland on protruding objects in their path, raising the tail high over the back in the process. No difference in marking frequency between the sexes. Excellent sense of smell, also sight and hearing. Insatiable curiosity.

Breeding. Mating in July–August (copulations 15–75 mins). Delayed implantation 165–210 days. Birth April–May. Sexual maturity: female 15 months; male 27 months. Gestation 28–30 days. Litter size 1–6 (usually 3). Litters per year 1. Mammae 4. YOUNG: Blind at birth. Emerge from den at *c.* 8 weeks, not fully independent until 6 months. Young of 2–3 months are observably more cautious in clinging onto branches. Young born with thin covering of white/grey fur; at 3 weeks greyish-brown. Juveniles light brown, moulting to dark brown in first spring. Weaning age 8–10 weeks. Parental care is exclusively by female. Some evidence of infanticide.

Lifespan. Max. recorded 17–18 years (in captivity). One study in Scotland indicates average lifespan 3–4 years, maximum 11 years, in wild. Man an important cause of mortality. Very susceptible to habitat disturbance. Trapping another major threat. Occasionally poisoning, especially in hill grazing areas (eg Ireland, where population was decimated by this). Road casualties. Direct predation by eagles, Eagle Owls and possibly Wolverine.

Measurements. Head-body length: 36–56cm (eg Germany male 48–53cm; female 40–45cm). Tail-length: 17–28cm. Hind-foot length: 8.5–

9.5cm. Shoulder-height: *c.* 15cm. Condylo-basal length: 77–88mm. Weight: 0.5–2.2kg, male *c.* 12% heavier than female (Eastern Europe male 670–1050g, female 484–850g). Regional variation (heaviest Denmark and Caucasus, lightest Eastern Europe) (birth *c.* 30g). Dental formula: 3/3, 1/1, 4/4,1/2 = 38.

General. Arboreal adaptations. Very agile, moving fast through treetops. Ascends by series of upward jerks while embracing tree, as does squirrel. Record leaps of 4m, using tail to balance. Important fur-bearing species in Russia – although current harvests 20% those of 1920s. Commercially unsuccessful attempts at captive breeding (failed due to late maturity, small litters, general reluctance to breed in captivity). Where squirrels migrate, Pine Martens reported to follow them.

BEECH MARTEN *Martes foina* Pl. 24

Name. Stone Marten, House Marten. *Foinos* = dark red (Gk.).

Recognition. Rich brown with pure white or greyish throat bib divided by dark stripe into left and right parts (cf Pine Marten). Also distinguished from Pine Marten by less hairy feet (pads exposed) and smaller, narrower ears, and shorter and broader muzzle. Variations include the population on Crete where Beech Martens have only a small greyish throat patch. Bounding gait. Third premolar tooth has convex external border (concave in Pine Marten). Faeces at latrine sites – unlike Pine Marten. Droppings 4–10 × 1cm, may contain fruits such as cherry stones and remains of food discarded by people.

Habitat. Depends on region. Deciduous woodland, wooded margins, open rocky hillsides, often near houses. In some regions (Switzerland, N. and E. France and S. Germany) very common in towns, and effectively a commensal of Man, frequently occupying houses (where damages insulation and electrical wiring). In some areas, rare in woods, common in towns. Up to 2400m (i.e. above tree-line) in Alps. Dens in hollow trees, heaps of stone, lofts, stables, barns, house attics, car engine spaces. Occasionally digs own burrow in ground.

Habits. Strictly nocturnal. FOOD: Largely carnivorous – principally rodents (mice, voles, squirrels), birds and berries. In some areas almost exclusively frugivorous/insectivorous and scavenging. HOME RANGE: *c.* 80ha, but extremely variable; town martens may have very small ranges which they cover several times each night. SOCIAL: Generally solitary, but town martens often forage in groups of 4–5 animals. COMMUNICATION: Voice similar to Pine Marten, but much noisier even outside of the breeding season. Shot as a nuisance in Switzerland on the grounds that they are noisy.

Breeding. Mating in mid-summer. Delayed implantation 230–275 days. Birth spring. Sexual maturity 1–2 years. Gestation 30 days. Litter size 1–8

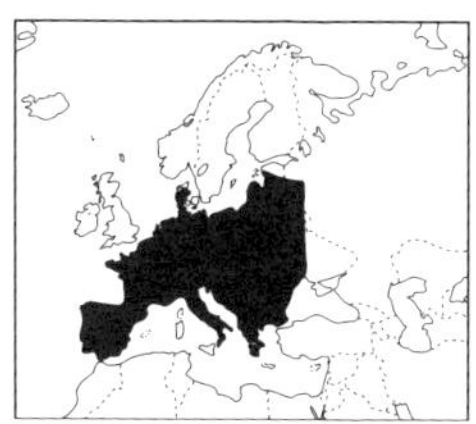

<*Beech Marten.* Throughout C&W Europe, and on Corsica and Balearic Islands.

>*Sable.* Probably exterminated W of 60°E, reintroduced in European Russia. Range shown is previous distribution.

(usually 3–4). Litters per year 1. Mammae 4. YOUNG: Blind at birth. Weaning age 8 weeks. Parental care exclusively by female. Young venture out of the den at 8–10 weeks, and accompany the female for 2–3weeks. Some hints at infanticide in towns where contacts and conflicts between martens are common. Very conspicuous vocal contact between mother and young.

Lifespan. 18 years (in captivity). Cause of death often related to human activities – trapping, shooting, high rate of road casualties, especially at onset of breeding season.

Measurements. Head-body length: 42–48cm. Tail-length: *c.* 26cm. Hind-foot length: 8–9cm. Shoulder-height: *c.* 12cm. Condylo-basal length: 77.6–84.6mm. Weight: 1.3–2.3kg (male larger than female) Dental formula: 3/3, 1/1, 4/4, 1/2 = 38.

General. Common in Continental Europe but absent from northern latitudes. Climbs well, but far more terrestrial than Pine Marten; very curious and extremely versatile: able to adapt to wide range of habitats/diets/climates. Habit of sleeping under bonnet of cars has spread as a cultural trait across Central Europe (40 cars per day are reported 'sabotaged' by Beech Martens in Switzeriand). Also bites cables in power stations and railway controls. They may do this as a consequence of great dietary versatility: the young bite almost anything new to sample its edibility – this may explain why damage declines in autumn when juveniles are more experienced.

SABLE *Martes zibellina* Pl. 24

Recognition. Pelage luxuriantly dense, soft brownish black with paler but indistinct throat bib. Similar to Pine Marten but longer legs and larger ears.

Habitat. Coniferous or mixed woodland (generally using undergrowth), often close to rivers and streams. Between rocks or tree roots.

Habits. Largely nocturnal, but frequently active by day. FOOD: Carnivorous. Birds, squirrels, small rodents. HOME RANGE: Population density one per 1.5km^2 in pine forest, one per 25km^2 in larch forest. SOCIAL: Solitary. COMMUNICATION: Reputedly less vocal than Pine Marten.

Breeding. Mating in June–July. Delayed implantation, 300–330 days. Births in April–May. Sexual maturity 15–16 months. Gestation 12–13

months (including delayed implantation). Litter size 1–5 (usually 3–4). Litters per year 1. YOUNG: Weaning age 7 weeks.
Lifespan. Max. recorded 15 years.
Measurements. Head-body length: 32–46cm. Tail-length: 14–18cm. Hind-foot length: 7–8.5cm. Condylo-basal length: 80–90cm. Weight: 0.9–1.8kg (birth 30–35g). Dental formula: 3/3, 1/1, 4/4, 1/2 = 38.
General. Widely extinct in Europe. Climbs and leaps with agility, but rarely swims. Reputedly hybridises with Pine Marten.

WOLVERINE *Gulo gulo* Pl. 21

Name. Glutton. *Gulo* = glutton (Lat.).
Status. Vulnerable.

Recognition. Short and heavy with a fairly long tail. Dark with light patches on head and yellowish stripes from shoulders along flanks to bushy tail. Dark brown, almost black or much paler (often bleached by sun). Powerful, short legs and big paws. Gallops, leaving paired or triple prints. Light patches on head extend forward over forehead. Largest terrestrial member of the weasel family; rather bear-like in stance and movement. Prints with 5 digits – if 1 digit fails to register then the print resembles that of Wolf. Bounds through snow, grouping its feet in fours, spaced 60–90cm apart. Droppings twisted and full of hairs: 15cm long × 2cm wide.
Habitat. Mountain woodland with rocky slopes, often near marshes. Bare mountains, mountain birch forests, submontane coniferous forests, open areas in wood: bogs and lakesides. In northern parts of range animals leave forest and wander on tundra in summer. Catches most food in coniferous forests. Dens usually deep down in a snowdrift, or in a shallow scrape on rocky slope in thick bushes or between rocks.
Habits. Active by night and day, with activity cycles of 3–4 hours. Mostly nocturnal in summer. FOOD: Rodents, birds, eggs, invertebrates, fruit, berries, carrion. Will occasionally attack Elk, Reindeer, foxes, Roe Deer and wild sheep if snow conditions right – but large mammals mainly taken as carrion in winter. Winter food of Wolverine in Finland is in the order of importance: carrion (encountered during long daily wanderings), Reindeer in Reindeer management area (especially if snow conditions allow, may also kill much more than can eat), Ptarmigen and other small game. Snowless season: vegetable matter especially berries also very important. Reputation for destructiveness due to raids on hunters' food caches and on animals caught in traps. HOME RANGE: 300,000ha per male, shared with 2–3 females. Female 50–350km^2; male 600–1000km^2. Female closely defends it against other females in spring, marking with scent glands. Male roams within ranges of several females. At least part of the range is territory, within which individuals of same sex not tolerated. Scandinavia – range size in winter up to 2000km^2. Migrations (70km or more) undertaken in winter. Up to 45km travelled per day.

SOCIAL: Usually solitary (except in breeding season). COMMUNICATION: Growls and hisses when angry; grunts and squeals at play.

Breeding. Mating April–August; delayed implantation until turn of year. Born February–March. Sexual maturity during second year, then breeds every year. Young stay in and around den until early May and within female's home range until late August, female young sometimes longer. Litter size 1–4. Litters per year 1. YOUNG: Nursed 8–10 weeks. Males disperse by onset of next breeding season; some female kits may remain in or near their mother's home range indefinitely.

Lifespan. Max. recorded 13 years (wild), 18 years (captivity).

Measurements. Head-body length: 62–67cm. Tail- length: 13–25cm. Hind-foot length: 14–18cm. Shoulder height: 40–45mm. Condylo-basal length: 130–140mm. Weight: 9–30kg. Dental formula: 3/3, 1/1, 2–4/2–4, 1–1/2 = 28–38.

General. Shot and trapped where still regarded as big game, as pest or furbearer. Alaska – 800 pelts per year, each worth U$150–200. 1973 Finland – 40 in wild. No effective protection in Finland and Scandinavia until 1978. Now partially protected in Norway, where 2 core areas have been defined (and a possible third proposed). Full protection in Sweden. The population is declining in Finland, mainly due to excessive persecution by hunters using snow-mobiles: supposedly protected, but special killing licences easily granted for Reindeer owners, also illegal hunting by using snow mobiles is common; in areas outside Reindeer management area population status stable or slightly increasing. Habit of removing bait from traps in regions where principal occupation is fur trapping makes it unpopular and consequently remorselessly hunted. Fur does not freeze and so used for coats or trimming round hoods by Eskimos. Often hunts by ambushing. Reputation for greediness (alternative common name is glutton) – caches food. Various lurid tales about its physical powers, cunning and appetite. Formerly found in area ranging from Scandinavia to Kamschatka, living up in the mountain forests, bogs and coniferous forests. Current population density in Scandinavia – one per 200–500km^2. Huge home range makes this species especially difficult to protect. Although apparently not yet threatened in large parts of its range, susceptible to human disturbance. Long-term survival open to doubt without better understanding of natural history and a sound conservation policy.

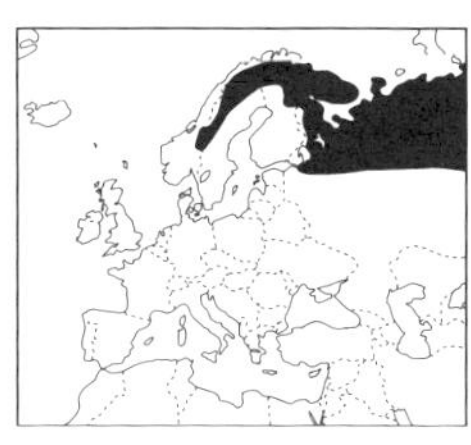

< *Wolverine.* Scandinavia, generally N of Trondheim.

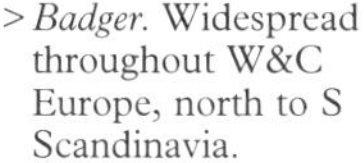

> *Badger.* Widespread throughout W&C Europe, north to S Scandinavia.

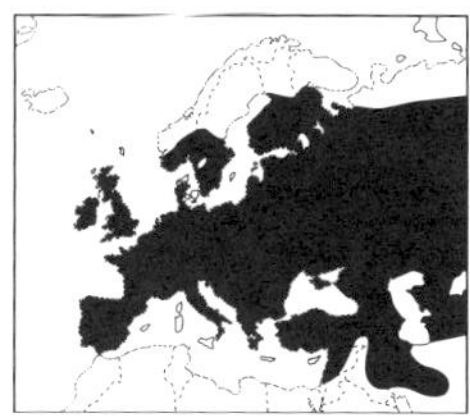

EURASIAN BADGER *Meles meles* Pl. 21

Name. Eurasian badger. *Meles* = badger (Lat.); 'badger' possibly from French *becheur* = a digger.

Recognition. Grey upperparts, black underside and legs. Heavily built, short legs and a short blunt tail with white tip, also white fringe to ears. Long snout. Hairs light at base and tip, black in middle. Black longitudinal stripe along either side of white head (covering eye region), suggested as warning coloration. Variation in pelage includes albino, semi- albino, melanistic, erythristic (reddish), which can be quite common in some areas. Wedge-shaped body, rather small head. Single, prolonged moult begins in spring. New guard hairs and underfur grows in autumn. Conspicuous fieldsign is the communal den or sett, consisting of 3–10 large entrances 10–20 metres apart. Exceptionally, entrances 100m apart may be linked. Chambers lined with bedding. Spoil heaps outside contain discarded bedding (bracken, hay) and soil. Paths leading from sett well marked. Faeces usually deposited in shallow, uncovered pits (latrines) – consistency varies with food eaten (muddy after eating earthworms). Latrines often linked by an obvious path. Prints register 5 toes and broad plantar pad – heel marks show in soft mud or snow. Fore-print larger than hind. Hindfoot length is 9–11cm. Forefoot width of adult is 4.5–6.5cm. Prominent claw marks from forefeet. Scratch marks on earth and signs of digging, especially when foraging for tipulid (cranefly) larvae, for example and for surface dwelling species of earthworms (*Allolobophora*; cf *Lumbricus* worms caught on surface). Grey, brindled hairs often snagged on barbed-wire fence.

Habitat. Most abundant in undulating, mosaic landscape of woods and pasture. Favours mixed deciduous woods with clearings. Also open pastures, railway embankments and large gardens. Not above tree-line on mountains. Sett typically has a number of entrance holes, several metres apart, leading into underground tunnels which usually stretch for 10–20m. A record sett had 180 entrances, 880m of tunnels, 50 nest chambers and numerous underground latrines. Sleeping chambers lined with bedding material of grass, leaves etc., which Badger gathers with forelegs and chin, dragging it backwards into a tunnel. Other oddments gathered in sett, including 250 golf balls in one sett. Diameter of tunnels at least 20cm, and often much more at entrances, which have large accumulations of soil at the front, sometimes together with old, discarded bedding. Setts used and enlarged by successive generations: usually one main sett with several 'outliers' or small setts scattered throughout the territory. One burrow system may be used for decades or centuries, by one generation of badgers after another, so sett continuously increases in complexity. Den sites include woodland, scrub, hedgerows, quarries, sea cliffs, moorland, fields, natural caves, tips, under buildings, embankments. Preferences for den sites in UK: 56% – deciduous woodland and copses; 13% – hedgerows and scrub; 9% – open fields.

Habits. Nocturnal, crepuscular except where undisturbed. Emerges before dusk May–August, usually after dusk rest of time. Emergence much less regular from November–February during period of winter lethargy when use deepest part of sett. Within the sett 2–3 animals share a nest chamber, but individuals tend to move chambers, and companions, every few days. Cubs (in particular) sometimes forage during daylight if food short in summer. FOOD: Omnivorous and opportunistic foragers: earthworms, plus other invertebrates and vegetation; insects, including beetles, caterpillars, wasps and bees' nests, plus birds' eggs (on occasion, also ground nesting or roosting birds), grubs, carrion (esp. winter), voles, moles, rabbits etc; South West UK: 75% stomach contents had earthworms present; 65% only earthworms. Also bulbs, fruit, hazel nuts, bilberries, raspberries, oats, wheat. In winter clover and grass. Some reputedly specialise on hedgehogs. Tend to be more carnivorous in spring and early summer, and take more vegetation in late summer and autumn – cereals and fruit being important items. HOME RANGE: Territories of 30–50ha in rich habitat, up to 150ha or more elsewhere. Population density 2–20 adults per 100ha, typically 10 per 100ha in good Badger areas in UK. Most prolonged foraging, up to 10 hours away from sett, occurs in autumn. Dispersal least likely in stable, high density areas. More males disperse than females, and the tendency to disperse is greater in sexually mature animals, i.e. older than 2 years. Emigrants sometimes join a neighbouring social group but may also disperse over several km. No difference in dispersal distances between sexes. Temporary excursions of over 1km by boars in late winter–early spring, probably to steal matings in neighbouring territories. SOCIAL: Territorial social groups (or clans), but feeding areas of neighbouring groups occasionally overlap. Each group member tends to forage alone (except when cubs). In some regions, eg Italian mountains, reported to have more typical mustelid spatial organisation, with single females maintaining territories, several of which may be overlapped by one male. Some groups have very large membership: 2–25 adults (averaging 5–8) plus yearlings. Most groups have more females than males. More than one individual of each sex may breed in a given year, and some litters may have multiple paternity. Females breeding in one year do not necessarily do so the next year. Mutual grooming common within group. COMMUNICATION: Numerous vocalisations have been described as snorts, scolds, moans, quavers, squalls, growls and a long drawn-out scream. Cubs make high-pitched whickers and puppy-like noises; also a screeching distress call. Adult makes a deep warning growl in aggressive encounters. Female said to make a call with one note reminiscent of the cry of a moorhen. Mating males make a throaty purr, and females and cubs make a quieter purr. Has paired anal glands, and a large subcaudal glandular pocket. Group members, especially dominant ones, wipe secretions of subcaudal pouches on each other. Frequently press rump to ground while foraging, thereby marking with subcaudal gland secretions. Some latrines near sett, but others (*c.* 70%) mark perimeter of territory using both droppings and secretions. Peak in latrine use in early spring, when bite wounds especially common in males. Dominant sense is smell, but hearing acute, sight poor – short-sighted and adapted for poor light conditions.

Breeding. Mating mainly Febrary–May, but known in every month (oestrus 4–6 days, mating 15–60 mins). Mating also quite common in July–September. Young born mid-January usually peaking early Feb (but births recorded mid-December–April). Sexual maturity: male 9–18 months; female 1–2 years (Southern England). First breeding age commonly at least 2 years old. Delayed implantation 3–10 months. True gestation 7 weeks. Litter size 1–5 (average 2.7). Litters per year 1. Mammae 6. YOUNG: Newborn is pink with grey silky fur. Eyes open at 5 weeks; milk teeth erupt at 4–6 weeks; permanent teeth appear *c.* 12 weeks. Weaning starts at 12 weeks, but if food short may be delayed for 4–6 months. Cubs remain below ground about 8 weeks. Live with, and are probably dependent upon, female until autumn, often over first winter. During weaning female may regurgitate semi-digested food.

Lifespan. Max. recorded 14 years (wild) – 16 (captivity). Only 50% cubs survive first year. Adult mortality averages 30% per year, with higher mortality rates in males, contributing to preponderance of females in adult population.

Measurements. Head-body length: male 68.6–80.3cm; female 67.3–78.7cm. Tail-length: male 12.7–17.8cm; female 11.4–19cm. Hind-foot length: 9–11cm. Shoulder-height: 30cm. Condylo-basal length: male 116–130mm; female 111–125mm. Weight: September–February average 12.2kg, March–May average 8.8kg. Male 12.3kg; female 10.9kg for SW England, with greater differences in north of range, where badgers store more fat. Females heavier than males in the autumn. Dental formula: 3/3, 1/1, 3/3, 1/2 = 34 (plus sometimes an additional minute vestigial premolar just behind canine teeth).

General. Widespread in Europe south of Arctic Circle including countries bordering the Mediterranean and some islands, eg Rhodes. Fairly common in hillier and less cultivated regions, but also occurs in some urban areas. Very common in south and south west Britain. Subject to persecution and road deaths. Highly adaptable species. Do not hibernate but become semi-dormant, although there is some evidence for reduced body temperature in winter. Weight in late autumn may be 3kg greater than in spring. Many subspecies named; improbable that more than 10 are valid. Fur of back and flanks traditionally used to make shaving brushes and sporrans. In UK (especially SW) and Ireland involved in epidemiology of bovine tuberculosis, and therefore trapped and killed where thought to be the source of infection in cattle. Numbers greatly reduced in many European countries when killed during the gassing of dens in an attempt to kill Red Foxes during anti-rabies campaigns. Traditionally the victim of badger-digging, where people dig into sett to catch a badger cornered by a terrier dog (the name of the German breed Dachshund translates as badger hound). This sport is now widely illegal. Black and white mask may be warning colouration.

OTTER *Lutra lutra* **Pl. 22**

Name. *Lutra* = otter (Lat.) from Sanskrit *udrah*, Gk. *enydris*, otter (Lat.).

Status. Vulnerable.

Recognition. Long slender body, short legs, long tail with thick base and tapering evenly. Broad muzzle, small ears. All four feet webbed. Swims smoothly (cf more jerky action of American Mink), leaving U-shaped wake. Watches from water with only eyes and nose above surface. Dives smoothly and silently from surface swimming position; back and base of tail arches above surface as drives itself downwards. Smooth sleek fur as emerges from water, soon dries to give 'spikey' appearance as guard hairs group together. Probably only one annual moult which progresses over a long period. 'Sprainting' (defecation) sites at traditional places set at intervals along water's edge, on otter paths, outside holts and at points of entrance to and exit from water. Often placed on prominent places such as large stones, logs at the water's edge or at the point of entry of tributaries or ditches to main river, or on ledges under bridges (accumulating into mounds or 'stools' along coast). Faeces have characteristic 'sweet' smell, more pleasant than that of mink scats, and usually lack the twisted end often composed of feather and fur in mink faeces. Droppings characteristic tarry black, but when leached by rain may appear like piles of cigar ash, containing fish bones and vertebrae. Runways (especially in areas of high population), rolling places and haul-out sites. Tracks 6–7cm length, width varying according to how well 5th toe prints and splay of foot; may show webbing and claws. Bounds *c.* 0.5m long. Broad furrows as slithers in snow or mud (mud slides on banks). When swimming underwater, leaves chain of bubbles at surface, which are trapped under ice. Vegetation may be scraped off outside holt entrances.

Habitat. Fresh water, where suitable cover (eg rivers, lakes, canals, marshes, sometimes ditches with only a few cm of water); also marine (eg coastal and estuaries), particularly where coast is rocky. Lives in holt: cavity in bank, hollow tree, between roots, rocky clefts or tunnels in peat. Entrance may be underwater with an air vent to the chamber. Lying up sites ('couches') are above ground in patches of *Phragmites* or other vegetation. One male known to use 37 different resting sites and holts within its home range (at intervals of 1320m), and a female 23 sites (at intervals of 970m). Chambers lined with dry vegetation.

Habits. Largely nocturnal (lie up by day in 'holts' – underground – or 'hovers' – above ground), with period of inactivity in middle of night. May be diurnal when feeding on nocturnal fish, eg NW Britain and Shetland. FOOD: Carnivorous. Principally fish (eg Eel, Perch, Pike, Burbot, Carp, Salmon, usually less than 20–25cm long) and occasionally other vertebrates (eg water-birds, Water Voles, rats, amphibians) and invertebrates (eg Crayfish, crabs, worms, insects). Invertebrates may be taken in quantity when available. HOME RANGE: Linear territories along streams and shorelines, eg Scotland: 39km long for male (57km^2), 16–22km long for female (14–30km^2); Swedish lakes: female 7km width, male 15km width, or irregularly shaped ones in marshes (eg up to 1km^2. male 15km diameter; female 7km diameter). Hunting trips 3–10km per night. SOCIAL: Lives in family groups of one or more females and cubs of the year occupying group territory, otherwise adults solitary and come together only for a few days at mating. COMMUNICATION: Not very vocal, soft shrill whistle, short chirp for contact between adults and cubs, growls, hisses, chatters. Jelly-like secretion used to mark faeces (spraints) themselves, placed at intervals to mark paths, holts. Occasional sign heaps: small mounds of sand or mud scraped up or twists of grass, sometimes with a spraint on top. Hunts most efficiently by sight, but vibrissae important in detecting fish movements in deep or dark water.

Breeding. Any time of year but highly seasonal in Shetland (summer). Approximately equal numbers of cubs born in every month in Britain. Earliest known breeding: male 1 year 5 months; female 1 year 10 months. Gestation 61–63 days. Litter size 1–5 (usually 2–3). Litters per year 1 (or once every two years), capable of producing cubs at 12–14 month intervals. Mammae 6. Contrast with Canadian Otter which in very similar but which has delayed implantation. YOUNG: Cubs blind up to 35 days. Weigh 1.4kg (male) or 1.0kg (female) at 2 months and grow at rate of 0.65kg per month (male) and 0.45kg per month (female) from 2–10 months. Taken to water and encouraged to swim around 3 months – appear very buoyant at this age. Weaning age *c.* 16 weeks. Parental care by female only. Cubs remain with mother for 10–12 months.

Lifespan. 11–15 years (in captivity), but much shorter in wild (mean 3–4 years). Causes of death include food shortage, road traffic, drowning in lobster creels and eel fyke nets, disease. Cubs occasionally killed by dogs. Hunting with hounds/trapping no longer legally permitted in Britain.

Measurements. Head-body length: male 60–90cm; female 59–70cm (Birth 15cm). Tail-length: male 36–47cm; female 35–42cm. Hind-foot length: 11–13.5cm. Shoulder-height: *c.* 30cm. Condylo-basal length: 10.5–12.5cm. Weight: 6–17kg (male *c.* 10kg (6–17kg); female *c.* 7kg (6–12kg)). Dental formula: 3/3, 1/1, 4/3–4, 1/2 = 36–38.

General. Amphibious adaptations eg 1.5–2km per hour swimming speed for up to 7–8 hours at a time. Normal dive times when hunting 10–40

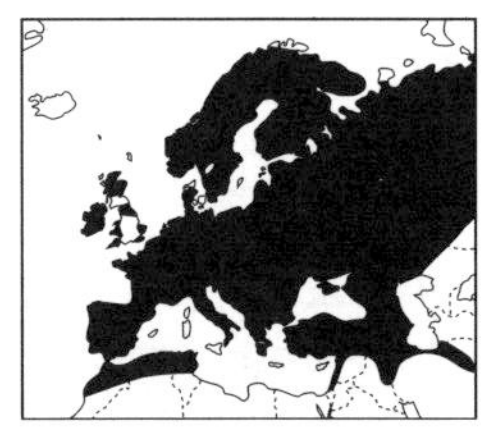

< *Otter.* Ubiquitous throughout W Europe, but not Mediterranean islands.

> *Common Genet.* Widespread through Iberian peninsula.

seconds for successful dives. Recent decline in numbers associated with pollution, habitat destruction and human disturbance. Overlap in diet with American Mink *c.* 60–70%. Very playful species (eg dives for pebbles).

COMMON GENET *Genetta genetta* Pl. 27

Name. European Genet, Small-spotted Genet. *Vivera* = ferret (Lat.).
Status. *G.g. isabelae*, the Ibiza Common Genet of Spain, is rare.
Recognition. Slender, elongate feline build, short legs, long tail with somewhat pointed tip. Pelage: tawny-grey with dark spots merging to form longitudinal stripes. Tail alternate dark and light rings. Semi-retractable claws. Forefoot tracks 3 × 2.5cm, hindfoot 3 × 3cm. 4 upper digits are spread evenly above plantar pad, 5th digit, small and eccentric, rarely leaves a mark. Leaves claw marks on tree trunks like those of squirrel, marten, Wildcat. Best fieldsign is droppings; very large relative to Genet's size: 10–24cm long, 1.5–2cm diameter. Always laid in a horseshoe shape and end in a tuft of blades of grass; black when fresh, whiten with age. Habitually defecates in the same place, the midden – on a shelf of rock, or in the fork of a branch overlooking the range; or sheds, rafters of buildings or hunting platforms etc. Several genets use the midden which can contain dozens of faeces.
Habitat. Quiet areas, far from human habitation: dense scrub and rocky terrain, woodland with streams. Up to 2000m in the Pyrenees. Nest lined with vegetation in trees or burrow, among rocks or under bushes. During the day the genet rests in the tops of bushy-topped trees, eg some conifers or chestnut.
Habits. Crepuscular and nocturnal. FOOD: Omnivorous. Principally rodents (especially mice), Rabbits, fruit, berries and insects. HOME RANGE: *c.* 5 km^2. Travel up to 3km per hour. SOCIAL: Solitary but seemingly tolerant in encounters with others. COMMUNICATION: Abrupt scream in alarm.
Breeding. Throughout year, but peaks in births April–May and August–September. Sexual maturity 2 years. Gestation 70 days. Litter size 1–4. Litters per year 2 (in captivity). Mammae 4. YOUNG: Begin to leave the nest at 8 weeks old. Weaning age 6 months. Young independent after 12 months.
Measurements. Head-body length: 47–60cm (birth: 13.5cm). Tail-length: 40–51cm. Hind-foot length: 7.5–8.5cm. Shoulder-Height: 18–20cm. Condylo-basal length: *c.* 90mm. Weight: 1–2.2kg. Dental formula: 3/3, 1/1, 3–4/3–4, 1–2/1–2 = 32–40.
General. Agile. Climbs, leaps and swims well.

< *Introduced Mongooses.* The *Egyptian Mongoose* has been introduced to Portugal and southern Spain, and recently to the island of Mljet, Yugoslavia. There have also been scattered introductions of the *Small Indian Mongoose* to Yugoslavia and Adriatic islands, Hvar, Korcula, Peljesac and Mljet. The *Indian Grey Mongoose* was introduced to Italy in the 1960s.

EGYPTIAN MONGOOSE *Herpestes ichneumon* Pl. 27

Name. Ichneumon mongoose. *Herpestes* = creeping (Gk.); *ichneumon* = mongoose (Gk.). Literally the tracker.

Recognition. Pelage uniform greyish-brown, coarse hair. Tail has broad base tapering to narrow tip. Small, broad ears. Non-retractable claws. Horizontal pupil (uncommon among carnivores). Footprints *c.* 3cm long, naked finger pads, end in long curved claws. Droppings *c.* 4.7 × 1cm.

Habitat. High ground, rocky scrub, heath but also in cultivated habitats, orange groves etc. Strictly terrestrial, though may live in marshy habitats near water. May excavate own burrow or den in dense shrubs or an existing hole of any kind.

Habits. Largely diurnal, but active by day and night. FOOD: Carnivorous. Young Rabbits, rodents, birds, snakes, eggs. Eggs broken by being clasped in forepaws and thrown backwards between hindlegs against rock (see vignette). Insects an important component of food. May feed on garbage dumps. SOCIAL: Family groups. Young of previous litter remain in association with mother after birth of subsequent litter. COMMUNICATION: occasional whistles.

Breeding. Gestation 84 days. Litter size 2–4.

Lifespan. 20 years (captivity).

Measurements. Head-body length: 50–55cm. Tail- length: 33–45cm. Hind-foot length: 8.5–9.5cm. Shoulder-height: 19–21cm. Condylo-basal length: *c.* 10cm. Weight: 2.3kg. Dental formula: 3/3, 1/1, 3–4/3–4, 1–2/1–2 = 32–40.

General. Mongoose figures on temple walls dating to 2800 BC. Considered sacred by ancient Egyptians. In Spain they are not protected and are harassed and hunted, so quite rare and shy. Contrasts with Israel, where they are quite tame and very common and can reach high densities, particularly around human habitations. In Israel an inverse relation has been found between mongoose density and the incidence of *Vipera palaestinae* (the most common venomous snake around agricultural settlements and elsewhere). The Egyptian Mongoose, along with several other mongooses, is resistant to snake poison.

INDIAN GREY MONGOOSE *Herpestes edwardsi* Pl. 27

Status. Introduced.
Recognition. Pelage brindled grey-brown, with paler tip to tail. Similar to Egyptian Mongoose but smaller.
Habitat. Scrub and rocky hillsides.
Habits. FOOD: Young Rabbits, rodents, birds, reptiles, eggs.
Breeding. One female reputed to produce 5 litters in 18 months.
Measurements. Head-body length: *c.* 45cm. Tail- length: *c.* 45cm. Condylo-basal length: male *c.* 82mm; female *c.* 73mm. Dental formula: 3/3, 1/1, 3–4/3–4, 1–2/1–2 = 32–40.
General. Introduced to Italy (just south of Rome, around Monte Circeo) in the 1960s. The Small Indian Mongoose *Herpestes auropunctatus* has been introduced to several Yugoslavian islands to control horned vipers. It is considerably smaller than the Indian Grey Mongoose: head-body length <39cm.

WILDCAT *Felis silvestris* Pl. 26

Name. *Felis* = cat (Lat.); *sylvestris* = of the woods.
Recognition. Similar to, but more robust than, striped tabby Domestic Cat. Wildcat has no blotchy markings on body (only stripes). Chief distinguishing feature is its bushy tail, which has 3–5 completely separate broad, black rings, and a rounded/blunt black tip. Tails of hybrids resemble Domestic Cat's being thinner with less distinct bands. Amber eyes, pink nose. Seasonal moults – fairly heavy moult in spring, lighter moult in late summer. Footprints in snow the most useful indicator of the presence of cats: four toes and tri- lobed main pad, 5th toe on front foot and metacarpal pad do not register. Generally larger than Domestic Cat (although some overlap may occur) and more rounded than that of fox. Faeces compact, cylindrical, often pointed ends, and traces of bone fragments, strong musty odour when fresh. Faeces sometimes distinguishable (especially from foxes') by measurements, but much overlap with those of feral Domestic Cats: Wildcat diameter 17mm (14–24mm), length 35mm (15–61mm); feral Domestic Cat diameter 15mm (6–24mm), length 34mm (14–90mm). Prey remains include chewed feathers; flesh eaten from larger prey but bones generally intact, and showing chewed scapulae and ribs; rarely decapitates prey, although young Rabbits and grouse may be almost entirely eaten, with the head sometimes removed. Rabbit skins characteristically turned inside-out and attached to leg bones/feet. Foxes may also invert lagomorph skins but generally consume animal from anterior end, eating most of the bones. Domestic or feral cats may leave similar feeding signs to Wildcats.

Characteristic scratch marks on trees and saplings (usually of hardwood species) – again, domestic/feral cats may leave similar traces.

Habitat. In general, occupies deciduous forests of the plains and lower hill regions, living mainly near natural clearings and in the peripheral zones of large forests. In Scotland favours border of forest and open hill (eg grouse moors), mixture of open land and scrub up to 2000m. Also, forestry plantations, especially in early stages, an important habitat. High moorlands mostly used in summer, and retreat to lower forested areas when snow cover prevails, though can withstand severe cold. Dens in hollow trees, rock crevices, rabbit burrows, disused Badger setts, under fallen debris. Dens used from November–February, but rarely goes underground during warmer weather.

Habits. Largely crepuscular and nocturnal, but found to be active from 16.00–02.00 hours in summer (Scotland) and active during other diurnal periods in winter. FOOD: Carnivorous. Principally small rodents and lagomorphs, and in some areas birds, with amphibians, fish, insects taken rarely, and exceptionally lamb and Roe Deer kids. Grass in small quantities probably important in diet to prevent hair balls (as with Domestic Cats). Generally prefers hunting in the open (eg fields, meadows and clearings) rather than deep forest. HOME RANGE: Females sedentary and exclusively territorial. Many males, particularly young animals, are nomadic, and movements overlap females' ranges. In winter and during mating season, males may restrict themselves to a forest home range. Range size in Europe varies from 60–350ha, depending on prey abundance. In good quality habitat, such as that found in NE Scotland, has a mean annual home range of 175ha. Densities also vary, ranging from one cat per 0.7–10km^2. Data from Lorraine (NE France): home ranges, female *c.* 200ha, no seasonal change; male 220–1270ha, smallest belong to old transient males and young males, largest to sedentary male. Speed of locomotion and daily distance travelled increase with size of home range. Male patrols ranges less often than does female. Resident males' home ranges overlap those of 3–6 females – both sexes sharing the same hunting/resting sites. Distribution of male dependent on that of adult females. Distribution of female dependent largely on prey. Substantial differences in food resources between areas, eg small rodents (France) versus Rabbits (Scotland). Young males disperse in autumn and winter of first year, most females remaining in birth areas. SOCIAL: Solitary, territorial (except when mating). Territorial marking by spraying urine over trees, vegetation, boulders etc and leaving faeces in prominent places. Mutual avoidance more common than aggression. COMMUNICATION: Vocalisation as Domestic Cat, purrs, growls, mews. Posture and expression as Domestic Cat. Scent marking includes sprayed urine, together with anal gland secretion; glandular areas of lips and cheeks rubbed on objects (and other cats). Some faeces placed on conspicuous landmarks along paths, other faeces buried. Scent glands: apocrine (sweat) and sebaceous (oil) scent glands on head, chin and base of tail; sebaceous glands between toes – possibly depositing scent partial function of scratching trees/saplings (also a visual marker). Good vision in poor light, due to presence of tapetum. Keen sense of smell, and exhibits Flehmen reaction. Vibrissae mostly white, majority on upper lips, few on cheeks/over eyes; shorter whiskers on chin. Ears can move through 180 to detect sound; hearing very sensitive. One female in a radio-tagging study suffered from snow-blindness for about 2 weeks in winter, yet still traversed most of her home range and survived in reasonable condition.

Breeding. Mating in late winter and spring, births in April–September (peaks in May). Second litters probably indicate hybridisation, as do litters outside usual breeding season, although both may be due to repeat litters where the first was lost. Sexual maturity: male 1 year; female 9–10 months, though not all young cats breed in first year, and not all adult cats every year. Gestation 63–69 days. Litter size averages 3.4 in captivity (range 1–8), in wild 3.7 (range 2–6). Litters per year 1 – occasionally a second litter in captivity. Mammae 8 (axillary pair sometimes absent). YOUNG: 100–163g at birth, blind but covered in fur, darker and more distinctly marked than adult, with less bushy tail. Pink pads darkening to almost black at 3 months. Eyes open 10–13 days, blue changing from 7 weeks to final amber colour at 5 months. Weaning age from 2–5.5 months. Only mother cares for young; family splits up at *c.* 5 months. Rudimentary den with no nesting material, but female cleans floor by scratching.

Lifespan. Max. recorded 15 years (captivity), 11 years (wild, Scotland). Persecution major cause of death in many areas of Scotland. Golden Eagles, foxes, Stoats and martens can all take kittens. Starvation during winter, especially during long periods of excessive snow cover which limits hunting. Road accidents. Disease (eye virus infection). Sex ratio at birth 1:1, but suggestion of higher mortality of females thereafter.

Measurements. Head-body length: 48–68cm (eg Scotland, male 51.5–65.3cm; female 49.5–59.5cm), means of 40.5cm (Germany) to 98.1cm (Romania). Tail-length: 21–38.5cm (eg Scotland, male 21–37cm; female 24–36cm). Hind-foot length: 10–16cm (eg Scotland, male 11.5–14.7cm; female 10.5–14cm). Shoulder-height: 35–40cm. Condylo-basal length: 82–93mm. Weight: 1.6–8kg (eg Scotland, male 3.0–7.1kg; female 2.5–5.6kg) (birth 100–163g). Dental formula: 3/3, 1/1, 3/2, 1/1 = 30.

General. Closely related (sub-species) to *Felis sylvestris lybica*, ancestor of Domestic Cats, found today in Middle East and N. Africa. Found throughout mainland Britain in historical times, but considerable reduction in range due to hunting, habitat destruction and intensive persecution as a predator of game birds. With diminution in persecution, and increased refuge provided by plantation forests, numbers and range have increased. However, it is still confined to Scotland north

< *Wildcat.* Populations in Scotland, Spain, Portugal, Germany, France and E Italy.
> *Lynx.* Scattered throughout Scandinavia, isolated pockets in Romania, Czechoslovakia, Yugoslavia, Greece and French Pyrenees.

of Glasgow and Edinburgh. Interbreeding with Domestic Cats leads to fertile hybrids and threatens the genetic integrity of the species. Formerly found throughout UK and wooded regions south of Netherlands, German, Polish and Russian lowlands. Climbs well but descends trees backwards. When threatened arches back and erects hair like Domestic Cat.

LYNX *Lynx lynx* **Pl. 26**

Status. Vulnerable.

Recognition. Long-legged, heavily built cat. Pelage yellowish-brown in summer, paler in winter, spotted heavily on legs, but more sparsely on back. Spots may be pale and joined into an almost roseate pattern, giving impression of being large. Tail short, with black tip. Ears have long tufts. Retractable claws. Pawprint rounded, *c.* 10cm long, similar to Wolf but pads smaller, forefeet turned slightly inward, and no claw marks. Scratches on tree bark. Droppings similar in shape to those of Red Foxes but two or three times bigger (length up to 25cm).

Habitat. Mature, high-timbered coniferous or mixed forest, with dense understorey. Steep, rocky, mountainous terrain. Usually 700–1000m, but up to 2000m. Dens in caves, Badgers' setts, dense thicket or beneath overhanging rock.

Habits. Crepuscular. FOOD: Carnivorous. Principally hares, also rodents, deer (especially Roe calves but also Reindeer or Chamois calves), ground birds (eg grouse, Woodcock, partridge). In Switzerland feeds mainly on Roe Deer followed by Chamois, which together amount to 85% of its food. Predation not confined to calves. In Yugoslavia also Red Deer and Roe Deer. In southern France 43% of diet is introduced White-tailed Deer. Requires *c.* 1kg meat per day. HOME RANGE: 2.5–1000km^2. SOCIAL: Solitary. COMMUNICATION: Male yowls in mating season. Vision very sharp (can spot buzzard in flight from 4km). Mothers and young rub heads together in greeting.

Breeding. Mating in January–March, births May–June. Sexual maturity: male 30 months; female 22 months. Gestation 74 days. Litter size 1–5 (usually 2–3). Litters per year 1. YOUNG: Eyes open 16–17 days, leave den at *c.* 4 months. Weaning age 2–5 months. Male expelled from lair

when young born, but brings food for mother for two months. Young stay with mother for *c.* 12 months.
Lifespan. 15 years (wild), 17 years (captivity).
Measurements. Head-body length: 80–130cm. Tail- length: 11–25cm. Hind-foot length: 19–22.5cm. Shoulder-height: 60–75cm. Condylo-basal length: 12–15cm. Weight: 18–25kg. Dental formula: 3/3, 1/1, 2–3/2, 1/1 = 28–30.
General. Climbs, but not high. Prey taken after explosive attack following stealthy stalk or ambush. Killing bite to throat of larger prey. Prey invariably dragged from point of capture (eg several hundred metres) before eaten or cached (covered in snow). Widely persecuted by hunters (seen as pest of game and domestic stock), and victim of deforestation. May be susceptible to fox mange. Until end of last century still living in wooded mountain regions of central and east France, south Germany, Switzerland, Austria and throughout Scandinavian Peninsula. Reintroduced into Switzerland, France (Voges), Yugoslavia (Slovenia), Czechoslovakia (Suminvar), Austria (Steinmark), and Bavaria (failed).

PARDEL LYNX *Lynx pardina* Pl. 26

Name. Iberian Lynx. *Pardus* = leopard (Lat.); *pardinus* = of a leopard.
Status. Endangered.
Recognition. Large, long-legged cat, very similar to Lynx. Distinguished from Lynx by smaller size, heavier spotting (but spots smaller) and more pronounced chin beard. Short tail with distinct black spots and black tip. Reputed to eat only part of prey, leaving legs, spine and entrails of rabbit; covers uneaten remains with a mound of sand and leaf litter.
Habitat. Open pine woodland on mountains, dense thickets of bramble, broom and gorse. Lowland scrub. Dens among rocks or hollow trees.
Habits. Crepuscular. FOOD: Carnivorous. Principally lagomorphs, also rodents, calves of Fallow and Red Deer, ground birds.
Breeding. Mating in January–March, births May. Gestation 63–74 days. Litter size 2–3. Litters per year 1.

<*Pardel Lynx.* Isolated pockets in Iberian peninsula.

Measurements. Head-body length: 80–110cm. Tail-length: 11–13cm. Hind-foot length: 17–19.5cm. Shoulder-height: 60–70cm. Condylo-basal length: *c.* 12cm. Weight: *c.* 13kg. Dental formula: 3/3, 1/1, 2/2, 1/1 = 28.
General. Formerly occurred in Portugal, Spain and part of France and possibly Sicily and S. Italy. Now restricted to small enclaves in the Iberian peninsular.

Order Pinnipedia – Walruses and Seals

Origins: Despite the similarities in their appearances, imposed by the demands of an aquatic life, the modern Pinnipedia are probably descendants of two stocks with separate origins. Both stem from the Carnivora as recently as 25 million years ago. The seals (phocids) have an otter-like ancestor, the eared seals and walrusses (otariids and odobenids) have a bear-like ancestor. The first phocid is found in the early Miocene, *c.* 20 mya in central France, and their radiation has taken them from the Arctic, through temperate marine habitats, to the tropics. There is even a remnant population in inland fresh water (Lake Baikal Seal). Eared seals and walrusses evolved slightly earlier in the late Oligocene. Although the dual origin of the Pinnipeds is generally accepted, one fossil, *Enaliarctos* (23 mya), shows signs of descent from bears and has both seal-like (thrusting hind flippers) and eared-seal (strokes of foreflippers) characteristics. Walruses appeared at the same time as fur seals, looking at first quite like modern sealions. From about 5–10 mya they were the most abundant and diverse Pinnipeds in the Pacific. By the end of the Miocene at least five genera of walrus migrated through the Central American Seaway (in an area close to where Panama is today) to the Atlantic and crossed to Europe in the early Pliocene. Their progenitors then became extinct in the Pacific, which was later re-populated when descendants of the Atlantic lineages travelled back via the Arctic ocean. At first walrusses ate fish, but some started to eat molluscs, as does the only remaining species, the Walrus.

Main features of the Pinnipedia: All Pinnipeds have many aquatic adaptations (blubber, large body, streamlining, oily fur, and adaptations of their circulatory and respiratory systems). They are tied to the sea for food, but must give birth on land (a dangerous period which they minimise by virtue of rich milk and rapid growth of the young). Pinnipeds have well developed eyes, no collar bone and short, stout forelimb bones. Eared seals (sea lions) retain the terrestrial art of wriggling their hind-flippers into a position under the hind end of the body to support themselves on land, and can use them to shove themselves forward. In contrast, true or hair seals' rear flippers contribute nothing to their movement on land, where they drag themselves forward with their forelimbs. The pinnipeds feet, which may be longer than the rest of the short limbs to which they are attached, are webbed.

Pinnipedia in Western Europe: The Pinnipedia include at least 14 extinct genera, 17 modern genera and 33 modern species. In Europe today they are represented by 8 species in 7 genera, from two of the three families (no eared-seals occur in Europe).

The Phocidae or hair seals have stiff fur and little underfur. They have well-developed eyebrow whiskers, and those in the moustache are often beaded. Many seals have spotted colour patterns and some, unlike other Pinnipeds, have a banded pattern which differs between the sexes. The young of some hair seals are born in lanugo, a dense, soft, woolly, and often white coat. They therefore have three distinct coats: newborn, sub-adult and adult. Hair seals carry plenty of insulating, energy-rich blubber. Males have a well-developed baculum, or penis bone, and females have 4 mammae. Their foreflippers are smaller than their hindflippers, and their flippers (which are flexible and have five digits each) are much smaller than those of eared-seals. They swim by moving the hindflippers up and down. The external ears of seals have no cartilage, and are visible as no more than a faint wrinkling of the skin. Dental formula: 2–3/1–2, 1/1, 4/4, 0–2/0–2. Upper incisors have simple pointed crowns, the canines are elongate, and post-canine teeth usually have 3 or more distinct cusps. Seals are good divers and can stay submerged for many minutes. They have poor hearing and good eyesight.

Odobenidae or walrusses are represented by only one species. The key feature of the family is that both sexes have tusks which are used in defence, to break through ice, to hook over ice for stability when sleeping in water, and to aid hauling out. A single tusk in an old male can weigh over 5kg. Walrusses have no external tail, small ears and small eyes. Their hindlimbs can turn forward to manoeuvre on land. Males have a large baculum, average 63cm. Walrusses eat molluscs.

WALRUS *Odobenus rosmarus* — Pl. 28

Name. *Odontes* = teeth, *baino* = to walk (Gk.). Scandinavian *valross* = whale horse. Rosmarus possibly connected with *ros maris* (Lat.) = rosemary (literally rose of the sea), possibly reference to pink tinge when cooling.

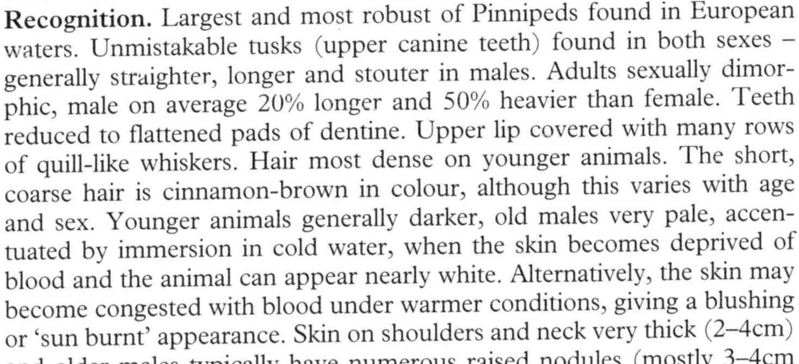

Status. Vulnerable.

Recognition. Largest and most robust of Pinnipeds found in European waters. Unmistakable tusks (upper canine teeth) found in both sexes – generally straighter, longer and stouter in males. Adults sexually dimorphic, male on average 20% longer and 50% heavier than female. Teeth reduced to flattened pads of dentine. Upper lip covered with many rows of quill-like whiskers. Hair most dense on younger animals. The short, coarse hair is cinnamon-brown in colour, although this varies with age and sex. Younger animals generally darker, old males very pale, accentuated by immersion in cold water, when the skin becomes deprived of blood and the animal can appear nearly white. Alternatively, the skin may become congested with blood under warmer conditions, giving a blushing or 'sun burnt' appearance. Skin on shoulders and neck very thick (2–4cm) and older males typically have numerous raised nodules (mostly 3–4cm diameter) in this region. Males of all ages tend to have scars on neck and body. Palms and soles of flippers rough and warty for traction on ice. Walrus uses quadrupedal locomotion (i.e. similar to sea lions) when on land. Head small in relation to body, small eyes, ears with no external pinnae. A pair of internal pharyngeal pouches (unknown in other seals except the Ribbon Seal) arise in the neck and pass along the back. Thought to act principally as resonance chambers for bell-like sound generally made underwater, but are also used as flotation devices when the animal is resting in the water. Hair shed June–July by male, giving naked appearance; replaced in July–August. Female apparently moults later or over a more prolonged period. Calves have prenatal moult 2–3 months before birth, and moult again in July of their first year.

Habitat. Moving pack ice in Arctic (seldom more than 15km offshore. Summer: haul-out on traditional coastal sites (headlands and small islands).

Habits. Diurnal (but active during arctic nights). FOOD: The pattern of abrasion on tusks and vibrissae suggests they feed by moving forwards with snout and tusks in contact with sea floor sediment, finding prey by touch. Upper edge of snout rather than tusks probably used for digging. Benthic (i.e. sea-floor) organisms, especially bivalve molluscs: only siphons and muscular feet of bivalves ingested, probably removed from shell by suction. Crustacea, echinoderms, sea cucumbers and fish also taken. Occasional young Ringed Seals and other marine mammals (which may be carrion). Adults can consume 45kg, or 6% of total body weight, per day. Also ingest and excrete a substantial amount of sand and gravel, acquired accidentally with prey. Most feeding done at depths of 10–50m. HOME RANGE: Most populations migratory, moving southward with advancing ice in autumn and northward with receding ice in summer. Remains close to edge of pack ice or in polynyas (areas of semi-permanent open water) in winter. Hauls out on land at traditional sites in summer when ice disperses. SOCIAL: Very gregarious, travelling almost always in small groups, hauling out in herds of up to several thousand animals. Tend to lie in close physical contact in any season. Uncommon to see solitary animals in water, single animals on ice invariably lone males; on land solitary individuals are usually sick or injured animals. Individual social status indicated by body and tusk size and aggressiveness. Polygynous mating equivalent to a 'mobile lek' in that males follow female herds and when latter stop to rest on ice, the males display incessantly in the water – a stereotyped sequence of underwater sounds followed by surface visual and vocal signals. Females leave resting colony to choose male, and mating takes place in water. Remarkable sensory abilities of moustacial vibrissae, which are used in feeding and can distinguish a mollusc from a stone when both measure less than 1cm across.

Breeding. Most matings in January–February. Births on ice between mid-April–mid-June, about 15 months after fertilisation. Sexual maturity: females at between 4–12 years, but most often at 5–6 years old; males mostly at 9–10 years, but generally do not breed until fully physically mature (14–16 years) and can compete for display positions in the lek. Gestation 11–12 months (15–16 months including 4–5 months delayed implantation). Litter size 1 (rarely twins). 2–3 year breeding cycle due

< *Walrus.* Coasts of north Norway, Baerents Sea and Iceland.

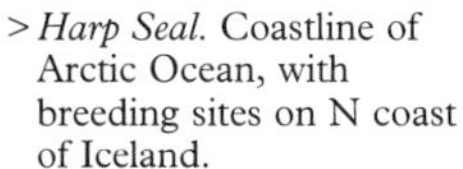

> *Harp Seal.* Coastline of Arctic Ocean, with breeding sites on N coast of Iceland.

to the lengthy gestation period. 4 abdominal mammae. YOUNG: Dependent on mother for milk and protection for up to 2 years, becomes fully independent by 4 years, at which point young females tend to remain with adult female groups, young males joining herds of other young, or older, males. At birth, appear ashen grey to grey-brown, becoming tawny brown with distinctive black flippers after one or two weeks. Cling to mother's neck while travelling pre-weaning or mother may dive with young between fore-flippers. Milk: 12% protein, 35% fat. Immature Walruses often ride backs of adults. Mother defends young vigorously.

Lifespan. Max. 16–30 years (max. reputedly 40 years). Causes of death largely unknown in Atlantic populations (but disease and predation by Polar Bears and Killer Whales amount to an annual mortality of *c.* 2% in Pacific populations). Harvesting by indigenous people may account for about 8% per year.

Measurements. Head-body length: males 280–360cm (average 320cm); females 230–310cm (average 270cm). Shoulder-height: *c.* 100cm. Condylo-basal length (total skull length): male 315cm; female 261cm. Weight: 700–2000kg (birth 60kg; 12 months 220kg; male *c.* 1270kg, female *c.* 850kg). Dental formula: 1/0, 1/1, 3/3, 0/0 = 18 (tusk length: male *c.* 75cm, female 60cm).

General. Thought to be two subspecies: *O.r. rosmarus* in the North Atlantic region, and *O.r. divergens* in the North Pacific region of the Northern Hemisphere. Differ principally in size, the latter being larger. Dives for 10 mins to at least 80m. Skin and blubber 5–7cm thick in adult. Tusks used for agonistic display, defence and levering ice. Tusks may also be hooked into ice in resting position (see vignette). Important role in Eskimo economy. Aggressive encounters with Polar Bears (for self-defence) and white whales (for mysterious reasons) have been observed.

HARP SEAL *Phoca groenlandicus* Pl. 29

Name. Saddleback (USA). *Phoca* = seal (Gk. and Lat.).

Recognition. Adult male Harp Seal generally light grey with a harp-shaped black band running along the flanks and across the back. The head to just behind the eyes is black. In females both facial and harp markings are usually paler and may be broken into spots. Tail short and

slightly flattened dorso-ventrally. Fieldsigns include afterbirth and blood on ice, blow holes, and turtle-like tracks in snow.

Habitat. Spend most of the year at sea. Otherwise, for mating, whelping and moulting climb onto sea ice. There are three breeding regions: (i) off Newfoundland, (ii) off Jan Mayen Island (iii) in the White Sea. It is believed that these three populations rarely mix. Young born among hummocks of ice on floe, within 1.5–2.5m of each other.

Habits. FOOD: Pelagic crustacea and fish, especially capelins, taken at surface and intermediate depths respectively (especially by weaned young and immature seals). Adults will dive to depths of 150–200m to feed on herring, cod and other ground fish. HOME RANGE: Highly migratory – cycle of spring migration, whelping, mating, moulting and autumn migration occurs yearly for adult seals. Young animals occasionally drift into other populations. SOCIAL: Gregarious. February-March: large breeding colonies, and later moulting group. Very old males may live alone or in small groups. COMMUNICATION: Barks, grunts, growls recognised as 15 distinct calls – produced only at the whelping and breeding grounds. No sounds made during the moult. The only air call known is by lactating females, or stressed captive animals. The pups cry and wail. Underwater hearing from 0.76–100kHz, and in air from 1–33kHz. Eyes adapted for dim light sensitivity – large eyes, sensitive (rod dominated) retina, tapetum, widely dilated pupil. Excellent visual acuity in both air and water.

Breeding. Mating in March. Delayed implantation: 11 weeks. Births in February–March. Sexual maturity: female 4–7 years, averaging 5.5; male 5.5 years, but density dependent and usually not sexually active until 8 years. Gestation 7.5 months (total including delayed implantation, 10.7 months). Litter size 1. Litters per year 1. Mammae 2. YOUNG: Pups have a white coat, and moult to juvenile coat covered in dark spots and blotches (see vignette). The dorsal areas and upper sides of yearlings lighten while the ventral surface darkens, causing the spots to become less obvious. The coat then starts to generally lighten, until by 3 years the background colour is pale grey, with indistinct markings. Adult coloration begins to develop at 4 years, though not complete until sexually mature, or even later. Male approaching maturity may have intervening 'sooty' coloration, lost at next moult. Weaning age 12 days. Pups deserted at approx. 12 days. Female will not defend young if disturbed in water but may do so vigorously on land. Feeds pup 2–3 times a day, on greyish white milk with a strong fish-like odour, and the consistency of thick cream. Contains 42.6% fat, 10.4% protein, 45.3% water and 0.8% ash.

Lifespan. Max. recorded 30 years. Mortality *c.* 10% annually.

Measurements. Head-body length: whitecoats (pups) 93–108cm (mean = 103cm), adult females 168–183cm (mean = 179cm), adult males 171–190 (mean =183cm). Condylo-basal length (skull length): *c.* 65cm. Weight: whitecoats average 11.8kg at birth, 22.8kg at 4 days and a max. of 33.3kg at weaning. Adult females average 119.7kg, and males 135kg.

Seasonal weight changes occur in all groups. Dental formula: 2–3/1–2, 1/1, 4/4, 0–2/0–2 = 26–38.

General. Dives to 280m for up to 30 mins. Very heavy hunting pressure, especially within breeding areas (eg Newfoundland breeding area, 180,000 young killed per annum during 1960s). Numbers declined (1800–1960s: total pop. perhaps *c.* 10 million to 2–3 million). In 1983 EEC stopped buying pelts and kill dropped to 30,000. Now back up to 70,000 per annum of all ages. Individuals reported occasionally from Scotland and Shetland Isles, and have been known as far south as the Bristol Channel and River Teign (UK).

RINGED SEAL *Phoca (Pusa) hispida* Pl. 29

Name. Common Seal. *hispida* = barbed or bristly (Lat.).

Status. Vulnerable in Europe, but extremely abundant in the Arctic, particularly off Arctic N. America.

Recognition. Smallest of pinnipeds. Distinguished from Harbour Seal by smaller size, and prominent grey-white rings on a generally dark grey back (adults). Belly usually silver and lacking dark markings (though these may be present on pups); brown whiskers, weaker dentition (mandibular teeth always aligned with axis of jaws). Inner surface of mandible concave between middle post-canine teeth (cf Harbour Seal where convex). Short 'cat-like' face, cf. Harbour Seal with more dog-like, longer snout. Moults on sea ice in late March–July, peak in June. Immatures moult earlier than breeding adults. Can be seen from air in spring when hauled out. Always positioned along narrow cracks with nose pointed to crack, or in centre of large open spaces on floes, never at edge. Enters water via hole in centre of floe, rather than running to edge of floe. Moves on land by reaching forward obliquely with fore-flippers (see vignette).

Habitat. Ice, on or near coast, in fjords, lakes and inland seas around Arctic and Baltic. Also far off-shore in arctic in polar pack-ice. Ice associated year round: winter – both shorefast (continuous cover) and pack ice; summer: remains with pack ice as it melts, receding north. Some, especially young, may remain in open water in summer. Young generally born on land-fast ice in cave ('lair'), but pack ice also used. Lair may be 2–7m long or more with multiple chambers and tunnels. Lairs located in snow drifts formed behind pressure ridges or other irregularities in ice. Both males and females use lairs to haul-out and rest. Each seal probably has 2–3 breathing holes and/or lairs. Breathing holes may be maintained through several metres of ice). Fieldsigns include haul-out lairs and the larger multi-chambered pupping lairs.

Habits. Diurnal, but active during Arctic nights. Haul-out once or twice during moult, peaks 10 am to 4 pm. FOOD: Small fish, especially arctic cod (*Boreogradus* 75% of diet (probably) in arctic, and almost exclusively

Boreogradus in winter and crustaceans (especially isopods and amphipods, March- May); rarely bivalves. Euphausiids (*Thysanoessa*) are a major food in late summer. Pelagic amphipods (*Parathemusto*) also. Fish are usually small (less than 20cm). Direct food competition with other marine mammals likely in some areas, as feed on the most abundant organisms at at least 2 levels of the food chain. Feeding at a minimum during the moult. HOME RANGE: Freshwater populations resident. In autumn, Arctic populations migrate south, with advancing ice, in spring go north with receding ice. SOCIAL: Mostly solitary in winter, or small groups. Adults may be territorial during breeding season. May be monogamous. Mating thought to take place under water. Social barriers break down during moult. Several (10 or more) seals may haul-out at same hole and more than 100 along the same crack spaced several metres apart. COMMUNICATION: Generally silent, occasional bleat or grunt.

Breeding. Mating in late April–early May. Delayed implantation: 14 weeks (mean implantation time (Alaska) late August). Births in late February–April (or late March–April in Arctic). Sexual maturity: male: 5–7 years; female 4–7 years, but varies with region. Litter size 1, though twins have been recorded. Litters per year 1. YOUNG: White woolly coat ('lanugo') begins to be shed 2–3 weeks after birth, completely moulted after 6–8 weeks; coat relatively longer and finer than that of adult, silver on belly and dark grey on back, sometimes with traces of adult ringed pattern. Newly moulted pups (termed 'silver jars' by Canadian hunters) much sought after for pelts. Enters water in first few weeks to move between lairs, avoiding predation. Female drags pup from one lair to next. Young can swim but seem to avoid the water. Weaning age 5–7 weeks, when weigh 9–12kg (i.e. 2–3 times birth weight). Only the female cares for young, returning to lair to nurse. Female makes lair, and pups may construct tunnels within. Female remains with pup until weaning at *c.* 6 weeks.

Lifespan. Max. recorded 43 years, although average probably 15–20. Main prey of polar bears, which eat on average one Ringed Seal every 6.5 days. Ringed Seals scan surroundings frequently when basking on ice in summer to avoid predation. Other significant predators reputed to be Arctic Foxes and Humans. Also Ravens, Red Foxes, Domestic Dogs, Wolves, Wolverines and, occasionally, Walruses (probably only pups or sick or disabled seals or carrion).

Measurements. Head-body length: averages of 121–135cm; 65cm long at birth (max. lengths: male 168cm; female 150cm). Condylo-basal length (skull length): *c.* 135cm. Weight: 4.5kg at birth. 36–113kg (max. wt. male: 113kg, female = 111kg). Dental formula: 2–3/1–2, 1/1, 4/4, 0–2/0–2 = 26–38.

General. Can dive for up to 20 min. Declining numbers in Europe, but abundant off N. America. Population density in arctic waters (fast ice) from 1–7 seals per nautical mile. Considerable annual variation in density on fast ice from about 50% below to 40% above the long-term means (according to 8 annual surveys covering 17 years).

COMMON SEAL *Phoca vitulina* **Pl. 28**

Name. Harbour Seal, Spotted Seal. *Vitulina* = of a calf (Lat.), *vitulus* = calf.
Recognition. Variable, mottled pattern of spots, ranging in colour from light grey to dark brown or black, on background of similar colour variability (from distance may appear silvery when dry). Nostrils close together and V-shaped: slightly concave muzzle, white whiskers: dog like appearance of head (cf the 'roman nose' of the Grey Seal). Top of head rounded and head appears small in relation to body. Post- canine teeth clearly tri-cusped and (excluding the first one) much longer than broad. Males often have heavily scarred necks and are more heavily built than females, who may be scarred on the back of the head as adults – from males biting during mating. Not sexually dimorphic – the male penile opening is usually fairly conspicuous, but so is the navel of the female, so the sexes are easily confused. Pups moult lanugo hair (the white coat) before birth. Moult is in August for females, late August–mid- September for males. Tracks visible in sand where haul-out at high tide and as follow water-edge as tide recedes (adult 55–75cm width, pups *c.* 30–45cm). New born pups leave scratch marks due to long nails. Droppings on sand banks, brown and like dog faeces, 2–3cm diameter.
Habitat. Sheltered coastlines of N. Pacific and N. Atlantic. Typically found on sandbanks in estuaries and sea locks but also regularly occurs on more rocky coasts. Pups born on inter-tidal rock or sand bank, occasionally in water.
Habits. Rhythm affected by availability of suitable nest sites. If these are unaffected by tides, then feeding usually by day. Otherwise, resting is confined to low tides. FOOD: Newly weaned pups feed primarily on bottom dwelling crustacea for 1.5–3 months. Older animals feed opportunistically on wide variety of fish, some cephalopods and crustaceans (daily requirement 1–3kg, or 5–6% of the body weight). In the Wash (England), where there are few fish, mainly whelks are eaten. In Holland the principal food is Common Flounder (and commercial fish overall comprise up to 75% of the diet). In Scotland gadoids and clupeoids are the main prey. Seals tend to eat one kind of fish each feeding period, with small fish eaten whole below the surface and larger fish consumed in pieces at the surface. They obtain most of their water requirements from their prey,

< *Ringed Seal.* Coastline of Arctic Ocean and Gulf of Bothnia, breeding sites on N coasts of Norway and Iceland.
> *Common Seal.* UK coastline (excluding S), and continental coasts of North, Baltic and Norwegian seas.

none from seawater. HOME RANGE: Extent of feeding range unknown, but apparently highly variable: some individuals return to same haul-out site day after day, others depart for extended periods. Largely sedentary, but juveniles tagged in the Wash recovered in France, Belgium and Holland, indicating travel distances of up to 300km. SOCIAL: Social system obscure: solitary in water but gregarious when hauled out, resting in groups of up to 1000 (15–500 in the Wash): usually mixed age and sex. In regions where haul-out sites are plentiful, different sites may be used for pupping, moulting and resting. Some evidence that males defend aquatic territories offshore of pupping sites. Cows defend area around calves, and will dive with pup in mouth or foreflippers if threatened: closest social bond exists between mother and calf during lactation. Courtship and (probably promiscuous) mating thought to take place in water. COMMUNICATION: Probably the least vocal Pinnepid: short barks, grunts, coughs (pups wail). Faint underwater clicks also recorded, though if echolocatory probably only of use during late stages of prey capture. More probably associated with a threat display. Indications of a highly sensitive retina probably permits good vision under water under low lighting conditions (such as great depth) but vision less effective in low light in air due to steropaic (elliptical) pupil. Auditory range to 180kHz (less sensitive in air than in water). Species very alert for danger (perhaps more so than other phocids). Vibrissae sensitive and may be used to find fish in murky water.

Breeding. Mating in July–early August (probably at sea), delayed implantation until November–December. Births in June–mid-July. Sexual maturity: males 3–6 (mostly 5) years; females 2 years, 20%; 3 years, 38%; 4 years, 34%; 5 years, 8%. Gestation 1.5–3 months delayed implantation, 8 months true pregnancy. Litter size 1. Litters per year 1. Mammae 2. YOUNG: Pups usually born with adult pelage (rarely foetal white coat). Can swim from birth, at 2–3 days can dive for up to 2 mins, and at 10 days for up to 8 mins. Young can be pulled along in mother's slip stream when swimming. Nurses for about a minute every 3–4 hours. Abandoned at weaning. Weaning age 2–6 weeks, when birth weight is doubled. Mother pushes pup under water if danger threatens.

Lifespan. Max. recorded: males 26 years, females 32 years. Causes of death mainly due to Man, and also Killer Whales and sharks.

Measurements. Head-body length: 120–195cm (male 130–195cm, female 120–155cm. Birth: 70–95cm). Condylo-basal length: *c.* 23cm. Weight: 45–130kg (birth 9–11kg). Dental formula: 3/2, 1/1, 5/5 = 34.

General. Large scale censusing difficult as herds are scattered, small and easily frightened. Estimated at 29,000–100,000, before viral infection of 1988–89. Locally regarded as pest of fisheries but less often than Grey Seal. One of the terminal hosts of the larval nematode *TerraNova decipiens* which infests the muscle of many fish, reducing their marketability. No pups culled in Wash since 1973 and none in Scotland since 1981. Protected throughout year in UK, Netherlands, Germany, Denmark and

Sweden (except around fishing gear). In Norway some populations have decreased markedly in recent years through hunting, pollution or disturbance.

GREY SEAL *Halichoerus grypus* Pl. 28

Name. *Halios* = sea, *khoiros* = little pig (Gk.); *grypus* = hook-nosed (Lat.).

Recognition. Distinguished from other seals by straight line profile from top of head to nose, the snout of the male being elongated, with a convex profile above the wide, heavy muzzle (hence 'hooknose' of name), and the female having a flatter profile with a more slender muzzle. Grey colour predominates although there is much individual variation – males may have an extensive darker tone forming a continuous background, with lighter patches or reticulations. Males have 3–4 conspicuous wrinkles around neck. Females have lighter continuous tones with the darker tone forming spots or blotches, usually denser on the back. Spots larger and less numerous than those of Harbour Seal. Immatures of both sexes have much less pronounced pattern. Sometimes there is a reddish tint on the head, neck, belly and flippers. Often has small external ear and long, slender claws on fore-flipper (cf Harbour Seal) but not a reliable feature due to wear. Nostrils clearly separated (cf. Harbour or Common Seal where the nostrils are inclined and nearly joined at the base). Post- canine teeth are large, strong and have one conical cusp. In the male the elongated snout is convex above the robust muzzle. The female has a slender muzzle with a flatter profile. Moult: cows January–March (peak at beginning of February) (during which they huddle together and make a 'song'); bulls March–May (peak in the middle of March). Do not usually use breeding sites for moulting. Dog-like brown or grey droppings 4–4.5cm diameter. Traces of hair catch on rocks at regular hauling-out sites. Tracks like common seal, but wider.

Habitat. Rocky or cliffy shores, but also estuaries, tidal flats and sandy shores. Breeding site termed 'rookery'. Haul-out sites often between tide-marks, eg Longone at the Farne Islands, and beach haul-outs in Wales. Breeding sites on uninhabited islands, caves and remote beaches – pups born above high water mark.

Habits. Rests by day at low tide and at sunset. Probably does not eat every day. Active throughout day and night at breeding sites. FOOD: General coastal feeder. Fish (21 species recorded), cephalopods and occasionally birds. In Scotland, salmon appears to be the most important prey item (both in and away from salmon nets), followed by cod and other fish. Require *c.* 5.7kg food per day: *c.* 4–4.5% of bodyweight. Frequently accused by fishermen of taking salmon and other commercially important fish from nets and damage most severe when set nets or other forms of fixed gear are used. Gill-netted salmon and cod particularly vulnerable. Neither sex feeds while breeding (cow fasts for 3 weeks, bull

for 6 weeks). HOME RANGE: Grey Seals not known to make definite and regular migratory movements and although major seasonal movements occur when seals congregate for the breeding season, and disperse when it is over, these do not appear to be directional. Breeding females usually return to same site every year, while young animals travel coastlines extensively and disperse widely; pups from Scotland have been recovered in Iceland. SOCIAL: Pups born in colonies of 50–70,000 cows. Bulls defend positions among groups of cows which are in or close to oestrus; most harems defended on land but sometimes the bulls defend females immediately offshore from the cows and their pups. There is no physically defined territory: the area dominated by each bull changes from day to day. Uses sexual activity rather than territorial fighting or boundary displays as reproductive strategy. Bulls pay constant sexual attention to cows once oestrous cows are present in the rookery. Non-breeding females may join cows with their pups. Cows with pups are very aggressive to each other, and to bulls. COMMUNICATION: Very vocal during pupping season and moult. Pups produce crying sound very like human baby. Females howl at each other in the breeding colonies. During the pupping season, males may produce a low frequency sound reminiscent of a steam train. 'Song' is a series of long drawn-out calls rising and falling in pitch.
Breeding. In NE Atlantic, births on land between September–December (peak in October); in Baltic on pack-ice, births in January–March; in NW Atlantic, on land or ice, births peak in January–February. Icelandic and Norwegian Grey Seals have a peak of pupping in October. Mating: *c.* 3 weeks after giving birth, taking place on land or at sea, lasting 15–45 minutes. Delayed implantation: 12 weeks, implantation February 10th–March 4th. Sexual maturity: female 4–5 years; male 6 years (but not socially in position to mate until 8–10 years). Gestation 8.5 months (total including delayed implantation 11.5 months). Litter size 1. Litters per year 1. YOUNG: Born in lanugo (white foetal coat), shed after 2–3 weeks (next moult after 15 months). Average daily weight gain while nursing: 1.8kg; milk contains over 53% fat. Weaning age 16–21 days, pups may remain on land for about 14 days before taking to the sea and feeding independently. Nursed at 5–6 hour intervals (milk: 67% solids, 53% fat). Mothers readily desert pups.
Lifespan. Max.: males 30 years; females 46+ years. 20% infant mortality common (due to starvation and desertion) but may be as high as 80% locally. Locally 35–65% mortality in first year.
Measurements. Head-body length: 210–330cm (max.: males 330cm, females 250cm; birth: 90–105cm). Condylo-basal length (skull length): *c.* 27cm. Weight: 125–315kg (eg males 170–310kg; females 105–186kg. birth: 14.5kg). Dental formula: 3/2, 1/1, 5/5 (often 6/5) = 34–36.
General. Largest member of the family Phocidae, and, apart from the Elephant Seal, shows greatest degree of sexual dimorphism – possibly correlated with the habits of land breeding and polygyny. Increased proportion of 'wormy' fish (nematode *PseudoterraNova decipiens*) has been

< *Grey Seal.* Coastlines of UK, Iceland, Faroes, W France and Artcic and Baltic coasts of Scandinavia.
> *Bearded Seal.* Coast of Baerents Sea, with breeding sites around Arctic coast of Norway and N Iceland.

associated with increasing numbers of Grey Seals (in whose stomach the adult codworm parasite lives), but no clear evidence. Superabundant in all localities previously inhabited except Baltic and Kattegat where they are endangered as a breeding species (but this is only a very small part of their range).

BEARDED SEAL *Erignathus barbatus* **Pl. 29**

Name. *Eri* = very (Gk.); *gnathos* = jaw (Gk.); barbatus = bearded (Lat.).
Recognition. Dentition very weak, loosely rooted post- canine teeth widely spaced. Dense moustache of long, flat vibrissae on upper lip, which are unique among North Atlantic seals being flat-sided as opposed to bearded in other seals (i.e. the Bearded Seal's whiskers have a smooth outline, which is confusing since it gets its name from the fact that the whiskers curl when dry). Third digit on fore-flipper slightly longer than others, giving a spade-like appearance (and Canadian name: Square Flipper) with strong nails. Toes of hind flippers are all of nearly equal length, again unlike any other North Atlantic seals. Distinguished from all other seals (except Monk Seals) by 4 mammae.
Habitat. Shallow coastal waters and offshore ice flow, apparently tending to stay with the pack ice whenever possible. Birth on pack-ice. On ice, tends to lie at edge of floe, facing the water or alongside it.
Habits. FOOD: Bottom-dwelling bivalves and crustaceans dug up with spade-like flippers, and fish (90% of diet is fish in Canadian Arctic) HOME RANGE: Considered non-migratory (except long movements of isolated stragglers in Scotland-N. France), but does advance and retreat with seasonal changes in extent of the pack ice. SOCIAL: Not colonial, but do congregate during the moult, for reasons as yet unknown – possibly co-operative vigilance against polar bear approach. COMMUNICATION: Males in breeding season make whistle that ends in loud howl. Whistle to pups under water but not, apparently, on land.
Breeding. Mating in mid-May. Delayed implantation 10–12 weeks, births in April–May. Sexual maturity: female 6 years, male 7 years. Gestation *c.* 8 months (total, including delayed implantation, 10.5–11 month-

s). Litter size 1. Litters per year variable – Alaska: interval between pups 1 year, in some parts of the North Atlantic mostly 2 years. Mammae 4. YOUNG: Born with woolly coat of grey-brown fur/lighter colour in mid-back). Moults at end of lactation to stiff adult bristly hairs, but many pups are born with woolly coat already partly moulted. Shed hair present as large (4–5cm diameter) 'hair balls' in amniotic fluid, and in colon of foetus/new born pup. Weaning age 12–18 days.

Lifespan. Max. recorded: male 25 yrs, female 31 yrs.

Measurements. Head-body length: 220–250cm (sexes nearly equal in size, though female tends to be slightly larger than the male). Condylobasal length (skull length): *c.* 210mm. Weight: 200–360kg (this latter only recorded for obese, pregnant females) (birth 34kg) Dental formula: 2–3/1–2, 1/1, 4/4, 0–2/0–2 = 26–38.

General. Dives to 30m, for 20 mins. On land moves forward by reaching ahead with both forelimbs at same time, holding with claws, lurching body along (see vignette). Liver so rich in vitamin A that it can be poisonous to humans in some localities, although liver is eaten by people around the Bering and Chukchi Seas (similarly, eating Polar Bear liver can cause hypervitaminosis-A). Parasitic disease 'trichinosis' can be caught from eating raw/incompletely cooked meat (1–5% Bearded Seals infected with the larvae); a disease of meat-eaters so may acquire it from eating scavenging crustaceans that have fed on the carcass of another infected mammal.

MEDITERRANEAN MONK SEAL *Monachus monachus* Pl. 28

Name. *Monachus* = monk (Med. Lat.).

Status. Endangered – about 500 remain scattered in small groups, with largest concentrations in the remoter parts of Greece and on the Atlantic coast of Mauritania.

Recognition. Adult pelage variable but usually dark dorsally with light patch ventrally. Variable colour patterns characteristic of the species, however, as they are constant among members of widely separated colonies. Annual moult.

Habitat. Subtropical coastlines, sheltered, small beaches (increasingly confined to remote spots). Pupping in grottoes and caves in sea cliffs, sometimes with only underwater entrances. This may be a result of disturbance, and they may prefer to pup on beaches under cliffs.

Habits. Diurnal, but increasingly nocturnal due to persecution. FOOD: Fish, including rays, and cephalopods. HOME RANGE: Thought to be generally sedentary, but known to have travelled 200–600km at edge of range. SOCIAL: Small herds, social organization unknown, but possibly polygynous.

Breeding. Births occur between May–November (peak September–October). Sexual maturity when approx 210cm. Litter size 1. Less than 1 litter per year. Mammae 2. YOUNG: Born with black, woolly coat, moulted

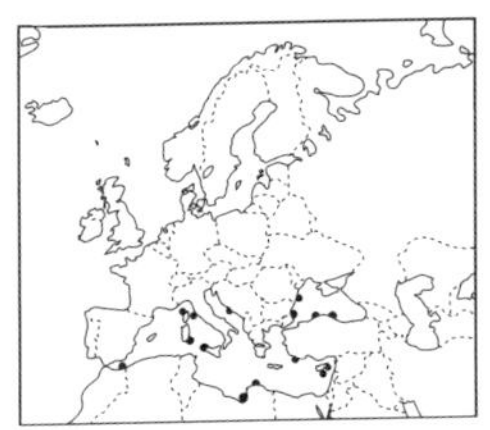

< *Mediterranean Monk Seal.* Found at less than twenty sites around the Mediterranean and Black Sea.

> *Hooded Seal.* Coastline of Arctic Ocean, with breeding refuges along coast of Norwegian Sea and N Iceland.

at 4–6 weeks to give juvenile pelage, silver grey dorsally and light ventrally. Weaning age 6 weeks. Young stay with female for 3 years.

Lifespan. Max. recorded 24 years (captivity). High rate of abortion, perhaps due to disturbance of pregnant females. Entanglement in fishing gear. Increased disturbance from fishing and recreational activities in even the most remote refuges.

Measurements. Head-body length: 230–278cm (birth: 91–92cm total length). Condylo-basal length (skull length): *c.* 25cm. Weight: on average 250–300kg; maximum recorded weight 400kg. Newborn pups *c.* 26kg. Dental formula: 2/2, 1/1, 5/5 (post canines) = 32.

General. The only seal in S. European waters. Once found on shores of Black and Adriatic Seas, the Mediterranean coasts of NW Africa as far as Cap Blanc, Madeira and the Canary Isles. Seriously imperilled by 1970s, but the creation of a marine park in the Greek Northern Sporades together with plans for other such reserves, attempts to get cooperation of local fishermen, and a rescue and information network throughout the Mediterranean to aid storm-washed/abandoned pups suggest there is now less danger of extinction. Many place names in Greece, Turkey and Yugoslavia derived from Greek word for seal (phoca), suggesting previously more widespread. Said to tame easily in captivity (due to their previous isolation, these seals are thought to be 'genetically' tame, i.e. with no inherent fear of man). Does not adapt well to captivity, rarely surviving for more than 1 year.

HOODED SEAL *Cystophora cristata* — Pl. 29

Name. Bladdernose Seal, Crested Seal. *kustis* = bladder, *phoros* = carrying, referring to nasal sac (Gk.); *cristata* = crested (Lat.).

Recognition. Mottled dark brown, brownish-black or black patches on a silvery grey background (more pronounced on back and upper sides). Appears more uniform dark grey in water. Sexual dimorphism, males being larger than females at maturity. Sac of skin (extension of nasal cavity), hangs like proboscis over mouth and can be inflated to size of football in males (much less pronounced in female). Male can also extrude the red nasal septum through one of the nostrils to form a bubble-gum-like

balloon when the hood deflates (see vignette). Post-canine teeth peg-like and widely spaced. Well-developed claws on all digits. Annual moult.
Habitat. Heavy, drifting ice in deep waters, especially for breeding and moulting, and open sea. Pelagic. Gives birth on old, heavy ice floes.
Habits. FOOD: Fish (including deep water species), molluscs (Octopus), and shrimps, starfish and mussels from sea bed. Hooded Seals fasts while breeding/moulting so dietary preferences and feeding behaviour not well known. Squid (eg *Gonatus fabricii*) and redfish may be important. HOME RANGE: Long migrations of mainly adults from breeding grounds to waters off Greenland, 'Denmark Strait' (summer moulting grounds) where virtually the entire active population appears to congregate: peak of abundance mid- June to mid-July. Disperse widely. Little is known of activities at this time. In February concentrate near thick sea ice around Nova Scotia, Labrador (Newfoundland) and Jan Mayan Island (breeding grounds). SOCIAL: Spring; family group of bull, cow and calf. Summer (mid-June–mid-July) assemble for moulting in large aggregations in Denmark Strait. Leave after moulting. Males probably opportunistically polygamous. Solitary throughout most of year, and little known of his behaviour. COMMUNICATION: Male roars in breeding season. Uses inflatable hood in aggressive display. Underwater calls include series of rapid (110 per second) clicks and low frequency pulsed sounds (males). Females and pups make variations on same kind of call.
Breeding. Births primarily in second half of March (mid-March–mid-April). Breeds Gulf of St Lawrence, NE Newfoundland, Davis Strait, NW Jan Mayen Island. Female courted by 1–7 males (who remain in water generally) while she attends pup on ice – referred to as 'families' though pair bond not particularly stable. Breeding behaviour commences at weaning. Female: on average first young born when 4.5 years (can mature at 2–9 years, but 50% mature at 3 and give birth at 4 years). Male: 5 years (but no opportunity to mate till older). Gestation *c.* 250 days from implantation to parturition, as show delayed implantation. Litter size 1. Litters per year 1. Mammae 2. YOUNG: Pale creamy-grey coat shed before birth: swallowed by foetus and defecated into amniotic fluid as discs measuring 4 × 0.5cm. Replaced by short slate-blue pelage, with cream ventral surface ('blueback' stage). Adult mottled patterns develops when yearling coat shed at first annual moults. Precocious – can swim and dive in first few days after birth. Sexual dimorphism apparent at birth. Combine large birth weight (averaging 22.0kg) with rapid neonatal weight gain (maximum of 42.6kg on day 4) to achieve a weaning weight comparable to other phocids in one third to one tenth the amount of time after birth. Weaning age 4 days average (range, 3–5 days) – the shortest lactation period known for any mammal. Thought to be adaptive for seals breeding on unstable pack ice. Female only nurses 3–4 times per day, milk having a fat content of 40–65%. Female defends pup aggressively, but deserts after weaning.

Lifespan. Max. recorded *c.* 35 yrs. Man is the most serious predator, but Polar Bears take many at whelping grounds, and Greenland sharks may take young/injured individuals. Natural mortality rates difficult to determine due to continuous exploitation. Total mortality estimates: 22% for adults (Newfoundland); 16% for females aged over 6 years, 23% for males aged 10 years or over. Intensive commercial fishing in some parts of range has lead to depleted stocks of some known prey spp. (redfish, capelin for example) but effect on seals unknown.
Measurements. Head-body length: on average, males 260cm (max. 350cm), females 200cm (max. 300cm). Pup: 87–115cm at birth. Condylo-basal length (Skull length): Male *c.* 26cm, Female 20cm (cranium very short, snout long and broad). Weight: on average, males 192–352kg, females 145–300kg (when adult) 10–30kg for newborn pup. Dental formula: 2/1, 1/1, 5/5 (post canines) = 30.
General. Dives to 300m for up to 20 mins. Trade in skins (eg 1977, Norway-Russia quota for Jan Mayen 46,000; Canada-Norway quota for NFL/Labrador 15,000; 3000 from Greenland). Blueback coat of pup regarded as most desirable seal pelt on European market. Capturing pup often requires killing of aggressive mother. Norway and Russia only two countries still engaged in commercial sealing, currently *c.* 10,000 animals per year. Greenland 'subsistence' hunt takes up to 5000 per year using rifles from motor boats. Trade bans on Hooded Seal products initiated by several countries in recent years (USA – Marine Mammals Protection Act 1972; New Zealand; Italy; Netherlands since 1969; Sweden since 1962; EEC since 1983). Populations probably at or above historic levels and increasing in Canada, where population well managed. Share much of range with Harp Seal and both species undertake almost parallel migrations. Competition unlikely, however, due to differences in many biological characteristics: Hooded Seals remain further offshore, occupy deeper water and thick drifting ice.

Order Cetacea – Whales and Dolphins

Origins: More than 50 million years ago there was an amazing group of hooved-predators called the mesonychids. They were members of the Condylarthra (the first ungulates, abundant in the Palaeocene and Eocene, 65–38 mya) which evolved from small insectivores, and probably gave rise to all later ungulates. Today, the idea of a hooved-predator seems a contradiction in terms, but the early mammalian predatory niche was occupied by the hooved descendents of these early herbivores. However, with the evolution of modern carnivores these mesonychids became extinct, but they bequeathed one remarkable line of decent: the whales. Early, otter-like mesonychids became progressively more aquatic, perhaps escaping competition from new land predators. In the early Eocene, 50 mya, *Pakicetus* was about 1.8m long, could not dive deep or hear well when submerged, but it was one of the first of the now extinct first whale suborder, the Archaeoceti. Today's suborders, Mysticeti (baleen whales) and Odontoceti (toothed whales) arose from the Archaeoceti and replaced it in the Oligocene. The ancient four-legged ancestry of whales was illustrated by the atavistic characteristics of a Humpback Whale stranded on Vancouver Island in Canada in 1919 which had internal legs 1.2m long. Today that ancestry is reflected in similarities between modern whales and Artiodactyles in blood composition, foetal blood sugar, chromosomes, insulin, uterine morphology and tooth enamel microstructure.

Main features of the Cetacea: Whales and dolphins are the apogee of mammalian adaptation to an aquatic lifestyle. With their system of tail-fluke propulsion, reduced forelimbs, absent hindlimbs, complex communication and use of echolocation, they are as adept under water as are fish, except for their need to surface for air. Nonetheless, even that constraint is reduced by their marvelously efficient breathing system: they use about 10% of the oxygen they breathe in, in contrast to about 4% used by terrestrial mammals. Whales and dolphins are characteristically large, and usually give birth to only a single young.

Cetacea have been hunted for many centuries and provide many useful products. Last century the first signs of over hunting appeared (with respect to sperm whaling). In belated response to this the International Whaling Commission was set up in 1946, at that time protecting right whales, grey whales and humpbacks in the Antarctic. However, this was largely ineffective until in 1982 a ban on all commercial whaling was agreed upon, to take effect in 1986. Subsistence whaling and 'scientific whaling' still continue, causing serious problems to some endangered species (eg Bowheads being taken by Inupiat Eskimos with new and very efficient methods). Another problem is that many of the small Cetaceans are caught incidentally in fishing nets – the notorious capture of several species of dolphins in the tuna purse-seine fishery in the eastern tropical Pacific is a major example, but dolphins are caught in all kinds of nets all over the world. Habitat destruction threatens whales which use the

Table 6. Call characteristics of selected cetacean species.

Species	Sound type	Frequency range (Hz)	Maximum energy (Hz)	Signal duration (s)	Comments
Mysticetes					
Blue Whale	AM moan	12.5–200	20–32	15–38	May comprise 1-3 parts. In latter case 1st part modulated at rate of 3.85 per sec, 2nd and 3rd at 7.7 per sec; brief intervals may occur between them, with 390Hz pulse of 0.5-1.0sec duration preceeding 3rd part
	Click	21,000-31,000	25,000		
Fin Whale	FM moan	?6-95	c.18-23	1	Pulses usually increase in amplitude in the first third of sound, then remain relaatively constant over middle period and then decline over final third of pulse. Repeated at relatively fixed intervals varying from 6-37+ sec between pulses, sometimes as doublets but otherwise repeated over 2-20 min period, with 1-3 min pause, extending over several hours. Variants include ragged pulses around 20Hz as short bursts of <0.1 sec to defined pulse of c. 1 sec; broadband rumbles of c. 20Hz
	FM moan		c. 40-75	0.3	Never repeated
	AM click	16,000-28,000		8.8	May comprise 2-3 parts, 3-3.4msec, with pulse intervals 250-336msec
Sei Whale	AM click		3000	0.7	7-10 pulese per burst, each of 4msec duration

Minke Whale	Grunt	80-140		0.165-0.32	Repeated at irregular/regular intervals at 2.1-2.3 pulses per sec
	FM thump train	<100-800+	100-200	0.05-0.07	Repeated over regular intervals over period of 1+min, but with individual variation in frequency composition and repitition rates
	AM click	3300-3800/5500-7200/10,200-12,00 (20,000)	4000-7500	0.5-1msec	
	Ratchet pulse		850	1-6msec/25-30msec	Single pulsed/multipulsed units
Humpback Whale	Moan/ groan	<4000		7-36min	Songs organised into themes composed of repeated phrases and syllables in fixed order, usually ended with a surface ratchet
	Snore				Includes short, long, 'elephant', 'lion' and two part snores
	Low grunt	120-250			
	Chirp/ whistle/ squeal	500-1650	1600	0.5	
	Click	2000-7000			
Bowhead Whale	Tonal FM moan	50-500	50-300	0.5-5	Frequency modulation descending, ascending or relativeely constant, sometimes accompanied by tonal FM sound of 400-1000kHz
	Pulsed tonal purr	100-800		1-3	30-75 pulses per sec (with 3-14 harmonics)
	Tonal AM call	150-375			Modulation frequency between 10-30Hz

	Complex pulsed call	100-3500		0.3-7.2	5-30 pulses per sec, each of 0.01-0.05 sec duration staccato pulse train produced in some, with a low 50-100Hz tonal component reproduced simultaneously
Montocetes					
Sperm Whale	Pulsed clicks	<100-30,000	10,000-16,000	1-10 secs to 20+ mins	1.5-3.0 pulses/burst, 2-30+ msec duration, but varying from 1 per 5-10 sec to 60+ per sec. Clicks repeated at very regular intervals, some with characteristic repitition rates (termed codas)
Narwhal	Whistle	300-10,000(18,000)		0.5-1.0 (0.05-6.0)	Some given at constant frequency but most with steadily decreasing or increasing pitch, and some fluctuating
	Pulsed tones		500-5000	0.56-1.34	Repeated at intervals from 1.2-10 sec; majority comprise single unbroken sinnals
	Pulsed clicks	500-24,000	500-5000/ 12,000-24,000	3.6(0.5-23.3)/ 0.7(0.2-1.8)	Mainly at 5-10/50-60 clicks per sec
White Whale	Whistle				
	Clicks		1200–1600 or 40,000/80,000/ 120,000	2-4 msec or 20-250 µsec	
Killer Whale	Clicks		250-500	0.8-25msec	
	Clicks	100-80,000	14,000-40,000	0.5-1.5msec	
	Clicks	to 35,000	12,000	0.1-0.5msec	
	Whistle	1500-18,000	6000-12,000	0.05-10/12	FM
	Pulsed call	1000-25,000	1000-6000	0.5-1.5(<0.05->10)	Harsh 'scream'-like sound may be unique to particular pod; calls may be discrete or less occasionally variable

	Pulsed clicks	to 35,000+		0.8-25msec	1->300 pulses per sec
Long-finned Pilot Whale	Whistle	2800-4700	3400-4700	0.65-1.0	Calls produced at 14.7-41.4 per min
Rough-toothed Dolphin	Whistle		3000-10,000	0.1-0.9	
	Pulsed clicks	100-20,000		50-250μsec	
Atlantic White-sided Dolphin	Whistle	8200-12,100		0.5	Series of distinct pulses, becoming a crackle at high repetition rates; also, rarely, a creak
	Clicks				
Common Dolphin	Whistle	4000-16,000			5 whistle types paired, with clicks and and whistles at the same time
	Clicks		4000-9000		
	Clicks	200-150,000	30,000-60,000	35-350μsec	
Bottle-nosed Dolphin	Bark	200-16,000		0.1	
	Whistle	4000-20,000		0.1-3.6	Narrow bands, 18 contours
	whistle	2000-20,000		0.8-0.9	Signature whistles, mostly pure tones
	Clicks	200-3000,000	15,000-130,000	10-200μsec	
	Clicks	100-300,000+		1-10msec	
Atlantic Spotted Dolphin	Whistle	6500-13,300		0.46	
	Pulses	to 150,000		0.075-0.2msec	
Harbour Porpoise	Pulses	41,000	2000	0.5-5msec	
	Pulses	<100,000-160,000	60,000/120,000/180,000	0.1msec	

Notes: This table was kindly assembled by Dr P.G.H. Evans. It is difficult to resolve data collected by different people and different sounds may be associated with different contexts. It is best regarded as a general and often preliminary description of sounds produced by selected cetacean species.
AM = Amplitude modulated; FM = Frequency modulated

coasts where most development is taking place. Increased boat traffic, pollution (eg heavy metals, pesticide and agro-chemical residues and oil) along with competition with Man for krill are all threats.

Cetacea in Western Europe: There are two modern suborders, Odontoceti and Mysticeti, which together include 9 families. The Cetacea include 150 extinct genera, 38 modern genera with 76 modern species. Today, 31 species from 20 genera and seven families of Cetacea are represented in Europe. Of the mysticetes these are the Balaenopteridae and Balaenidae. Of the odontocetes they are the Delphinidae, Phocoenidae, Monodontidae, Physeteridae and Ziphiidae. The mysticetes together with the Physeteridae are referred to as the great whales.

The Odontoceti or toothed whales comprise 66 out of the 76 extant Cetacean species. They feed mainly on fish and squid, using their teeth, and are the only group of mammals to have a single nostril. In most Odontoceti the jaws form a beak-like snout, behind which the forehead rises in a rounded curve, the 'melon'. The toothed whales arose in the late Eocene, *c.* 40 mya, and radiated fast in two ways, either they produced many simple teeth (up to 300) with single roots or, like the beaked whales or Ziphiidae, they lost most of their teeth.

The Mysticeti or baleen or whalebone whales include most of the great whales. They arose in the mid-Oligocene, coinciding with the circum-Antarctic current and the cooling of the oceans that caused an upsurge in plankton production. The baleen whales are named after the baleen plates through which these toothless whales sieve seawater. The baleen plates are sheets of stiff fibrous tissue derived from the upper palate, that shred as they are worn to form a filter plate for plankton. The whales catch the plankton on the straining-hairs and suck them off with their tongues. Baleen whales include the Blue Whale, the largest animal alive. The Blue Whale can afford to be so big because much of its huge weight is supported by the water. The great whales migrate from krill-rich high latitudes to warm tropics and back each year. They generally give birth every alternate year.

The Balaenopteridae or rorquals are characterised by their streamlined appearance, a straightish jawline, short baleen plates and many throat grooves (folds of skin from chin to under the belly).

The Balaenidae or right whales are the oldest of the 4 living mysticete families. They have an arched rostrum giving a deeply curved jawline, very long slender baleen plates and no throat groves. They are quite rotund with a relatively large head (*c.* ⅓ of body length), and a narrow upper jawbone, and have tactile hairs on their heads.

The Delphinidae, Phocoenidae and Monodontidae are all quite closely related (and probably diverged in the mid-Miocene *c.* 15 mya). The Delphinidae are the true dolphins (NB dolphins are called porpoises in the USA). They usually have functional teeth in both jaws, a melon with a distinct beak and a dorsal fin, a markedly asymmetrical skull and a single crescent-shaped blowhole. They include the killer whales. The Phocoenidae, the porpoises, are all quite uniform: they are small, lack a beak and

The Colour Plates

Each species entry has a cross-reference to the relevant Plate. Each plate entry has a cross reference to the relevant species entry. Species are grouped together on each plate, with similar species for ease of identification, in the following order:

53. Striped Mice
54. Rats
55. Aquatic Rodents
56. Aquatic Rodents and Porcupine
57. Red-backed Voles
58. Grass Voles
59. Grass Voles
60. Pine Voles
61. Balkan Snow Vole and Lemmings
62. Hamster
63. Hares and Rabbit
64. Primates

Plate 1

MARSUPIALS

RED-NECKED WALLABY *Macropus rufogriseus* **p. 20**

Key features: Characteristic bipedal gait. Coat grizzled greyish-brown above, white below. Silvery tail with a black tip. Ears, feet and muzzle become darker towards the tips.

Illustrated signs: *a,* Skull. *b,* one hind print; at slow grazing speed, fore-print and tail drag may register. *c,* Dropping ovoid shaped, usually 5–6 pellets in a loose chain, each 15 x 20mm.

Left, Marsupials give birth to undeveloped infants after a short pregnancy, and thereafter continue the 'pregnancy' externally, carrying the baby in a pouch.

Plate 2

HEDGEHOGS

Spines make hedgehogs unmistakable, and all species roll into a ball when alarmed. The Western and Eastern species have non-overlapping distributions.

EASTERN HEDGEHOG *Erinaceus concolor* **p. 24**
Key features: Differs from Western Hedgehog in having throat and chest paler than belly. *a*, The skull has a postero-dorsal process on the maxilla extending behind the lachrymal foramen.

WESTERN HEDGEHOG *Erinaceus europaeus* **p. 22**
Key features: Western Hedgehog has uniform colouring on the underside. *b*, Skull has a short postero-dorsal process on the maxilla, which does not extend behind the lachrymal foramen.

ALGERIAN HEDGEHOG *Atelerix algirus* **p. 25**
Key features: Paler underside than Western and Eastern Hedgehogs, with wider 'parting' free of spines on crown of head. External ears larger than in the Western and Eastern species.

Common features: *c*, Skulls similar. *d*, Droppings, *c.* 30–40mm × *c.* 10mm, with fragments of beetle wing-cases visible. *e*, Tracks from fore- and hind foot about the same size, 2.5cm long, 2.8cm broad, tracks with five toes on each foot. *f*, Typically nest under brushwood.

Below, Baby hedgehogs are born with short spines beneath their skin.

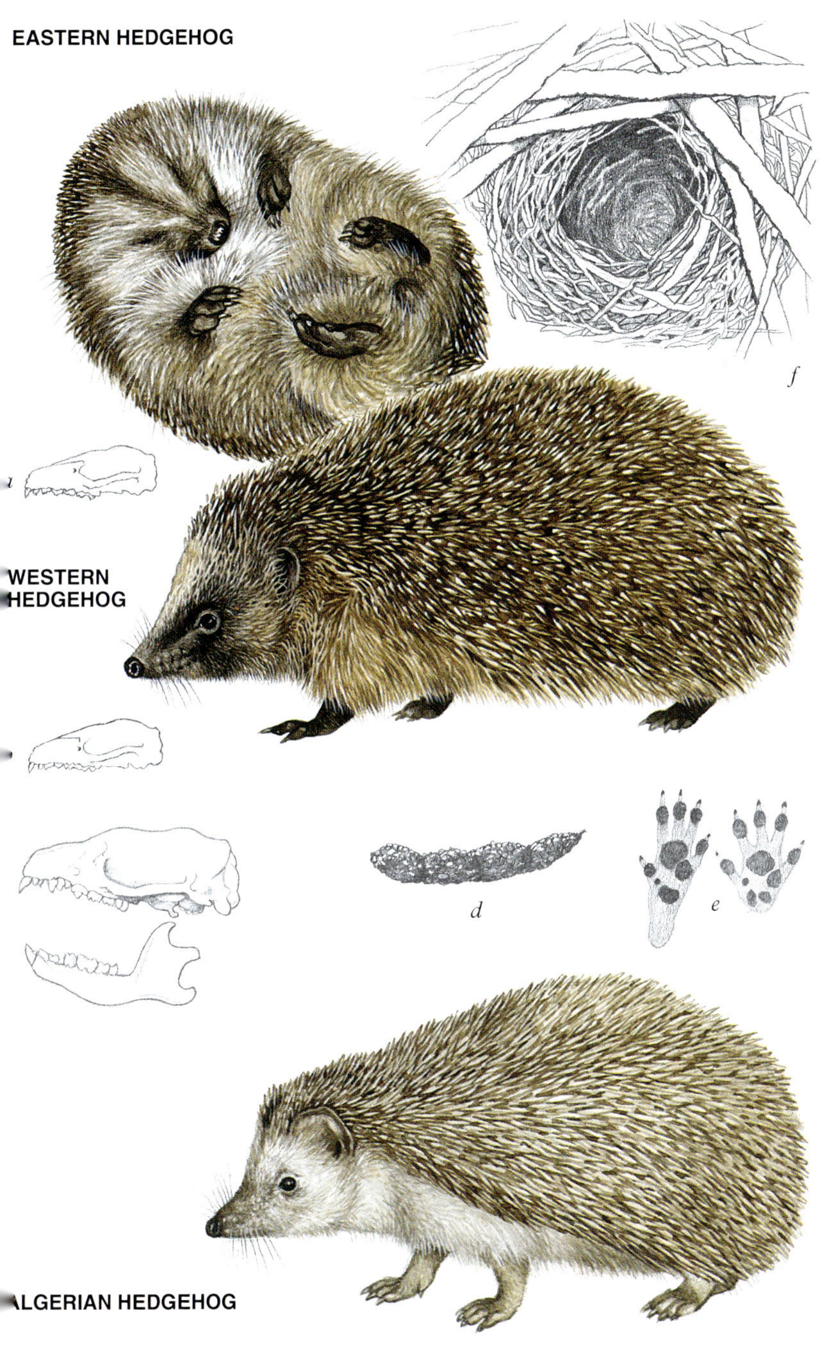

EASTERN HEDGEHOG
f
WESTERN
HEDGEHOG
d
e
ALGERIAN HEDGEHOG

Plate 3

WHITE-TOOTHED SHREWS

All have prominent ears, dense velvety fur, and long sparse hairs on tail.

PYGMY WHITE-TOOTHED SHREW *Suncus etruscus* p. 37
Key features: Extremely small, total head-body length less than 45mm. *a*, Four unicuspid teeth on each side of upper jaw.

BI-COLOURED WHITE-TOOTHED SHREW *Crocidura leucodon* p. 34
Key features: Clearer delineation between dark upperside and pale underside than in Greater White-toothed Shrew. Tail also bi-coloured. The rostrum of the skull is wider than in the Greater White-toothed Shrew. *b*, Three upper unicuspid teeth on each side of jaw, no marked difference in size between the 2nd and 3rd.

LESSER WHITE-TOOTHED SHREW *Crocidura suaveolens* p. 35
Key features: Only distinguished from Greater White-toothed Shrew by smaller size and teeth. *c*, Middle tooth of the three upper unicuspid teeth is much smaller than 1st and 3rd.

GREATER WHITE-TOOTHED SHREW *Crocidura russula* p. 36
Key features: Blurred border between dark upperside and pale underside (cf Bi-coloured White-toothed Shrew). *d*, Middle tooth of the three upper unicuspid teeth is slightly smaller than 1st and 3rd (cf Lesser White-toothed Shrew).

Common features: *e*, Skulls similar. *f*, Droppings blackish, *c.* 2–4mm long, containing insect wing cases, and may accumulate in piles. *g*, Tracks with five toes on each foot; footprints in snow vary from 0.5 to 1.0cm in length. *h*, nest a sphere of grasses and leaves. *i*, Narrow runway through dry grass tend to be flat in cross-section.

Below, Once the young of the Lesser White-toothed Shrew leave the nest they may hang on to their mother in a nose-to-tail procession known as caravanning.

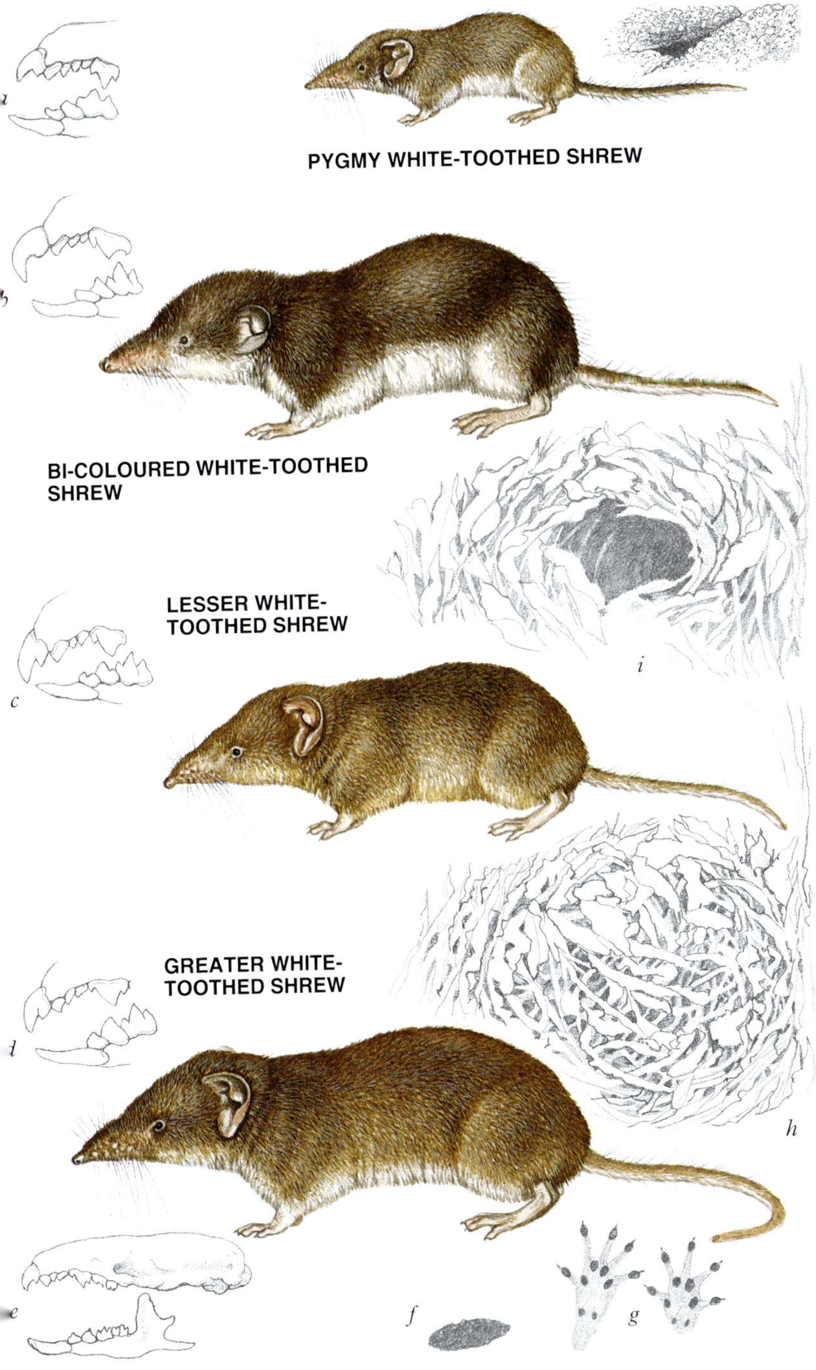

PYGMY WHITE-TOOTHED SHREW

BI-COLOURED WHITE-TOOTHED SHREW

LESSER WHITE-TOOTHED SHREW

GREATER WHITE-TOOTHED SHREW

a *b* *c* *d* *e* *f* *g* *h* *i*

Plate 4

RED-TOOTHED SHREWS

No protruding tactile tail hairs, in contrast to white-toothed shrews. Fur glossy.

LEAST SHREW *Sorex minutissimus* **p. 25**
Key features: Very small, with short hind feet. Back is dark brown with lighter flanks and underside. *a*, Second upper unicuspid tooth slightly smaller than 1st and 3rd.

APPENINE SHREW *Sorex samniticus* **p. 26**
Key features: Similar to Common Shrew but tail shorter. *b*, cusps of the upper incisors divided by a rounded notch; 2nd upper unicuspid tooth is larger than the 1st and 3rd.

COMMON SHREW *Sorex araneus* **p. 26**
Key features: Adults have lighter brown band along their flanks, separating dark upperside from pale underside; juveniles lack this band and are lighter brown above. Tail shorter relative to body length, when compared to the Pygmy Shrew. Sexually mature male with lateral scent glands illustrated. *c*, First three of five unicuspid teeth in each side of upper jaw are all of similar size. The concavity between the two cusps in the upper incisors has a V-shaped notch.

PYGMY SHREW *Sorex minutus* **p. 28**
Key features: Very small, but with relatively long hairy tail. Does not have the lighter brown flank band of Common Shrew. *d*, Third of five unicuspid teeth in upper jaw is slightly larger than the second.

Common features: *e*, Skulls similar. *f*, Droppings blackish, *c.* 2–4mm long, containing insect wing-cases and found in no particular pattern. *g*, Tracks with five toes on each foot. *h*, nest is a sphere of grasses and leaves. *i*, Narrow runways through dry grass tend to be flat in cross-section.

Right, Shrews engage in refection: they produce a special faecal pellet that is eaten and passed through the digestive system for a second time. This female Common Shrew also exhibits a nape bite, caused by the male's grasp during mating.

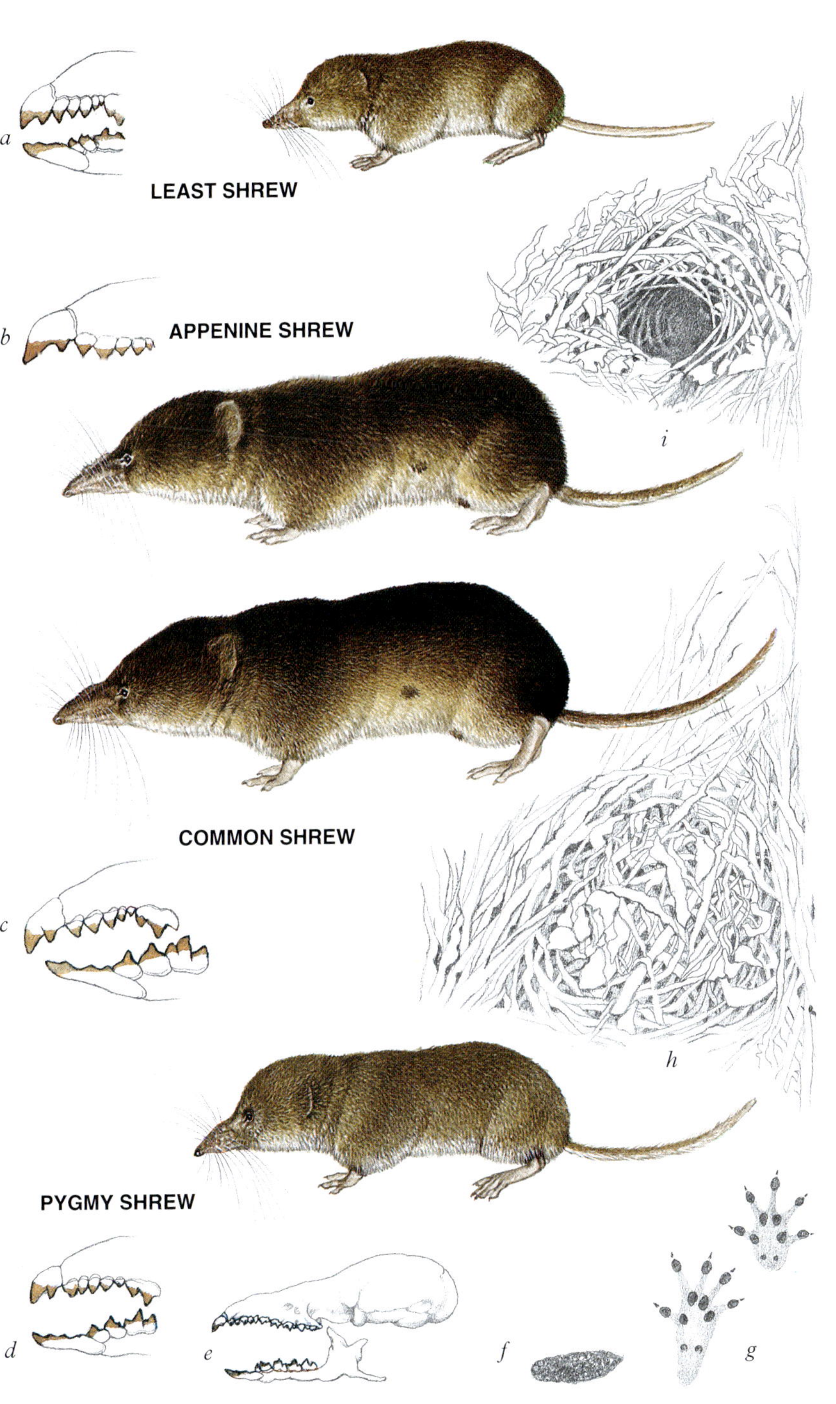
a
LEAST SHREW
b
APPENINE SHREW
i
COMMON SHREW
c
h
PYGMY SHREW
d
e
f
g

Plate 5

RED-TOOTHED SHREWS

DUSKY SHREW *Sorex isodon* **p. 29**
Key features: Relatively large shrew, with head and body longer than tail. Coat bi-coloured, but underside almost as dark as upperside. *a*, Five unicuspid teeth in upper jaw get progressively smaller towards the back.

MASKED SHREW *Sorex caecutiens* **p. 30**
Key features: Size intermediate between the Common and the Pygmy shrew. Noticeable contrast between the dark back and the lighter underside. Feet are white with silvery, shiny hairs. *b*, Five upper unicuspid teeth with no marked difference in size.

ALPINE SHREW *Sorex alpinus* **p. 30**
Key features: A greyish shrew with tail roughly same length as head and body. Fur uniformly dark with the feet and underside of the tail lighter. *c*, Fourth and fifth of the five upper unicuspid teeth are of equal size.

MILLET'S SHREW *Sorex coronatus* (Not Illustrated) **p. 31**
Key features: Slightly smaller jaws than Common Shrew but only reliably distinguishable by detailed examination of chromosomes.

SPANISH SHREW *Sorex granarius* (Not Illustrated) **p. 31**
Key features: Slightly smaller than Common Shrew. Only identifiable by detailed skull measurements and chromosome counts.

Common features: *d*, Skulls similar. *e*, Droppings blackish, *c.* 2–4mm long, containing insect wing-cases and found in no particular pattern. *f*, Tracks with five toes on each foot. *g*, Nest is a sphere of grasses and leaves. *h*, Narrow runways through dry grass are flat in cross section.

Below, Shrews are distasteful to most mammalian predators, probably due to the secretions from their lateral scent glands. However, predators such as Red Foxes may nonetheless kill and discard them (often squashed by a pounce), perhaps because they kill the shrew before recognising what it is. Owls, in contrast, regularly eat shrews and their undigested remains are obvious in regurgitated owl pellets.

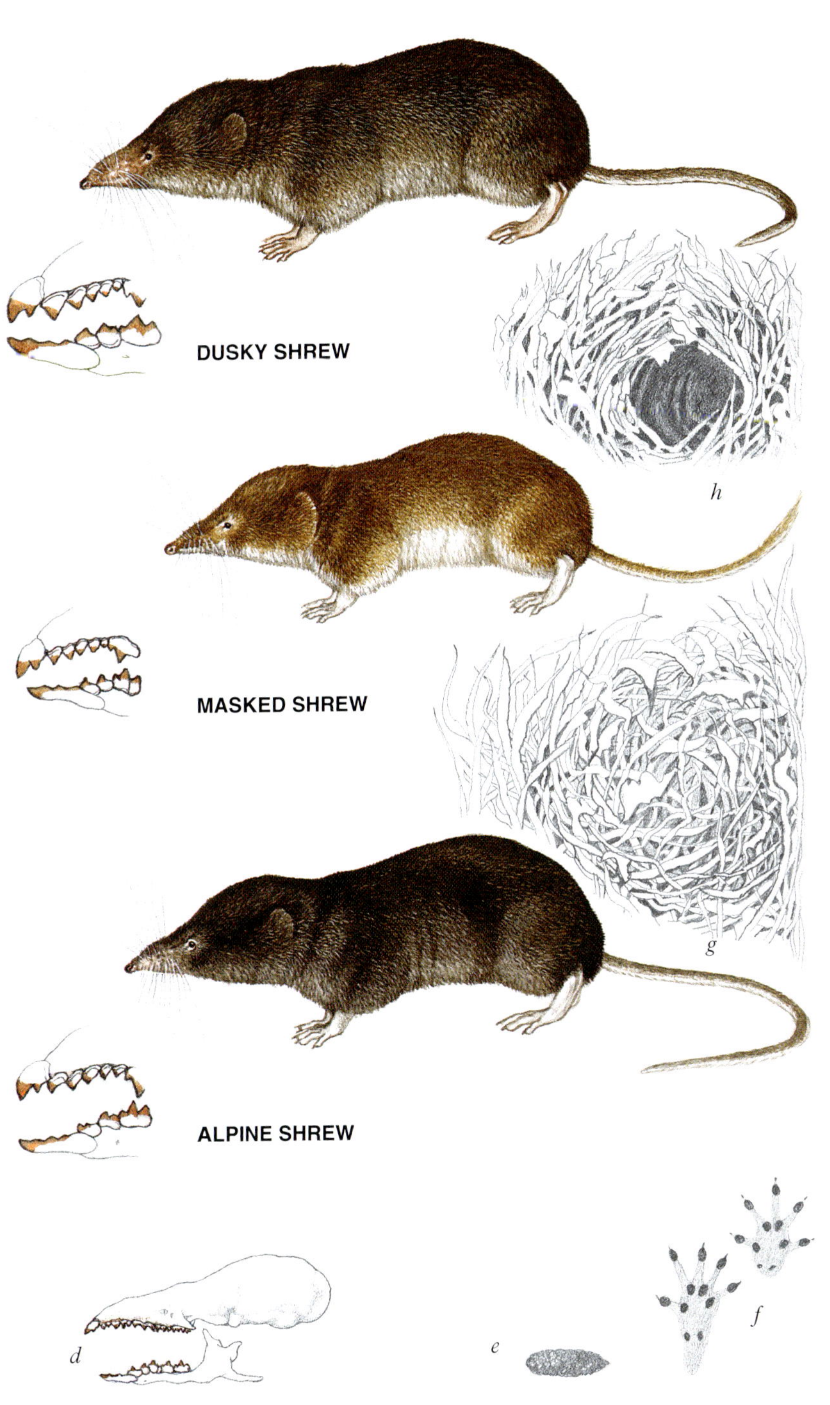
DUSKY SHREW
h
MASKED SHREW
g
ALPINE SHREW
f
d
e

Plate 6

RED-TOOTHED WATER SHREWS

WATER SHREW *Neomys fodiens* **p. 32**

Key features: Largest European shrew, blackish above and pale below with distinct demarcation. Obvious fringe of silver hairs along underside of tail and similar fringe of bristles on outer edge of all feet. *a*, Four unicuspid teeth in each side of upper jaw.

MILLER'S WATER SHREW *Neomys anomalus* **p.33**

Key features: Distinguished from Water Shrew by less developed bristles on feet and tail. Bristles on tail may be confined to last third. Also dark above and whitish below with distinct demarcation. *b*, The four upper unicuspid teeth on each side of the jaw tend to have rather longer bases and smaller cusps than those of the Water Shrew.

Features common to both water shrews: *c*, Skulls similar. *d*, Droppings *c.* 5mm long, found in latrine sites on rocks near water. *e*, Tracks with five toes on each foot. *f*, Burrow entrance, *c.* 2 cm diameter, in bank near water. *g*, Runways through grass, not associated with grazed grass (cf Water Voles).

DESMAN

PYRENEAN DESMAN *Galemys pyrenaicus* **p.38**

Key features: Extremely long snout, long tail flattened at tip. Coat dark brown to black above and grey to light brownish below. The large hind feet are partly webbed and fringed with silvery bristles. The root of the tail is rather constricted and then broadens due to the presence of an enlarged scent gland.

Illustrated signs: *h*, Skull. *i*, Tracks with five toes on each foot. *j*, Burrow in bank near water.

Below, In Water Shrews the underside is very variable in colour, ranging from silvery-grey to pale brown, occasionally black.

WATER SHREW
a
g
c
d
e
f
MILLER'S WATER SHREW
j
h
PYRENEAN DESMAN
i

Plate 7

MOLES

Velvety black fur and cylindrical body. Huge forefeet adapted for digging.

BLIND MOLE *Talpa caeca* **p. 39**

Key features: Slightly smaller than Common Mole with longer, more slender muzzle (*b*), and whitish hair on legs and tail. The measurement of the hindfoot is probably the most reliable feature: usually less than 17mm long, excluding the claws. *a*, Central incisors almost twice size of peripheral (3rd) incisors. Upper incisor teeth form a V-shape in sideview (not illustrated).

COMMON MOLE *Talpa europaea* **p. 40**

Key features: This species has a narrower muzzle than the Roman Mole (*g*), with a skull width of 11–13mm, measured across the cheek bones (*d*). Also distinguished from Roman Mole by shape of the three cusps on each of the three upper molars of each jaw: in the Common Mole these are simple cones (*e*), but in the Roman Mole each central cusp has an additional notch (*f*).

ROMAN MOLE *Talpa romana* **p. 42**

Key features: Longer body and hind feet than Blind Mole. Wider muzzle than in the Common Mole, width of skull across cheek-bones 13.3–15mm (*g*). Also distinguished from Common Mole by shape of the three cusps on each of the three upper molars of each jaw: in the Common Mole these are simple cones (*e*), but in the Roman Mole each central cusp has an additional notch (*f*).

Common features: *h*, Molehills pushed up to remove soil debris as mole digs tunnels. *i*, Runs close to the surface raise arched ridges with somewhat displaced grass. *j*, Tracks in soft soil. Due to the modifications for digging, the forelegs cannot be used in the normal walking position: the mole steps on the front edge of the forefoot, leaving the track of the five claws in a slightly curved row. Hindfoot track with five toes. Body drag present. *k*, Skulls similar.

Right, Moles paralyse worms by damaging their nerve cord with a bite. These worms are cached alive in large numbers for future use when food is short.

BLIND MOLE
a
b
j
k
c
d
COMMON MOLE
e
h
i
f
g
ROMAN MOLE

Plate 8

SMALL AND MEDIUM HORSESHOE BATS

LESSER HORSESHOE BAT *Rhinolophus hipposideros* **p. 47**
Key features: Smokey brown above, grey to grey-white below. *a*, Very delicate noseleaf with short, rounded upper connecting process and longer, pointed lower connecting process. Roosts in caves and buildings.

MEDITERRANEAN HORSESHOE BAT *Rhinolophus euryale* **p. 49**
Key features: The fur is grey-brown and fluffy with a reddish or lilac tinge above, grey-white to yellowish-white below. *b*, Noseleaf with upper connecting process of sella pointed and delicately bent downwards, the lower connecting process looks rounded when viewed from below. Roosts in caves.

BLASIUS' HORSESHOE BAT *Rhinolophus blasii* **p. 50**
Key features: The fur is grey-brown above and pale (almost white or yellowish) below. The border between the two is well defined. *c*, Upper connecting process of the noseleaf is straight and longer than the lower connecting process, which looks narrow and rounded from the front. Roosts in caves.

Common features: *d*, Skulls similar. *e*, Droppings indistinguishable.

Right, Both Greater and Lesser Horseshoe Bats wrap themselves up completely in their wing membranes when hibernating or torpid. Mediterranean and Mehely's Horseshoe Bats do not envelope their bodies completely.

a LESSER HORSESHOE BAT

b MEDITERRANEAN HORSESHOE BAT

c BLASIUS' HORSESHOE BAT

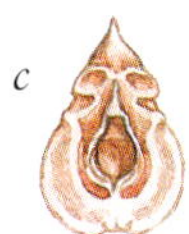

d

e

Plate 9

HORSESHOE BATS AND VAGRANTS

GREATER HORSESHOE BAT *Rhinolophus ferrumequinum* **p. 50**
Key features: Grey-brown or smoke grey fur with reddish tinge above and greyish-white to yellow-white below. Noseleaf (*b*) with short upper connecting process. Lower connecting process pointed in profile. Roosts in caves and buildings.

MEHELY'S HORSESHOE BAT *Rhinolophus mehelyi* **p. 52**
Key features: Relatively thick fur, grey-brown on the back and almost white underneath. Dark grey-brown 'spectacles' around the eyes. Noseleaf (*c*) has a relatively blunt upper connecting process in profile, only slightly longer than the rounded (in front view) lower connecting process. Roosts in caves.

Common features: *a*, Greater and Mehely's Horseshoe Bats have similar skulls. *d*, Droppings of all European Horseshoe bats are indistinguishable.

HOARY BAT *Lasiurus cinereus* **p. 80**
Key features: Rare in Europe. The fur has a mottled 'frosted' appearance. Characteristic dense covering of fur on the dorsal surface of the tail membrane.

EGYPTIAN SLIT-FACED BAT *Nycteris thebaica* **p. 46**
Key features: Very rare in Europe, but clearly identified by the slit-like nasal apperture.

Left, All bats have nipples on their chests, but female horseshoe bats, like this Greater Horseshoe Bat, possess two additional 'false nipples' near the genitals, to which the young can cling.

GREATER HORSESHOE BAT
a
b
c
MEHELY'S HORSESHOE BAT
d
HOARY BAT
EGYPTIAN SLIT-FACED BAT

Plate 10

SMALL BATS: PIPISTRELLES

COMMON PIPISTRELLE *Pipistrellus pipistrellus* **p. 66**
Key features: Smallest European bat. The fur is uniformly coloured, with individual coloration varying from orange to dark brown above, and yellow-brown to grey-brown below. Thumb short (equal to width of wrist), lower legs and tail membrane hairless. First upper premolar tooth is small and partially covered by the canine tooth when viewed from the side (*a*).

NATHUSIUS' PIPISTRELLE *Pipistrellus nathusii* **p. 69**
Key features: Slightly larger than the Common Pipistrelle, fur on upperside less uniformly coloured. In summer the upperside is red to chestnut brown becoming darker brown with grey tips after moult. Underside light brown to yellow-brown. The thumb is longer than the width of the wrist. Also distinguished by the presence of hairs on the upper surface of the tail membrane and on the underside of the tail membrane along the lower legs. The first upper premolar tooth is visible from the side (*b*).

KUHL'S PIPISTRELLE *Pipistrellus kuhlii* **p. 70**
Key features: Fur on upperside varies from brown to cinnamon or yellow-brown; underside light grey to whitish-grey. White edge to wing membrane between foot and fifth finger. The thumb is short. The first upper incisor tooth has one point, the second upper incisor is very small, and the first upper premolar is situated inside the tooth row and not visible when viewed from the side (*c*).

SAVI'S PIPISTRELLE *Pipistrellus savii* **p. 72**
Key features: Fur is relatively long and varies from pale yellow-brown to dark brown on the upperside, yellowish-white to grey-white on the underside, with a clear demarcation between the two. The first upper premolar tooth is displaced and not visible when viewed from the side (*d*); the first upper incisor is bicuspid. Roosts in mines and caves, as well as in tree holes and buildings.

Common features: Roost in tree holes or buildings. *e*, Skulls similar. *f*, Droppings similar to those produced by mice but exclusively of very fine insect remains; dark brown to black, frequently found in abundance in roosts, but break up quickly.

Left, An adult Common Pipistrelle on an adult human's thumb.

a
COMMON
PIPISTRELLE
b
NATHUSIUS'
PIPISTRELLE
KUHL'S
PIPISTRELLE
SAVI'S
PIPISTRELLE
e
f

Plate 11

LARGE AND MEDIUM VESPER BATS

SERONTINE BAT *Eptesicus serotinus* **p. 76**

Key features: Size large. Fur dark brown above and yellow-brown below. Ears are about twice as long as wide; muzzle very bulbous. Tragus up to ⅓ length of ear with a blunt, concave front edge and convex rear edge. Canine teeth large.

Illustrated signs: *a*, Skull. *e*, Droppings 3.5–4.0mm in diameter, more bulbous than those of Northern and Parti-coloured Bats. Commonly roosts in buildings.

NORTHERN BAT *Eptesicus nilssoni* **p. 78**

Key features: Size medium. Fur is shaggy above; each hair has dark brown base and light glossy or sandy coloured tip. This species is yellowish-brown below. The ears are relatively short. The tragus is short, broad and delicately bent inwards with a rounded tip. The tip of the tail protrudes 3–4mm from the edge of the interfemoral membrane.

Illustrated signs: *b*, Skull. *d*, Droppings indistinguishable from those of Parti-coloured Bat, but less bulbous than those of Serotine. Roosts in tree holes, caves and buildings.

PARTI-COLOURED BAT *Vespertilio murinus* **p. 79**

Key features: Size medium. The frosted appearance of the long dense fur is due to bi-coloured hairs on the back which have black-brown bases and silvery white tips. The underside is whitish. The ears are slightly rounded and short with an almost mushroom-shaped tragus.

Illustrated signs: *c*, Skull. *d*, Droppings indistinguishable from those of Northern Bat, but less bulbous than those of Serotine. Roosts in caves, tree holes and buildings.

Left, Insect remains accumulate below a Serotine's feeding site. Long-eared and Horseshoe Bats probably use regular feeding sites even more than do Serotines.

a
SEROTINE BAT
b
NORTHERN BAT
c
PARTI-COLOURED BAT
d
e

Plate 12

MEDIUM TO LARGE VESPER BATS: NOCTULE AND ALLIES

LEISLER'S BAT *Nyctalus leisleri* **p. 72**

Key features: Medium-sized. The fur is long and composed of bi-coloured hairs which are darker towards the roots than at the tips. The upperside is rufous-brown, the underside yellow-brown. The ear has a mushroom shaped tragus. Lateral membrane starts at heel..

Illustrated signs: *a*, Skull small (condylo-basal length: 15–16mm). Roosts in tree holes and buildings.

NOCTULE *Nyctalus noctula* **p. 74**

Key features: One of Europe's largest bats. Forearm more than 48mm long. The fur short, smooth and uniformly coloured with the upperside rufous-brown and the underside light brown. The tragus is mushroom-shaped.

Illustrated signs: *b*, Large skull, distance between canine and third upper molar of 7.0–7.4mm. Roosts in tree holes.

GREATER NOCTULE *Nyctalus lasiopterus* **p. 75**

Key features: Identical to Noctule, but larger.

Illustrated signs: *c*, Skull. Roosts in tree holes.

Common features: Droppings of all *Nyctalus* species are similar (*d*).

Above, Tree holes heavily used by Noctules or other bats may become discoloured by urine and faeces.

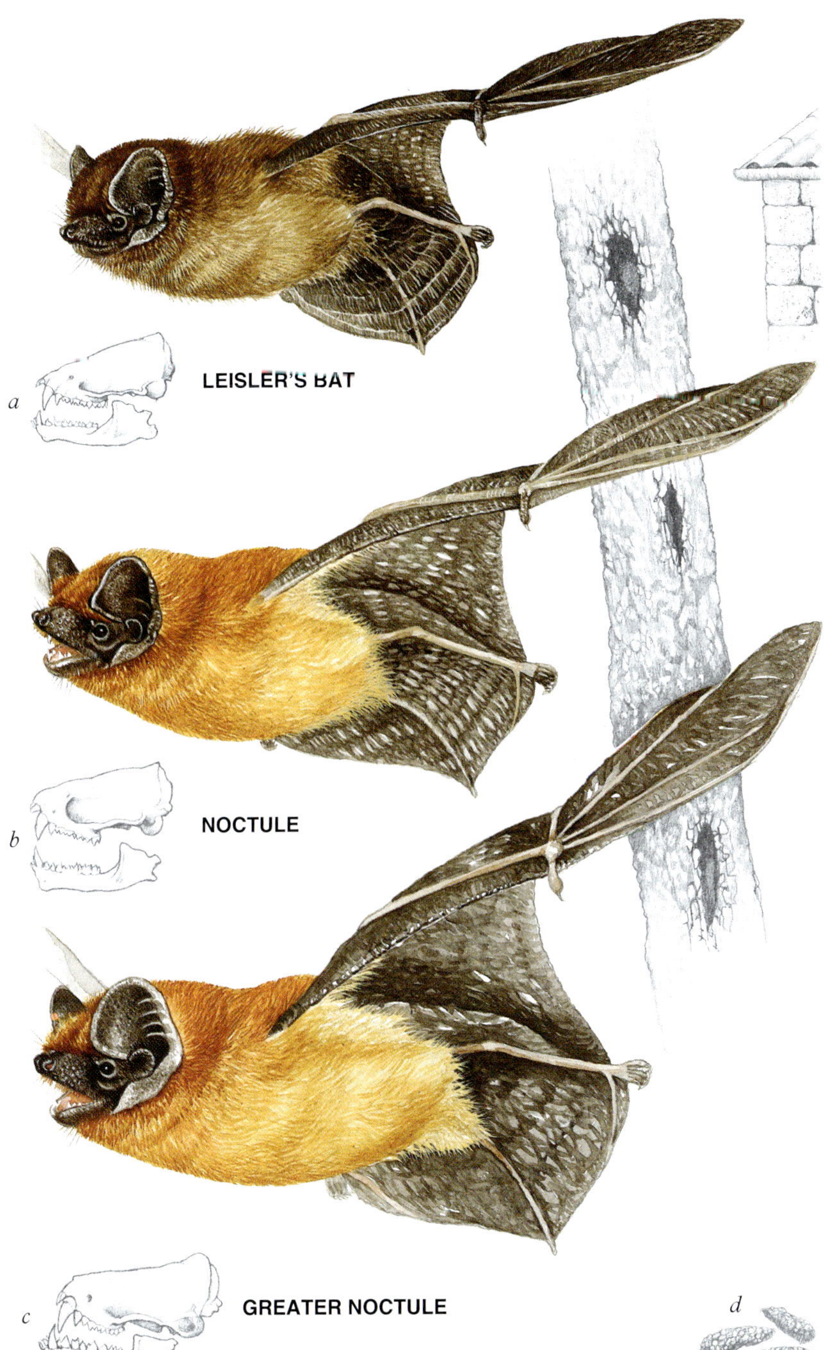

a
LEISLER'S BAT
b
NOCTULE
c
GREATER NOCTULE
d

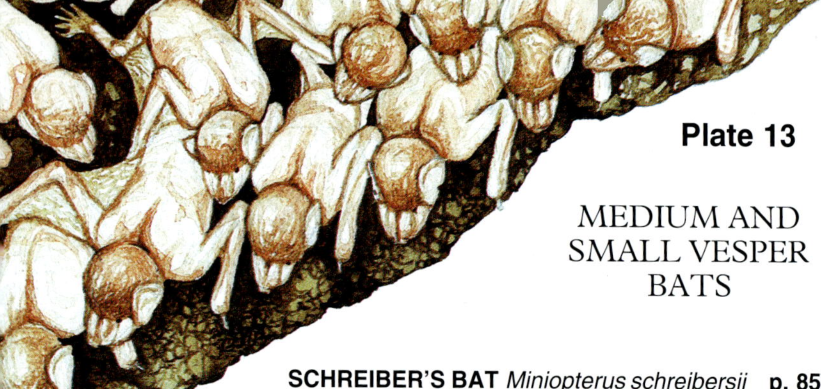

Plate 13

MEDIUM AND SMALL VESPER BATS

SCHREIBER'S BAT *Miniopterus schreibersii* **p. 85**
Key features: Medium sized. The short, dense fur of the head stands erect. Fur on the upperside is grey-brown to ash-grey, the underside is lighter grey. Ears also grey-brown with a yellowish-white or greyish tinge. Tragus short and bent inwards with a rounded tip. At rest bends fingers differently to all other bats. Lateral membrane starts at heel. No post-calcarial lobe.
Illustrated signs: *a*, Skull. Roosts in caves or buildings.

WHISKERED BAT *Myotis mystacinus* **p. 53**
Key features: Small. The long shaggy fur shows some variation in colour from dark brown to grey brown above and dark to light grey below. Lateral membrane starts at base of toes. The tragus is straight or concave on the outer edge and relatively longer than in Brandt's. Penis (*c*) is not club shaped, cf Brandt's Bat (*e*). Also differs from Brandt's in the lack of a conspicuous cusp on the anterior inner angle of the third upper premolar tooth (when viewed from inside the mouth, *d*). Second lower premolar tooth significantly smaller than first (cf Brandt's. Not illustrated).
Illustrated signs: *b*, Skull similar to Brandt's. Roosts in tree holes, caves and buildings.

BRANDT'S BAT *Myotis brandtii* **p. 55**
Key features: Small. Fur long, light brown with golden sheen. The inner edge of the ears and the base of tragus are pale. The tragus is almost half the length of the ear and has a more or less convex posterior margin. Penis (*e*) is club-shaped, cf Whiskered Bat's (*c*). Distinguished from Whiskered Bat by a conspicuous cusp on anterior inner angle of upper third premolar (when viewed from inside the mouth, f). Second upper premolar tooth not significantly smaller than first (cf Whiskered Bat, not illustrated).
Illustrated signs: Roosts in caves, tree holes or buildings. *b*, Skull similar to Whiskered Bat.

Common features: *g*, Droppings similar for these species.

Top, Infant Schreiber's Bats, like the young of other bats, cluster together for warmth while their mothers are away foraging.

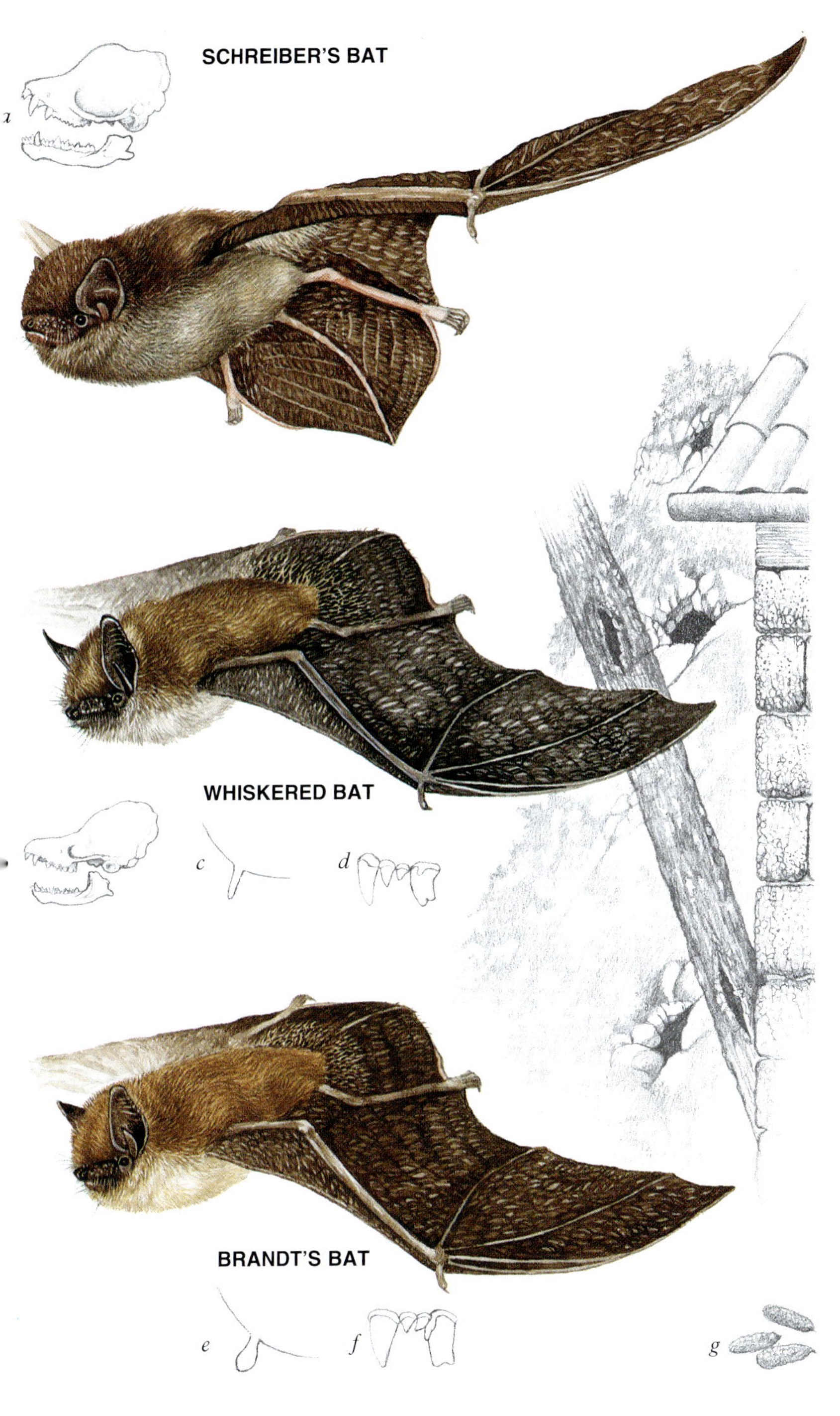

SCHREIBER'S BAT
a
WHISKERED BAT
c
d
BRANDT'S BAT
e
f
g

Plate 14

MEDIUM VESPER BATS

DAUBENTON'S BAT *Myotis daubentonii* **p. 56**
Key features: Medium sized. Fur is fluffy, brown-grey to dark bronze above, silvery-grey below with a sharp border between. The face is pinkish and the nose red-brown. Bases of hairs dark grey-brown. Ears short. Tragus less than half the length of the ear. Posterior edge of tragus convex, front edge straight. Feet large with long bristles.
Illustrated signs: *a*, Skull. Roosts in tree-holes, caves and buildings.

LONG-FINGERED BAT *Myotis capaccinii* **p. 58**
Key features: Medium sized. Fur smoke grey with slight yellowish hue above, light grey below; the border between the two is indistinct. The nostrils are more prominent than in other *Myotis* species. The tragus is half the length of the ear, pointed, with a convex inner edge and serrated outer edge. Thick brown hairs cover both sides of the wing membrane. Feet large and bristly.
Illustrated signs: *b*, Skull. Roosts in caves.

POND BAT *Myotis dasycneme* **p. 59**
Key features: Medium sized. Fur long and thick, brownish or pale grey-brown with a silky sheen above and white-grey below. The border between the two is fairly sharp. Fine white hairs are present on the underside of the tail membrane and lower part of tail.
Illustrated signs: *c*, Skull. Roosts in tree-holes, caves and buildings

Common features: *d*, Droppings are similar.

Left: Hibernating Daubenton's Bat covered with drops of dew caused by the high humidity in the hibernaculum.

a
DAUBENTON'S BAT
b
LONG-FINGERED BAT
POND BAT
d

Plate 15

MEDIUM VESPER BATS

NATTERER'S BAT *Myotis nattereri* **p. 60**

Key features: Medium sized. The upperside is light buff-brown, contrasting with the white underside. The face is red-pink and more or less bare. Light grey-brown wing, lateral membrane starts at base of toes. Characteristic conspicuous dense fringe of stiff downwardly pointing bristles along the outer edge of the interfemoral membrane. These characteristics and the slender tragus help to differentiate the species from the smaller Whiskered and Brandt's Bats. **Illustrated signs**: *a*, Skull. Roosts in tree-holes, caves and buildings. This is one of the species most likely to be found in any cave-like site.

GEOFFROY'S BAT *Myotis emarginatus* **p. 61**

Key features: Medium sized. Fur woolly. Hairs on back are grey at base, yellowish in the middle and rufous at the tips. Underside yellow-grey. The tragus is lancet-shaped with notches on the outer edge. The extreme tip of the tail projects beyond the tail membrane. Free edge of tail membrane has sparse short, straight, soft hair (much sparser than Natterer's Bat). **Illustrated signs**: *b*, Skull. Roosts in tree-holes, caves and buildings.

BECHSTEIN'S BAT *Myotis bechsteinii* **p. 62**

Key features: Medium sized. Fur light to reddish-brown above, greyish-white below. Bases of hairs are darker than the tips. The face is bare and pink. The noticeably long ears are fairly broad and, when folded forward, extend beyond the nose; the tragus is lancet-shaped and long. There are no hairs on the interfemoral membrane. **Illustrated signs**: *c*, Skull. Roosts in tree-holes and caves.

Common features: *d*, Droppings similar.

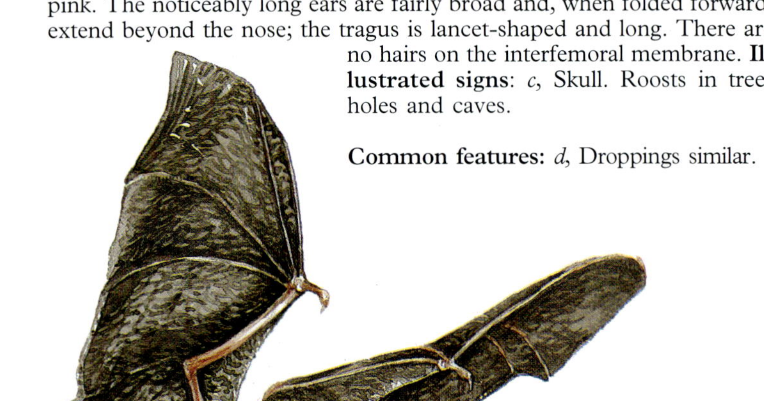

Left, Bats, like this Bechstein's, typically drink in flight by skimming low over water.

NATTERER'S BAT
GEOFFROY'S BAT
BECHSTEIN'S BAT
d

Plate 16

LARGE VESPER BATS

GREATER MOUSE-EARED BAT *Myotis myotis* **p. 64**
Key features: Large. Short dense fur, light grey-brown above (the bases of hairs are brown) and grey-white below. The face is almost bare and pinky-brown with a short broad nose of same colour. The ears are thick and long with a broad-based, pointed tragus that reaches almost half way up the ear. The last vertebrae of the tail projects beyond the membrane.
Illustrated signs: *a*, Skull. Roosts in tree-holes, caves and buildings.

LESSER MOUSE-EARED BAT *Myotis blythi (oxygnathus)* **p. 65**
Key features: Smaller than the Greater Mouse-eared Bat. The total length of the skull (*b*) is under 23mm and the distance from the canine to the third upper molar is under 9.5mm. The fur is light brownish-grey above and grey-white below. Compared to the Greater Mouse-Eared Bat the muzzle is narrower and more pointed and the ears shorter. The narrow-based, lancet-shaped tragus reaches about half way up the ear.
Illustrated signs: *b*, Skull. Roosts in caves, treeholes and buildings

EUROPEAN FREE-TAILED BAT *Tadarida teniotis* **p. 86**
Key features: Very large with short, soft, mole-like fur. Black-grey to smoke grey with brownish sheen above, lighter grey below. The ears are long, broad and project forwards. The tail protrudes from the edge of the membrane.
Illustrated signs: *c*, Skull. Roosts in caves and buildings.

Common features: *d*, Droppings similar.

Below, Some species, like this European Free-tailed Bat, are strong and active on all fours: others, like the horseshoe bats, are less adept on the ground.

GREATER MOUSE-EARED BAT
LESSER MOUSE-EARED BAT
b
EUROPEAN FREE-TAILED BAT
d

Plate 17

MEDIUM BATS

BROWN LONG-EARED BAT *Plecotus auritus* **p. 83**
Key features: Medium sized. Fur long and fluffy. Yellow to white below, light brown above, with indistinct demarcation on neck. Ears are especially long (up to 41mm long, ¾ of head-body length), with a long, pale pink, lancet-shaped tragus with light grey pigment at tip. The thumb is shorter than that of the Grey Long-eared Bat, under 6 mm long. **Illustrated signs:** *a*, Skull similar to Brown Long-eared Bat. Roosts in trees, caves and buildings.

GREY LONG-EARED BAT *P. austriacus* **p. 82**
Key features: Medium sized. Long fur, generally grey above, white below. Hairs on underside have black bases. Grey mask around the eyes and the nose. Upper lips are dark brown. This species is easily distinguished from all except the Brown Long-eared Bat by the huge length of its grey ears. The tragus is 5.5–6mm wide at its widest point. The length of the thumb usually over 6mm. **Illustrated signs:** *a*, Skull similar to Brown Long-eared Bat. Roosts in caves and buildings.

BARBASTELLE BAT *Barbastellus barbastellus* **p. 81**
Key features: Black-brown fur with whitish tips give upperside frosted appearance; underside dark grey. The naked parts of the face and ears are black. The muzzle is particulary short. Ears are shorter than those of the previous two species and join on top of the head. The tragus is triangular towards the base and rounded at the tip. **Illustrated signs:** *a*, Skull. Roosts in trees, caves and buildings

Common features: *c*, Droppings are indistinguishable.

Top, When hanging up Brown Long-eared Bats fold their ears out of the way to prevent damage and presumably to minimise heat loss.

BROWN LONG-EARED BAT
EY LONG-EARED BAT
BARBASTELLE BAT
c

Plate 18

FOXES

RED FOX *Vulpes vulpes* **p. 96**

Key features: Coat colour variable, usually reddish-brown. Slender muzzle, with white on upper and lower jaws. Pointed ears with black backs. Longer ears than the Arctic Fox. Long thick furry tail often white tipped. Darker variety called Cross Fox. 4 week old cub (*a*) typically has dark brown coat. Characteristic trotting gait. **Illustrated signs:** *b*, Skull. *c*, Droppings sometimes end in curl of undigested prey's fur and are often sited on tussocks of grass, stones or molehills. *d*, Tracks more oval than those of dogs *c*. 5cm long and 3-4cm wide, with the hindfeet slightly smaller than the forefeet, both with four toes. *e*, Den with freshly dug earth, remains of prey in spoil heaps. *f*, Reddish guard hairs or crinkly woolly underfur caught on barbed wire fences; narrow paths through tall grass. Also well-trodden paths around outer furrows of ploughed fields (not illustrated).

ARCTIC FOX *Alopex lagopus* **p. 94**

Key features: Two colour phases, one greyish-brown in summer (*h*) turning white in winter (*g*), the other, the blue phase, has brown fur in summer (*i*), blue-grey in winter (not illustrated). Muzzle shorter than in the Red Fox, ears shorter and more rounded. Body covered by a very dense coat. Characteristic cantering gait. **Illustrated signs:** Fieldsigns similar to those of Red Fox. *j*, Shull. *k*, Droppings often on conspicuous site. *l*, Tracks. *m*, Den often used over many years so accumulations of prey remains and excreta fertilise ground and produce lusher vegetation than surroundings.

Below, This partially moulted white morph Artic Fox is changing from summer grey to winter white. This fox is feeling the heat, and is stretched out, panting (foxes, like other members of the dog family, do not sweat and so can only lose heat effectively through their foot pads and by panting).

RED FOX

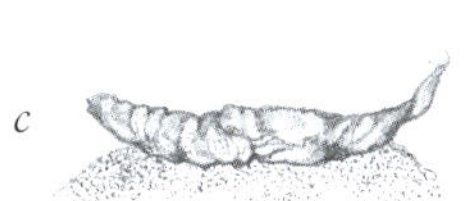

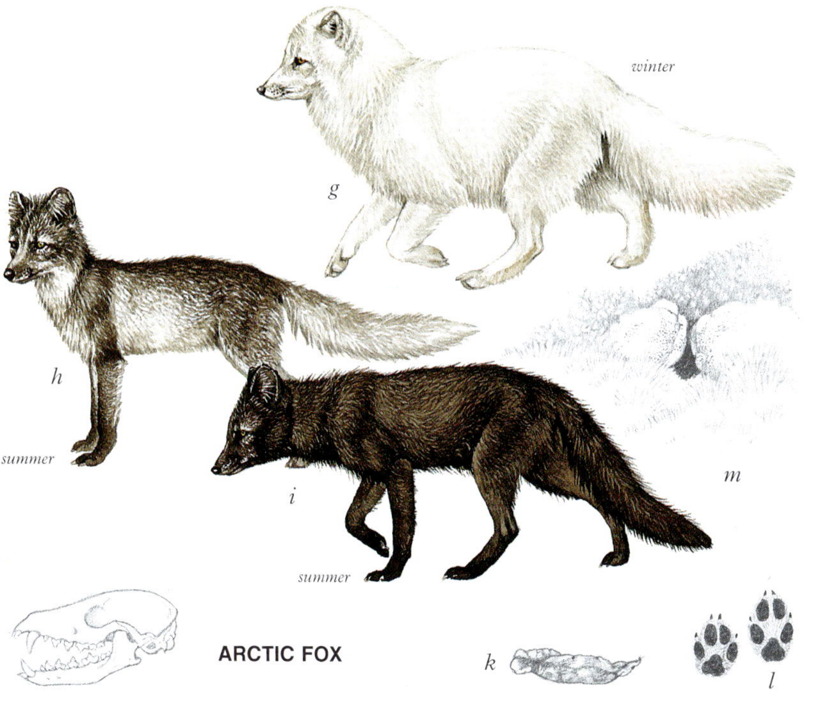

winter

g

h

summer

m

i

summer

ARCTIC FOX

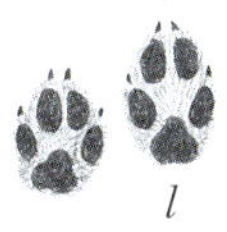

Plate 19

WILD DOGS

WOLF *Canis lupus* **p. 92**

Key features: Grey to greyish-fawn coat with reddish-brown on head and ears, and on back in southern specimens. General appearance resembles a German Shepherd dog, but shallower chest and broader head with a conspicuous ruff of hair around cheeks. Neck shorter and thicker than most domestic dogs. **Illustrated signs:** *a*, 4 week old pup. *b*, Skull. *c*, Droppings in prominent sites, especially at the junctions of well-travelled trails. *d*, Tracks difficult to distinguish from those of same size dog; four toes on each foot. *e*, Den in ground burrow or cave.

GOLDEN JACKAL *Canis aureus* **p. 93**

Key features: Smaller and more slender than Wolf with relatively larger ears and shorter legs. Grey coat with reddish tinge. **Illustrated signs:** *f*, Skull. *g*, Droppings generally left singly on prominent sites such as tussocks, bushes or boulders, but occasionally large middens of many droppings accumulate, generally near territorial border. *h*, Tracks with four toes on each foot, smaller than those of Wolf. *i*, Den in burrow or cave.

RACCOON DOG *Nyctereutes procyonoides* **p. 101**

Key features: Similar to Raccoon but larger in size, uniformly coloured, with shorter tail and ears. A distinctive black mask on face. **Illustrated signs:** *j*, Skull. *k*, Droppings left in pits at latrine sites. *l*, Tracks similar to fox but with toes widely spread. Forefoot track is 4–5cm long and 5-6cm wide, hindfoot slightly smaller. *m*, Den in ground burrow.

Below, Body posture and expression communicate social rank within the complicated stucture of Wolf packs. High-ranking animals threaten with an upright stance and a shortened snarling mouth, while subordinates threaten with flexed legs and a widely snarling mouth.

e
a
c
d
WOLF
i
g
h
GOLDEN JACKAL
m
k
l
RACCOON DOG

Plate 20

BEARS

POLAR BEAR *Ursus maritimus* **p. 101**

Key features: Coat whitish-yellow or grey but varies according to season and light conditions. Male has more aquiline nose than female. Very small ears. **Illustrated signs:** *a*, Skull. *b*, Droppings firm when eating mainland diet, but oily, black and liquid droppings when found on ice (not illustrated). *c*, Track with five digits on each foot. Den in snow cave (*d*), with slide marks descending from entrance (*e*); lots of trample marks in snow below den at foot of slope (*f*); platform or pit nearby where female nurses the cubs (*g*).

BROWN BEAR *Ursus arctos* **p. 105**

Key features: Coat fawn to dark brown. Large, heavily built body, no tail, ears short. Young bears, like the 6 week old cub (*i*), frequently have a whitish collar. **Illustrated signs:** *h*, Claw scratch marks reaching up tree trunks. *j*, Skull. *k*, Droppings cylindrical, diameter about 6cm, containing vegetable matter. *l*, Track with five toes on each foot with long claw marks; hindfoot *c*. 28cm long by 21cm wide; track of forefoot slightly shorter and broader. *m*, Den in ground burrow; snow or soil flattened within 1–2 m radius around den entrance.

Below, Polar Bears can swim for many hours between ice floes. They can smell seals one kilometre away and trek 30km a day through shifting pack ice in search of good hunting grounds.

d
g
e
f
a
POLAR BEAR
b
c
h
m
i
j
BROWN BEAR
k
l

Plate 21

RACCOON, BADGER AND WOLVERINE

RACCOON *Procyon lotor* **p. 107**

Key features: Easily recognizable by the black mask around the eyes and the bushy tail banded with four to six rings. Greyish-brown coat. **Illustrated signs:** *a*, Skull. *b*, Dropping similar in size and shape to those of a medium-sized dog. *c*, Tracks with five toes, with large claws. The forefoot track is about 7cm long and 7cm wide with widely separated toes; the hindfoot has the toes closer together and is about 9cm long and 6–7cm broad. *d*, Often dens in tree-hole, with scratch marks on bark around entrance to den.

EURASIAN BADGER *Meles meles* **p. 126**

Key features: Coat grey above with black underside and legs. Head white with two prominent black stripes over eyes and ears. The body is robust with short legs and short, greyish-tipped tail. **Illustrated signs:** *e*, Scratch marks on tree bark up to 1m above ground. *f*, Skull. *h*, Droppings often in shallow pits at latrine sites, especially near sett and near territorial boundary; often have muddy consistency (*g*). *i*, Tracks with five toes on each foot; claws leave deep gouge marks in mud. Lives in communal den or sett, commonly with 3–8 entrances, each associated with substantial spoil heap and well-trodden interlinking paths (*j*). Paths also radiate out from sett and criss-cross the territory.

WOLVERINE *Gulo gulo* **p. 124**

Key features: Coat dark brown to black with yellowish band from shoulders along flanks to tail, with a lighter patch around cheeks. Heavy build, with short, powerful legs and big claws. **Illustrated signs:** *k*, Skull. *l*, Droppings twisted, about 15cm long and 2cm diameter, containing hair. *m*, Tracks with five toes on each foot. *n*, Den in small cave.

Below, Scent marking is important in Eurasian Badgers, as it is in all Carnivores and the great majority of mammals. This badger is scent marking with its subcaudal scent pouch.

d
RACCOON
b
c
j
RASIAN BADGER
g
h
i
n
WOLVERINE
l
m

Plate 22

AQUATIC CARNIVORES

OTTER *Lutra lutra* **p. 129**

Key features: Long slim body with short legs; tail has thick base, tapering. Toes on both front and back feet are webbed. **Illustrated signs:** *a*, Floats low in the water. *b*, Skull. *c*, Droppings with scent jelly from anal sac (*h*) often on boulders, promontories, waterside ledges. *d*, Tracks with five toes on each foot. Forefoot track almost circular: 6.5–7cm long and 6cm broad. Hindfoot track is longer: varies from 6 to 9cm. Tail drag may show if mud soft. Length of stride about 70–80cm. *e*, Dens in burrow with clear path leading down bank into water. *f*, Muddy slides down steep banks for play and fast access to water. *g*, Aquatic vegetation may be flattened at site of day bed on promontory at waterside.

AMERICAN MINK *Mustela vison* **p. 114**

Key features: Very similar to the European Mink. Slender body, short legs and slightly bushy tail approximately half body length. Pattern of white hairs on chin, throat and undersided are unique to each individual. Little or no white on upper lip (cf. European Mink).

EUROPEAN MINK *Mustela lutreola* **p. 113**

Key features: Uniformly shiny brownish-black fur, with white upper and lower lips and chin. Slender body, short legs and slightly bushy tail which is approximately half body length.

Features common to the two species of mink: *i*, Float high in water. *j*, Skull. *k*, Droppings long and cyclindrical, 6-8cm long and 0.9cm wide, twisted and pointed at one end, normally contain fish scales, fur, feathers, pieces of bone and remains of berries. Because droppings vary with diet, it is difficult to differentiate between those of mink and otter. *l*, Tracks show distinctly only four of the five toes on each foot. *m*, Dens in thick hollow log, ground burrow or vegetation.

Left, Although feral American Mink are now so well-established in the UK that most have the same coloration as the wild population, some farmed colour varieties are still found – the pink nose is conspicuous. American Mink are now found throughout UK and are effectively impossible to eradicate. Contrary to earlier fears, it seems that they are generally not a threat to native wildlife, although they may threaten ground nesting birds on offshore islands.

g f e

h

a

b

OTTER

c d

AMERICAN MINK

m

i

EUROPEAN MINK

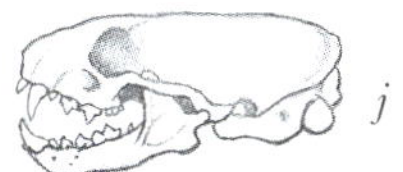

j

k

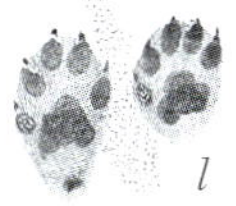

l

Plate 23

POLECATS

STEPPE POLECAT *Mustela eversmanni* **p. 116**
Key features: Light beige on back and sides with almost white head; legs, feet and part of the underside are dark.

WESTERN POLECAT *Mustela putorius* **p. 117**
Key features: Darker than the Steppe Polecat with white at sides of nose, but dark on top of muzzle. Body fur darker in summer (winter coat illustrated). The belly fur is yellowish with a glossy lustre. Flanks are lighter brown and the fringes of the ears pale.

Features distinguishing Steppe from Western Polecat: The Steppe Polecat's skull, like the domestic ferret's (*a*), has a waist less than 15mm across; in the Western Polecat (*b*) it is over 15mm. *c*, Top of Steppe Polecat skull shows constriction where post-orbital region meets the brain case; this is level in the Western Polecat (*d*)

MARBLED POLECAT *Vormela peregusna* **p. 118**
Key features: Mottled brownish and yellowish-white above, uniform dark brown below. Wide white band behind the eyes from cheek to cheek. Short legs and large ears with white edges. Bushy tail. Dominant males in breeding season more vividly orange coloured. **Illustrated signs:** *h*, Skull

Common polecat features: Mustelids have characteristic bounding gait (shown in Steppe and Marbled Polecats) and upright scanning posture (as in Western Polecat). Den in ground burrow or rocky crevice (*g*, *k*). Tracks (*f*, *j*) show five toes on each foot. Droppings are cylindrical, twisted and drawn out to a point (*e*, *i*).

Left, Feral ferrets *Mustela furo* are common, as they often escape while being used to bolt Rabbits from warrens. Some are albino, but others have the same coloration as wild polecats and can easily be confused with them. Ferrets and Western Polecats hybridise readily.

TEPPE POLECAT
WESTERN POLECAT
MARBLED POLECAT
a
b
c
d
e
f
g
h
i
j
k

Plate 24

MARTENS

PINE MARTEN *Martes martes* **p. 119**

Key features: Dark brown with a creamy yellow patch on throat. Long bushy tail. A long-legged appearance distinguishes this species from most other mustelids (but not Sable). Ears longer and broader than those of the Beech Marten. Feet densely furred. **Illustrated signs:** *a*, Skull. Droppings often on top of conspicuous sites; soft and shapeless if eating berries (*b*), twisted if mammalian diet (*c*). *d*, Tracks with five toes on each foot. *e*, Dens in squirrel dreys, tree-holes and rock crevices.

BEECH MARTEN *Martes foina* **p. 122**

Key features: Brown coat, more greyish then Pine Marten, with a white patch on the throat divided in two by a dark stripe. Feet less furred, ears smaller and muzzle shorter than those of the Pine Marten. **Illustrated signs:** *f*, Skull. *g*, Droppings accumulate at latrine sites. *h*, Tracks with five toes on each foot. *i*, Commonly dens in buildings, also in tree-holes or rock crevices.

SABLE *Martes zibellina* **p. 123**

Key features: Extremely thick coat, brownish-black with a lighter but ill-defined throat patch. Legs are longer and ears larger than in the Pine Marten. Droppings and tracks similar to Beech Marten.

Common features: All three species show the typical mustelid bounding gait.

Right, In parts of Germany and Switzerland Beech Martens are closely associated with human dwellings and have recently taken to denning under the bonnets of cars. This habit is spreading and, as the martens often chew cables, can be expensive.

PINE MARTEN
a
b
c
d
e
BEECH MARTEN
f
g
h
i
SABLE
j
k

Plate 25

WEASELS

STOAT *Mustela erminea* **p. 109**

Key features: Long slim body with a long black-tipped tail. Summer coat is chestnut brown above, yellowish-white below (straight line demarcation between). Edge of ears white. In winter partly or completely white (according to location). Male much larger than female. Female in scanning position.

Illustrated signs: *a*, Skull. *b*, Droppings *c*. 4–8cm long and 0.5cm in diameter, twisted and drawn out to a point, often prominently placed. *c*, Tracks register five toes on each foot: forefoot about 2cm long and 1.5cm broad, hind foot about 3.5 by 1.3cm. *d*, Dens in hollow logs, rock crevices, ground burrows.

WEASEL *Mustela nivalis* **p. 111**

Key features: Smaller than the Stoat and with shorter tail, lacking the black tip. Coat chestnut-brown above and white below, with the border between the two clearly defined but jagged (except in Least Weasel in which the the demarcation is a straight line). Cheek spot in Common Weasel, missing in Least. Tendency to moult to white winter coat varies with location.

Illustrated signs: *e*, Skull. *f*, Dropping similar to those of the Stoat but may be only 0.2cm in diameter. *g*, Five toes on all feet, tracks about 1.4cm long and 1.0cm broad. *h*, Dens in ground burrows, under tree roots or stone wall crevices.

Below, A Stoat can hunt prey much larger then itself, in this case an adult rabbit.

d
male, summer
female, winter
STOAT
b
c
Common Weasel
h
summer
winter
Least Weasel
f
WEASEL
g

Plate 26

CATS

LYNX *Lynx lynx* **p. 136**
Key features: Large cat with long legs. Yellowish-brown coat closely spotted on the legs and more sparsely on back. Long hairy tufts on ears. Short tail, black-tipped. Appears more thick-set in winter due to bulkier coat.

PARDEL LYNX *Lynx pardina* **p. 137**
Key features: Very similar to Lynx, distinguished by smaller size, and more numerous, darker spots on the back.

Features common to both species of lynx: *a*, Skull. *b*, Droppings are cylindrical, usually 6–8cm long and 1–1.5cm wide, deposited in a small hole scraped in the soil. *c*, Tracks with four toes on each foot, with no claw marks. *d*, Dens in rocky crevices and small caves. *e*, Scratch marks in the bark of trees.

WILDCAT *Felis silvestris* **p. 133**
Key features: Stockily-built medium-sized cat, resembling a domestic tabby cat but with a more robust body. The tail is diagnostic, being bushy, with 3–5 separate bands and a thick tip. Coat striped but never mottled. Eyes amber, nose pink. **Illustrated signs:** *f*, Skull. *g*, Droppings sometimes buried (especially near den) but otherwise may be on conspicuous objects and along trails (similar to Red Foxes). *h*, Track shows four toes on each foot, with no claw marks.

Right, A northern Lynx showing the winter coat.

NX
b
c
e
d
RDEL
NX
WILDCAT
g
h

Plate 27

GENET AND MONGOOSES

COMMON GENET *Genetta genetta* **p. 131**

Key features: Dark spots forming long stripes. Tail with dark and light rings.

Illustrated signs: *a*, Skull. *b*, Large dropping 10–24cm long and 1.5–2cm wide, often horseshoe-shaped with a tuft of grass at one end. *c*, Tracks with five toes on each foot, fifth toe registers only slightly and without claw. Forefoot tracks 3cm wide by 2.5cm long and hindfoot 3cm by 3cm. *d*, Dens in dense vegetation, hollow logs and rock crevices.

EGYPTIAN MONGOOSE *Herpestes ichneumon* **p. 132**

Key features: Uniform greyish-brown coat. Tail broad at the base tapering to a narrow, black tip. Small, wide ears.

INDIAN GREY MONGOOSE *Herpestes edwardsi* **p. 133**

Key features: Smaller than the Egyptian mongoose. Tawny grey-brown, tip of the tail paler.

Features common to both species of mongoose: *e*, Skull. *f*, Droppings 4–7cm in length and 1cm diameter. *g*, Track shows five toes on forefoot, four on hindfoot, claws register only faintly. *h*, Dens in ground burrows.

Right, The scent glands of most members of the family Viverridae are highly developed. Situated between the scrotum and penis, the male Genet has a complex three-chambered pouch containing the perineal glands (the female's scent pouch is single chambered).

d
COMMON GENET
b
c
EGYPTIAN
MONGOOSE
e
f
g
h
DIAN GREY
ONGOOSE

Plate 28

WALRUS AND TEMPERATE SEALS

WALRUS *Odobenus rosmarus* **p. 140**
Key features: Largest pinniped in European waters. Conspicuous tusks, larger and straighter in males (*a*). Males much heavier than females. Both sexes with relatively small head. Very short, inconspicuous, coarse, light-brown hair. At less than two weeks old the pup is dark grey (thereafter with distinctive black flippers and tawny-brown fur).

GREY SEAL *Halichoerus grypus* **p. 148**
Key features: In profile the top of the head forms a straight line with the nose. Great variation in colour: males usually grey with lighter patches on a darker background; females are lighter with darker spots, concentrated on the back. The pups' fur is white until they are two weeks old.

COMMON SEAL *Phoca vitulina* **p. 146**
Key features: Mottled with spots varying from light grey to dark brown or black. Small rounded head, slightly concave profile with white whiskers and a dog-like appearance. Pups born with adult colouring.

MEDITERRANEAN MONK SEAL *Monachus monachus* **p. 151**
Key features: Uniform dark brown coat with white patch below. Newborn pup has a woolly black coat.

Below, Walruses often change from grey to pink after emerging from cold water: when cold they can stop the flow of blood to their skin (a condition called aeschemia), thus preventing heat loss, but on returning to the surface the blood supply is restored to the skin. They also use their tusks to rest on the edge of ice.

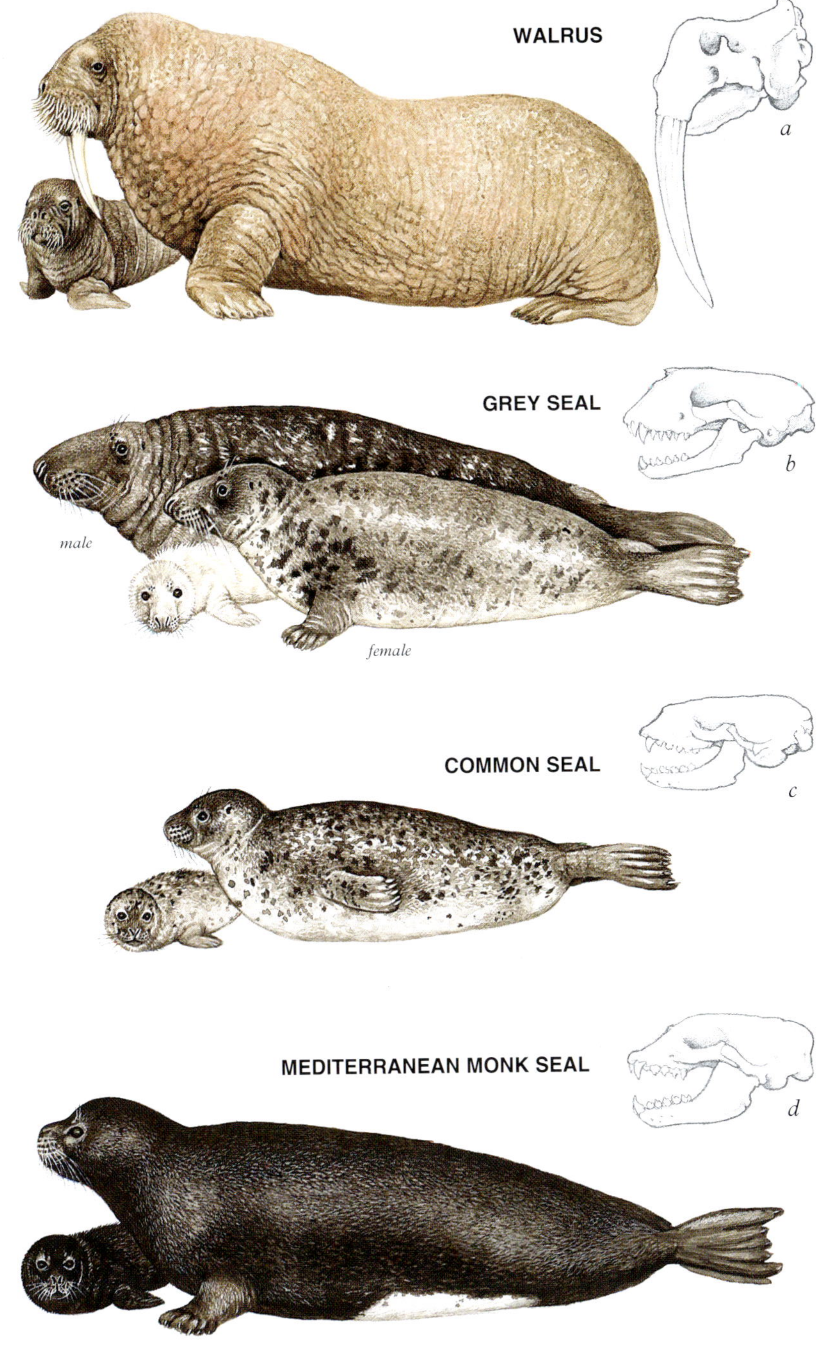
WALRUS
a
GREY SEAL
b
male
female
COMMON SEAL
c
MEDITERRANEAN MONK SEAL
d

Plate 29

ARCTIC SEALS

HOODED SEAL *Cystophora cristata* **p. 152**
Key features: Dark brown or black patches on silvery grey background. Coat of newborn bluish above and cream below. Male larger than female and with a much more pronounced proboscis hanging over the mouth. **Illustrated signs:** *a*, Skull with widely spaced peg-like teeth.

BEARDED SEAL *Erignathus barbatus* **p. 150**
Key features: Large, sausage-shaped body with relatively small head. Long vibrissae on upper lip which are crinkly when dry. All toes on hind flipper of similar length. Newborn with woolly coat of grey-brown fur, mid-back lighter. **Illustrated signs:** *b*, Skull has weak, widely-spaced teeth with loose roots.

HARP SEAL *Phoca groenlandicus* **p. 142**
Key features: Males light grey with a harp-shaped black band along flanks and back. Females acquire harp marking when older, although usually less distinct. Both sexes have black head. Female's head markings are paler. Newborn pup white.
Illustrated signs: *c*, Skull.

RINGED SEAL *Phoca hispida* **p. 144**
Key features: Smaller than the Harp Seal and with conspicuous grey-white rings on a dark blackish back, paler below. Brown whiskers. Short cat-like face. Newborn pup with white woolly fur, often in snow lair.
Illustrated signs: *d*, Skull.

Left, A male Hooded Seal showing proboscis inflated (top) and nasal septum everted and inflated (bottom).

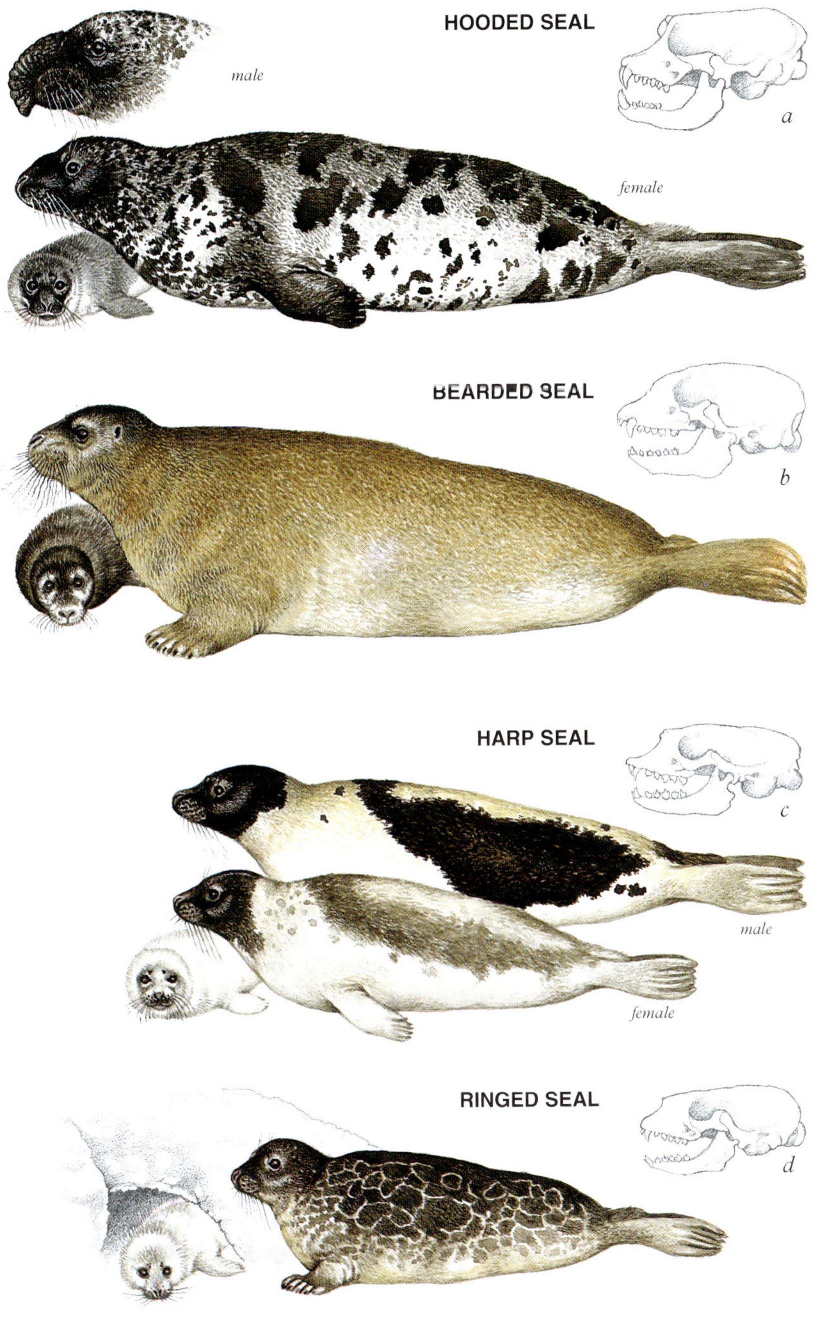
HOODED SEAL
male
a
female
BEARDED SEAL
b
HARP SEAL
c
male
female
RINGED SEAL
d

Plate 30

DOLPHINS

RISSO'S DOLPHIN *Grampus griseus* **p. 169**
Key features: Dark grey, with light grey patches on chest and belly. Dorsal fin, flipper and flukes all dark. Numerous parallel scars on flanks. Tall, sickle-shaped dorsal fin. Blunt head (cf. Bottle-nose Dolphin), prominent eyes, squarish melon bisected by deep crease. Long pointed flippers. Lighter colour of adults distinguishes them from Bottle-nosed Dolphins and False Killer Whales. **Illustrated signs:** *a*, Skull with few teeth (usually four, and no more than seven, pairs) at tip of lower jaw. Generally there are no upper teeth.

BOTTLE-NOSED DOLPHIN *Tursiops truncatus* **p. 165**
Key features: Brown to dark grey on back and upper flanks, paler lower flanks and belly. Short beak with protuberant lower jaw. The dorsal fin is tall, slender, sickle-shaped and located in middle of back. Robust head and body. Rough-toothed Dolphin has long, conical head and narrower cape markings on back. Risso's Dolphin has blunt, beakless head and taller, more sickle-shaped dorsal fin. The Spotted Dolphin has a longer, slimmer beak and a complex pattern of spots. **Illustrated signs:** *b*, Skull with 20–26 teeth in each jaw.

ROUGH-TOOTHED DOLPHIN *Steno bredanensis* **p. 161**
Key features: Back and flanks dark grey, tinged with purplish-black, underside spotted with white. Lips and tip of snout white, eyes dark. Head conical; beak long and slender with no crease separating it from the forehead (cf. Bottle-nosed Dolphin and Spotted Dolphin). Sickle-shaped dorsal fin, in middle of back. Tail stock with keel above and below. **Illustrated signs:** Skull with small teeth in both jaws (*c*).

Below, Whales and dolphins are born backwards, breach births allowing the infant to swim to the surface to breath as soon as it has completed its exit from the womb. In Bottle-nosed Dolphin society another female often assists at the birth.

RISSO'S DOLPHIN
a
BOTTLE-NOSED DOLPHIN
ROUGH-TOOTHED DOLPHIN
c

Plate 31

DOLPHINS

COMMON DOLPHIN *Delphinus delphis* **p. 163**

Key features: Black back and upper flanks, creamy white chest and belly. Conspicuous yellow hourglass pattern on sides, yellow in front of dorsal fin, becoming less conspicuous and paler grey behind it. Black stripe from the lower jaw to the flipper. Lips black. Dark grey to black tail flukes with distinct median notch. Slender, stream-lined shape. Long, well-defined beak generally black, but often tipped with white. Tapering flippers and slender, sickle-shaped to erect dorsal fin. Striped Dolphin lacks the hourglass below the dorsal fin and is larger, with a more complex pattern of stripes. **Illustrated signs:** *a*, Skull with 40–55 pairs of small sharp pointed teeth in both jaws.

STRIPED DOLPHIN *Stenella coeruleoalba* **p. 162**

Key features: Black lateral stripes from eye to flipper and eye to anus; distinctive pale V-shaped shoulder-blaze on flanks originates above and behind the eye, and narrows to a point below and behind the dorsal fin. Elongated dark beak. The forehead is not prominent but is separated from the beak by a distinctive crease. Tapering black flippers. Dorsal fin slender, sickle-shaped, centrally placed. Caudal peduncle narrow, lacking strong keel. Common Dolphin differs in the detail of colour pattern. **Illustrated signs:** *b*, Skull with 45-50 pairs of sharp, slightly curved teeth.

ATLANTIC SPOTTED DOLPHIN *Stenella frontalis* **p. 162**

Key features: Black with white spots above, light below. Black around the eye forms a 'bridle' extending from beak to eye; a dark line from the beak to the pectoral fin. Dark spots on lower flanks and belly in adult animals only. Young are less spotted, with a more complex colour pattern and a more obvious cape-like pattern on back. Sickle-shaped dorsal fin in middle of back. Long, slim beak. Body form differs slightly with region, coastal animals tending to be larger and more robust. **Illustrated signs:** *e*, Skull.

Below, Up to several thousand Common Dolphins may congregate in schools.

COMMON DOLPHIN
STRIPED DOLPHIN
ATLANTIC SPOTTED DOLPHIN

Plate 32

DOLPHINS

WHITE-BEAKED DOLPHIN *Lagenorhynchus albirostris* **p. 166**
Key features: Large dolphin with short white beak. Black back with pale grey to white on flanks behind dorsal fin. Oblique grey to white stripe on flanks in front of dorsal fin. Dorsal fin sickle-shaped, large, mid-back. Beak thick and rounded. Flippers broad at the base, curving backwards and narrowing to a point. Caudal peduncle thickened to a strong keel above and below. Tail has a concave trailing edge and a shallow notch. White-sided Dolphin has a different pattern of white or light patches on the flanks. **Illustrated signs:** *a*, Skull with fewer and larger teeth than those of White-sided Dolphin.

ATLANTIC WHITE-SIDED DOLPHIN *Lagenorhynchus acutus* **p. 166**
Key features: Similar to the White-beaked dolphin. Large robust body with short, usually bi-coloured beak. Dark back and distinctive long white patch on flanks extending backwards into a narrow yellow band. Dark, narrow stripe from the corners of the mouth to the front of each flipper. Black eye patch, from which a thin line extends forwards to the dark beak. Sickle-shaped dorsal fin in middle of back. Pointed, sickle-shaped flippers curve backwards. The caudal peduncle has a thick keel above and below and does not narrow laterally until very near the flukes, which have a shallow notch. White-beaked Dolphin similar but has two pale patches on each flank. **Illustrated signs:** *b*, Skull with numerous smaller teeth than the White-beaked Dolphin.

HARBOUR PORPOISE *Phocoena phocoena* **p. 171**
Key features: Back dark grey, flanks paler, belly white but no sharply defined pattern. Grey stripe from jaw line to flippers. Body small and chunky. No prominent forehead or beak; short, straight mouthline tilts slightly upwards. Dorsal fin low and triangular, mid-back. Flippers short and blunt. Tail stock flattened laterally into a noticeable keel; flukes have median notch. Dolphins are larger with tall, sickle-shaped dorsal fins and prominent beaks. **Illustrated signs:** *c*, Skull with flattened spade-shaped teeth (*d*) in both jaws.

Left, Many dolphins are noted for their acrobatics. Here a White-beaked Dolphin leaps clear of the water while upsidedown.

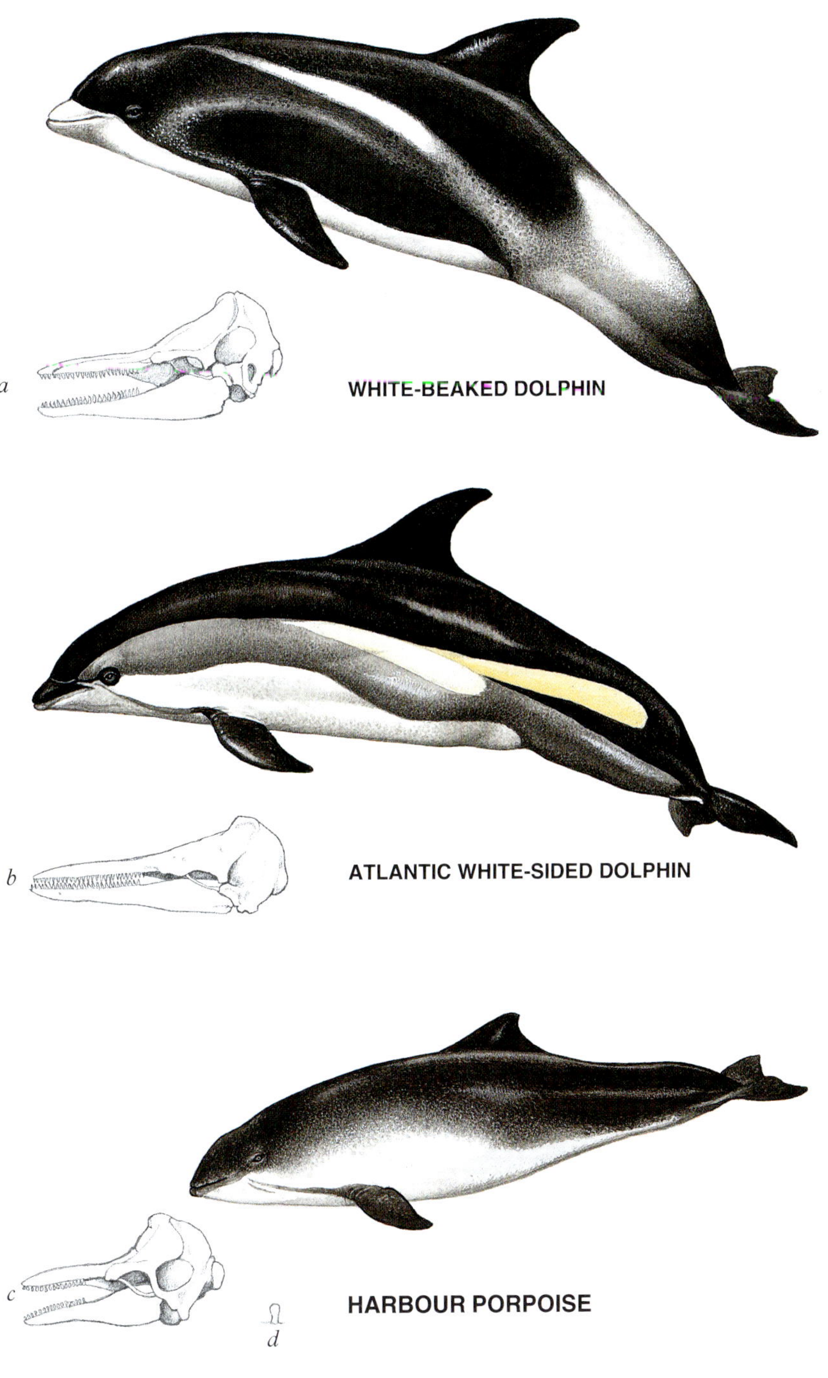
a
WHITE-BEAKED DOLPHIN
b
ATLANTIC WHITE-SIDED DOLPHIN
c
HARBOUR PORPOISE
d

Plate 33

MEDIUM WHALES WITH DORSAL FINS

KILLER WHALE *Orcinus orca* **p. 168**
Key features: Striking pattern of black with a grey saddle behind the dorsal fin, large white patch extending from underside up to flanks. White, oval patch situated behind and above each eye. The chin, throat and underside of the flukes are white and the white of the throat continues along the ventral midline narrowing between the flippers. The dorsal fin is tall and triangular in males (*b*), shorter and sickle-shaped in females (*a*). Flippers broad and paddle-shaped. Head broad. False Killer Whale is smaller, more slender and without striking white markings. **Illustrated signs:** *c*, Skull with 10–13 pairs of large conical teeth.

FALSE KILLER WHALE *Pseudorca crassidens* **p. 167**
Key features: Generally black, slightly paler on throat and neck. Slender body with an elongated tapered head. Tall, sickle-shaped dorsal fin in middle of back. Flippers narrow, short and pointed. Killer Whale has a tall dorsal fin and striking coloration, a chunkier body, broader, more rounded head; Long-finned Pilot Whale has a rounded, bulbous head and long dorsal fin, low in profile, located further forwards on the back. **Illustrated signs:** *d*, Skull with 8–11 pairs of large teeth (25mm in diameter).

LONG-FINNED PILOT WHALE *Globicephala melaena* **p. 170**
Key features: Black or dark grey with bulbous forehead and short almost imperceptible beak. Light markings on throat and belly and sometimes behind dorsal fin and eye. Long, sickle-shaped flippers 15–20% of total body length. Sickle-shaped dorsal fin is low in profile, with long base, located relatively far forward on back. In the southern parts of its range can be confused with the False Killer Whale, but latter has more tapered head, no beak, longer mouth, more slender and erect dorsal fin situated farther back. **Illustrated signs:** *e*, Skull reveals similarity with that of the Killer and False Killer Whales but has smaller teeth (8–12 pairs).

Below, Long-finned Pilot Whale spy-hopping.

a

b

KILLER WHALE

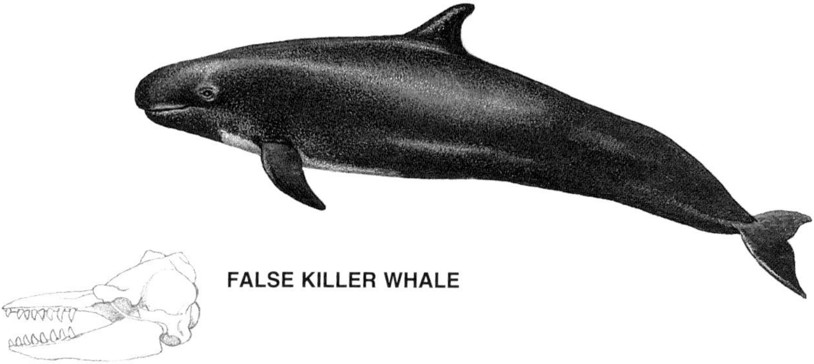

c

FALSE KILLER WHALE

LONG-FINNED PILOT WHALE

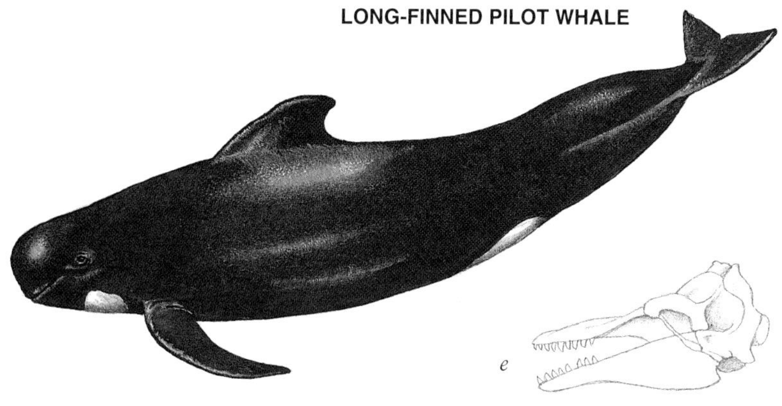

Plate 34

SMALL WHALES WITH DORSAL FINS

GRAY'S BEAKED WHALE *Mesoplodon grayi* **p. 181**

Key features: Dark brownish-grey to black above, flanks often mottled grey; lighter belly with conspicuous spots in genital and anal region. Throat often white or white flecked. Small head. Straight mouth line (cf. Blainville's Beaked Whale). **Illustrated signs:** *a*, Skull with one pair of large teeth, 10cm high and 8cm wide situated approximately 20cm from the tip of the jaw.

PYGMY SPERM WHALE *Kogia breviceps* **p. 174**

Key features: Head shark-like in shape (underslung lower jaw, false gill on side of head). Blowhole on the top of the head but left of the midline. Flippers wide at base, tapering to a blunt point, situated below and behind the false gill. Low, strongly sickle-shaped dorsal fin more than half-way along back. Larger than the similar Dwarf Sperm Whale, with more numerous teeth, rounder head and more sickle-shaped dorsal fin. **Illustrated signs:** *b*, Skull with 12–16 pairs of thin, sharp teeth on the lower jaw only.

MEDIUM WHALES WITHOUT DORSAL FINS

NARWHAL *Monodon monocerus* **p. 173**

Key features: Adult mottled bluish grey. Newborn white. Long tusk unmistakable in male (*c*). No dorsal fin, head rounded with slight hint of beak. Toothless, except for spiralled tusk of males. Trailing edge of flukes strongly convex in adults. Distinguished from the white adult Beluga by the dark blotches on the head and back.

BELUGA *Delphinapterus leucus* **p. 172**

Key features: Adults milky white. Juvenile shows typical slate grey to reddish brown coloration. Robust body and proportionally small head. The beak is short, broad and often overhung by a noticeable melon. Flippers short and rounded; dorsal fin absent. Narrow ridge along the spine just behind the midpoint of the back, often darkly pigmented. The blowhole is a transverse slit located just in front of the neck crease. - **Illustrated signs:** *d*, Skull with 8 to 11 pairs of irregular, often curved teeth in upper jaw, 8 to 9 pairs in lower jaw.

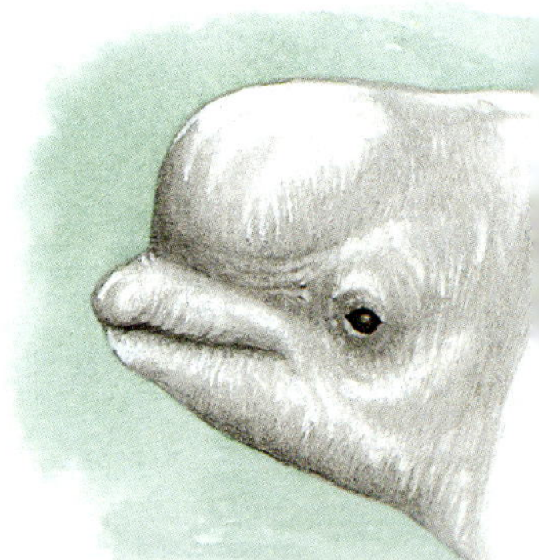

Right, The Beluga's melon changes shape as the whale produces sounds. Although the details of its function are obscure, one theory is that the oil inside the melon focuses the sound waves in the same way as a lens focuses light.

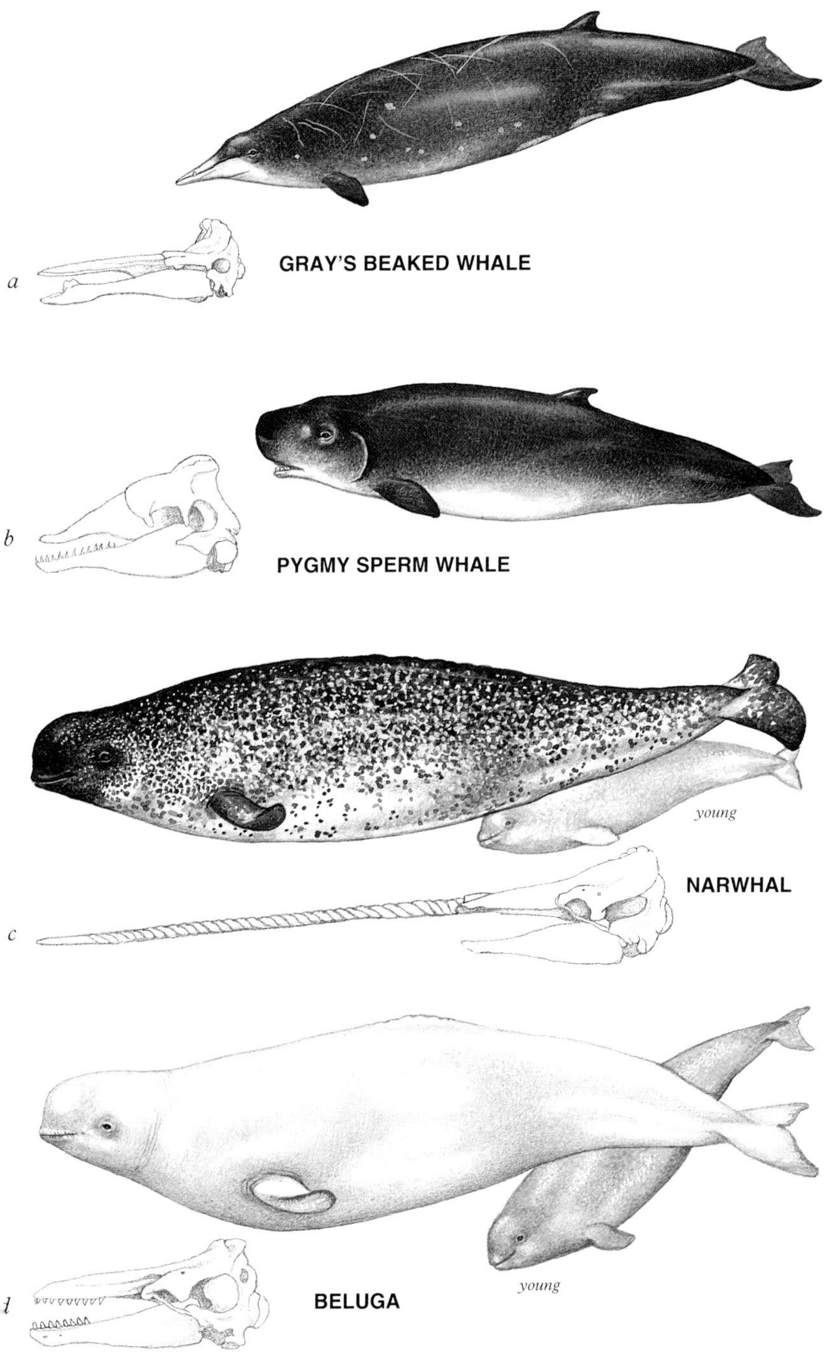
a
GRAY'S BEAKED WHALE
b
PYGMY SPERM WHALE
young
NARWHAL
c
young
BELUGA
d

Plate 35

MEDIUM WHALES WITH DORSAL FINS

MINKE WHALE *Balaenoptera acutorostrata* **p. 187**
Key features: Top of head appears flattened in profile (*c*) and has marked beak ridge and pointed snout. From above, head is dark, ridged along midline and sharply pointed. 50–70 ventral throat grooves. Fairly prominent, tall, dorsal fin, set farther back on body than that of the Pilot Whale (Plate 41). Much more pointed rostrum than Sei and lacks the white right lower lip of the Fin Whale. In northern hemisphere flippers have transverse white band. Tail flukes not raised before diving. **Illustrated signs:** *a*, Baleen plates up to 30 cm, yellowish white anteriorly becoming grey to brown black posteriorly. *b*, Indistinct blow, up to 1.8m. *c*, Skull.

NORTHERN BOTTLE-NOSED WHALE *Hyperoodon ampullatus* **p. 183**
Key features: Large bulbous forehead and short dolphin-like beak. Much larger than any dolphin. Single crescent-shaped blowhole in an indentation behind the bulging forehead. Chocolate brown to greyish-brown above, lighter on flanks with irregular blotches on belly. Moderately high dorsal fin, often strongly hooked and located about ⅔ down back. Flippers short and tapered. Broad unnotched tail flukes with deeply concave trailing edge. Cuvier's Beaked Whale has less bulbous forehead and shorter, less well- defined beak; Sowerby's Beaked Whale is much smaller, with less bulbous forehead (Plate 39).

CUVIER'S BEAKED WHALE *Ziphius cavirostris* **p. 182**
Key features: Heavily built with slightly concave head, forehead sloping to a short poorly defined beak. Two conical teeth at tip of lower jaw, only exposed in males (see skull, *e*). Grey to blue-grey body, head usually white. Linear scars often present on back and flanks. Pale spots on sides and belly. V-shaped pair of throat grooves characteristic of all beaked whales. Small, triangular sickle-shaped dorsal fin, up to 38cm high, well behind the midpoint of the back. Flippers small, with an angular outer edge. Tail un-notched. Less bulbous forehead, shorter and less well-defined beak than Bottle-nosed Whale.

Right, Minke Whale lunge feeding.

MINKE WHALE

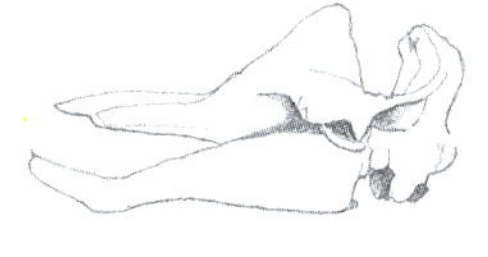

NORTHERN BOTTLE-NOSED WHALE

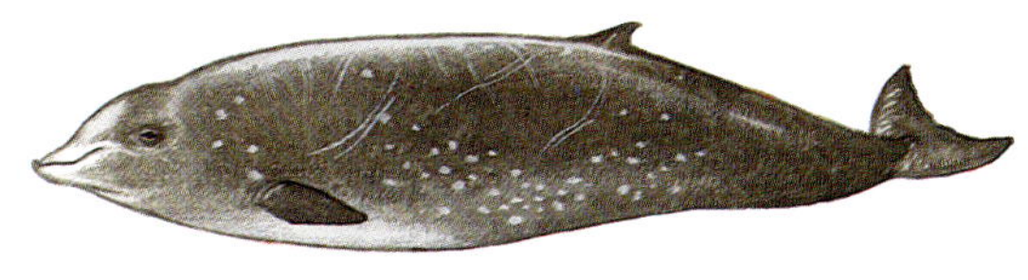

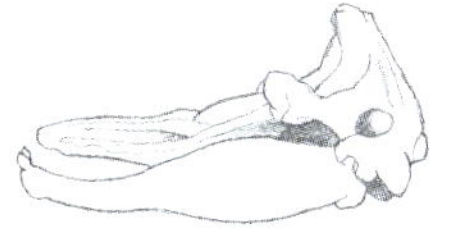

CUVIER'S BEAKED WHALE

Plate 36

MEDIUM WHALES WITH DORSAL FINS

GERVAIS' BEAKED WHALE *Mesoplodon europaeus* **p. 179**
Key features: Body uniformly coloured, but individuals variable from greyish-blue to black with lighter underside. Characteristic laterally compressed body. Flippers positioned low on the sides. The small dorsal fin can be sickle-shaped, blunt or triangular. Tail fluke without notch. Mandibular teeth about one-third of the way from tip of jaw to corners of mouth (see skull (*a*); cf. True's Beaked Whale which has teeth at tip of mandibles).

BLAINVILLE'S-BEAKED WHALE *Mesoplodon densirostris* **p. 179**
Key features: Black or dark grey coloration with paler blotches on flanks and slightly lighter on the abdomen. General appearance similar to other *Mesoplodon* species. Male has a huge tooth dominating the middle third of each mandible (*b*). The tooth rises up above the beak and is enveloped and supported by a massive bony protuberance. Short flippers. Relatively large sickle-shaped and pointed dorsal fin. Body frequently scarred.

TRUE'S BEAKED WHALE *Mesoplodon mirus* **p. 180**
Key features: Dark grey to grey-black back, lighter slate grey sides and grey belly with light spots, especially in the genital and anal region. Body shaped like that of Cuvier's Beaked Whale: chunky in the middle and narrowing rapidly towards the tail. The dorsal fin is triangular or slightly sickle-shaped. Unnotched tail with slightly concave edge. Two teeth at the tip of mandibles, exposed outside mouth in adult males (*c*). Pronounced beak.

SOWERBY'S BEAKED WHALE *Mesoplodon bidens* **p. 181**
Key features: Dark body, pale grey below, back and belly spotted with lighter blotches. Small flippers, one-seventh to one-ninth of total body length. Dorsal fin triangular or sickle-shaped. Unnotched tail with slightly concave edge. Prominent bulge on forehead and moderately long beak. Mandibular teeth appear near the middle of beak (see skull (*d*); cf. tooth position in Gervais', Blainville's, and True's Beaked Whales).

Left, Adult female Dense-beaked Whale, viewed from the front, showing that the high curve of the lower mandible makes it virtually impossible for the animal to see forwards.

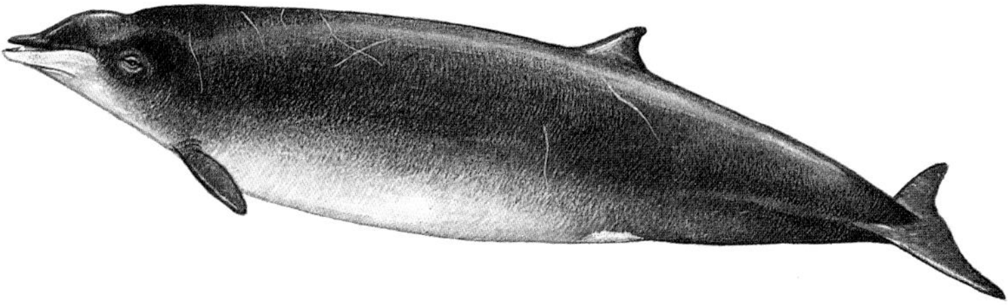

GERVAIS' BEAKED WHALE

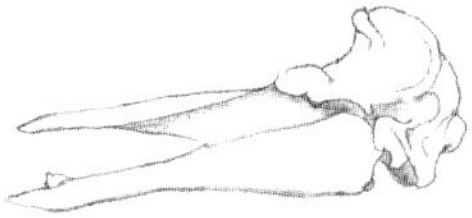

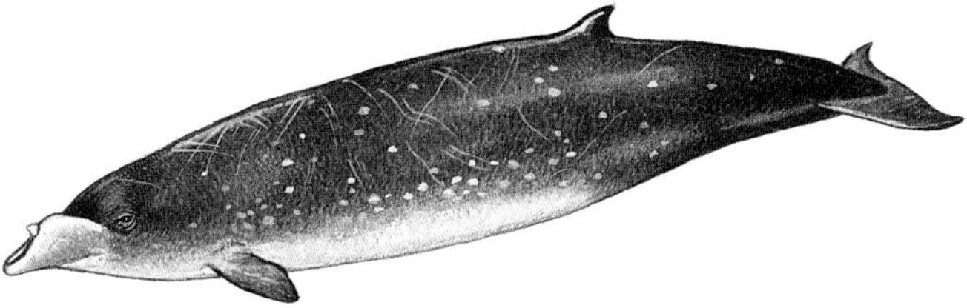

a

BLAINVILLE'S-BEAKED WHALE

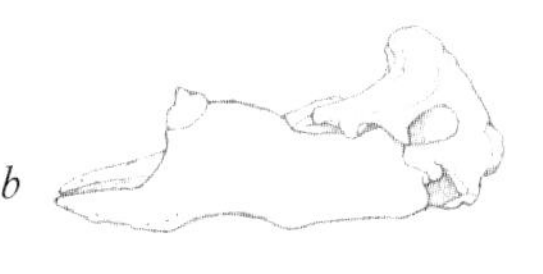

b

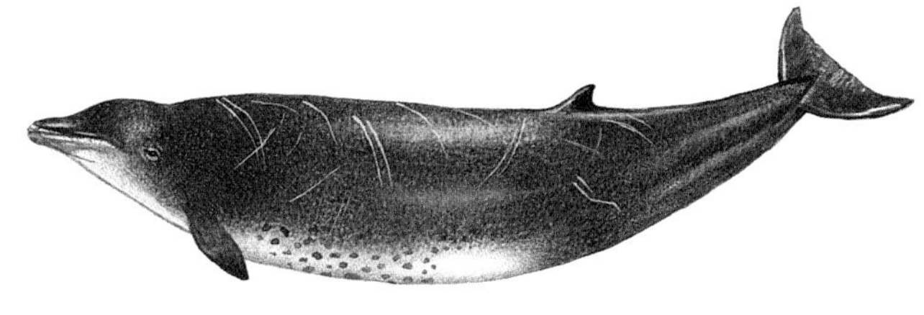

TRUE'S BEAKED WHALE

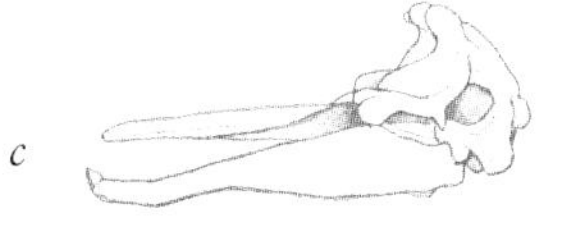

c

SOWERBY'S BEAKED WHALE

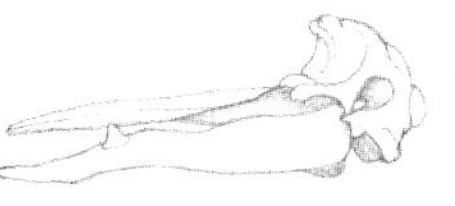

d

Plate 37

LARGE WHALES WITH DORSAL FINS

BLUE WHALE *Balaenoptera musculus* **p. 184**

Key features: Body long, bluish-grey, mottled with grey or greyish-white. Flat in profile in front of blowholes. From above, body broad, head U-shaped, longer and less blunt than that of Sperm Whale. Small dorsal fin situated more than ¾ down back. Long, pointed pectoral fins, white below. **Illustrated signs:** *a*, Blow up to 9m. *b*, Baleen plate short (up to 80cm), broad and black. *c*, Skull. *d*, Tail flukes raised only slightly above water before diving.

FIN WHALE *Balaenoptera physalus* **p. 185**

Key features: Uniform slate grey above, white below with asymmetrical head coloration. Frequently has greyish-white chevron on back behind head. White on right lower lip and palate diagnostic (*g*). Slightly sickle-shaped dorsal fin, more prominent than Blue Whale's. Head narrower than Blue Whale's and more V-shaped from above. Usually rises obliquely, so the top of head breaks the surface first. **Illustrated signs:** *e*, Blow up to 6m. Tail flukes not raised above water when diving (cf. Blue Whale). *f*, Baleen plates up to 72cm in length and 30cm in width. *h*, Skull. *i*, Arched back and fin on surface before dive.

SEI WHALE *Balaenoptera borealis* **p. 186**

Key features: Steely grey back, often with round, grey or white scars. Dorsal fin large; almost ⅔ down back (more sickle-shaped and further forward than other large baleen whales). **Illustrated signs:** *j*, Blow up to 3m during which dorsal fin often visible. *k*, Baleen plates up to 75cm, about 300 to 400 per side, uniformly ash-black with fine white fringes. Tail flukes not raised above water before diving (cf. Blue Whale).

HUMPBACK WHALE *Megaptera novaeangliae* **p. 188**

Key features: Dark grey body, head and snout. Very long pectoral fins, nearly one-third of length, mainly white and lumpy, with knobs on leading edge. Underside of tail often white. Profile of head in front of blowholes is straight but knobbly. Can be confused with Blue, Fin or Sei whales. Dorsal fin usually small and indistinct, but shape varies: resembles Blue Whale, but is further forward and generally more prominent. **Illustrated signs:** *m*, Blow up to 3m. *n*, Baleen plates. *o*, Skull. Before diving back is strongly arched (*p*) and tail flukes (*q*) often raised above surface (cf. Fin and Sei Whales).

Left, A Humpback Whale breaching.

a
d
BLUE WHALE
c
i
g
h
FIN WHALE
j
SEI WHALE
q
n
p
o
HUMPBACK WHALE

Plate 38

LARGE WHALES WITHOUT DORSAL FINS

BOWHEAD WHALE *Balaena mysticetus* **p. 191**

Key features: In profile deep depression divides triangular head from rounded back; broad, strongly bowed lower jaws. Narrow, arched rostrum, see skull (*a*). White on chin often with string of black spots. Large, broad pectoral fins; very broad tail. No dorsal fin or ridge. Big head, 33–40% of total length. Affinity for ice, rarely seen near other large whales. **Illustrated signs:** *b*, Baleen plates up to 450cm. *c*, V-shaped blow, up to 3.9m. *d*, Tail flukes raised before diving.

NORTHERN RIGHT WHALE *Eubalaena glacialis* **p. 191**

Key features: Large head, arched upper jaws and bowed lower jaws (see skull, *h*), and narrow rostrum. Callosities near the blowholes, on chin and lower lips. No dorsal fin or ridge. Broad tail flukes deeply notched with concave trailing edge. In the north can be distinguished from Bowhead Whale by the callosities. **Illustrated signs:** *e*, Baleen plates up to 240cm, usually blackish, but at sea can look pale. *f*, V-shaped blow, up to 4.8m. *g*, Tail flukes often raised before diving.

SPERM WHALE *Physeter macrocephalus* **p. 176**

Key features: Head extremely large (25–35% of total body length) barrel-shaped with single blow hole on the left side at front. Small dorsal hump, with spinal ridge between hump and tail. The skin has shrivelled appearance due to corrugations. Body dark greyish-brown to brown. Flukes dark, nearly straight, rear margin even, but deeply notched (cf. Humpback's flukes, Plate 36). **Illustrated signs:** *i*, Blow angled forward from front of head, up to 5m. *j*, Tail flukes raised before diving. *k*, Skull with teeth in lower jaw.

Below, An entire pod of Sperm Whales often strands together. Perhaps the distress calls of the first one to strand lure the others to the same fate.

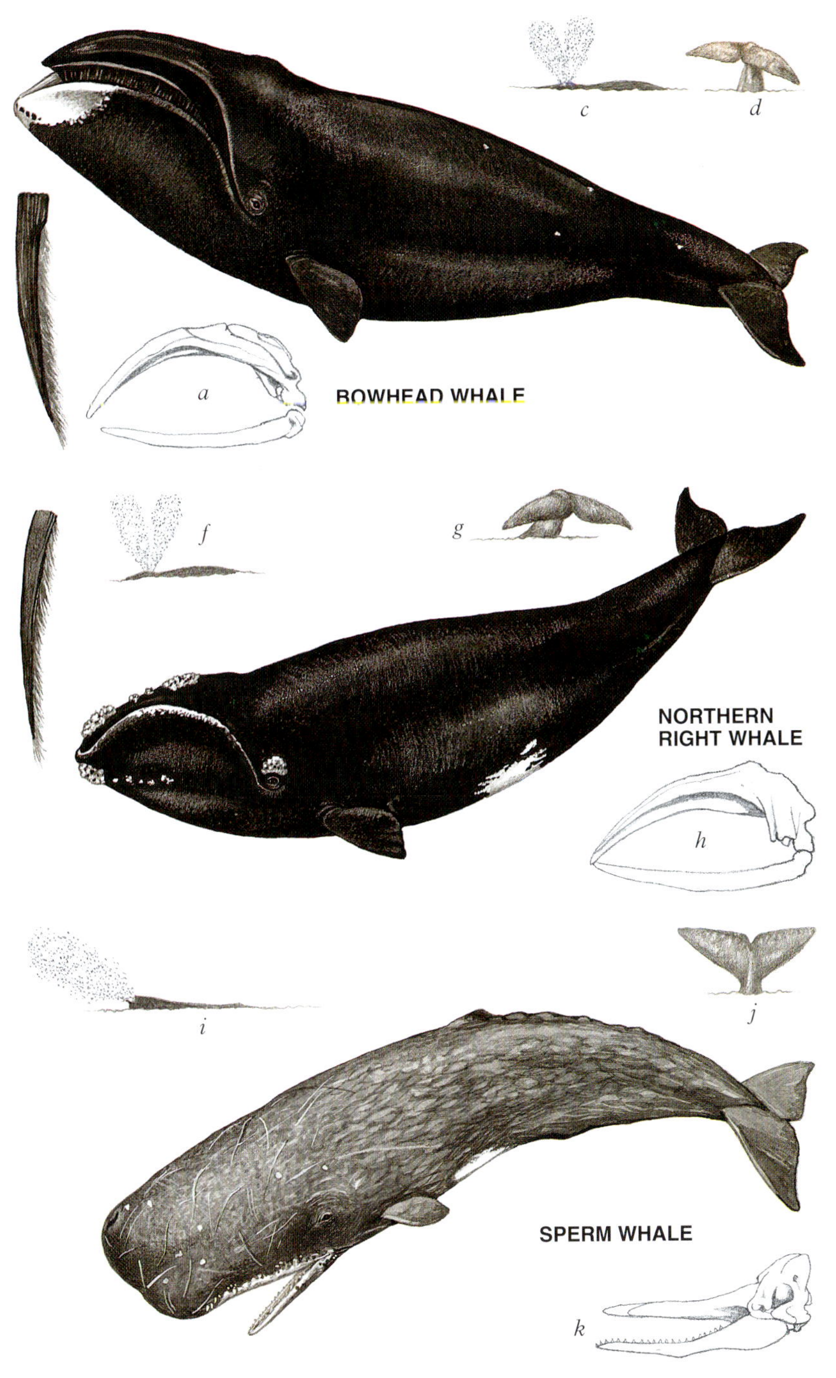
c
d
a
BOWHEAD WHALE
f
g
NORTHERN
RIGHT WHALE
h
i
j
SPERM WHALE
k

Plate 39

SMALL, UNSPOTTED DEER

ROE DEER *Capreolus capreolus* **p. 211**
Key features: Adult coat sandy to red-brown in summer and grey-brown to blackish in winter. Black nose and 'moustache', white chin. Appears tail-less. Three points on antlers. Antler size increases with age: buck illustrated is mature (7 years), antlers (*a*) from yearling buck from same population. White to buff patch on rump (in winter, inverted heart-shape in females, kidney-shape in males, *g*). In winter females have a tuft of white hairs projecting backwards between the hindlegs (resembling a tail). Fawn spotted.
Illustrated signs: *c*, Male skull. *d*, Droppings are shiny black, cylindrical pellets pointed at one end (*c.* 14 by 18mm). *e*, Tracks characterised by small size and narrow pointed shape of the hoof (4.5cm long and 3cm wide). *f*, Bark frayed on saplings, sometimes associated with scent marks from ant-orbital glands (*b*). *h*, Well trodden rutting ring.

CHINESE WATER DEER *Hydropotes inermis* **p. 198**
Key features: In summer reddish and sleek, in winter coat is grey brown or pale fawn and very thick (not illustrated). Black nose and conspicuous dark eyes. No antlers. Large ears. Fawn spotted. **Illustrated signs:** *m*, Arched path through tall vegetation and hedgerow boundaries. *n*, Skull shows male's elongated upper canines, *c.* 8cm forming a tusk. *o*, Dropping more elongated than in the Muntjac. *p*, Track.

REEVE'S MUNTJAC *Muntiacus reevesi* **p. 199**
Key features: Upper canines form small tusks, obvious in male's skull (*n*). Coat chestnut above and white below; tail is ginger, erected when alarmed. Antlers (*c.* 6–8cm) pointing backwards with single spikes.
Illustrated signs: *m*, Arched path through tall vegetation and hedgerow boundaries. *o*, Droppings are black, rounded or cylindrical, sometimes pointed at one or both ends. *p*, Small tracks *c.* 3cm long and with a tendency for the inner half of the hoof to be less distinct than the outer half.

Right, Alarmed Roe Deer with tail erected to show white rump patch.

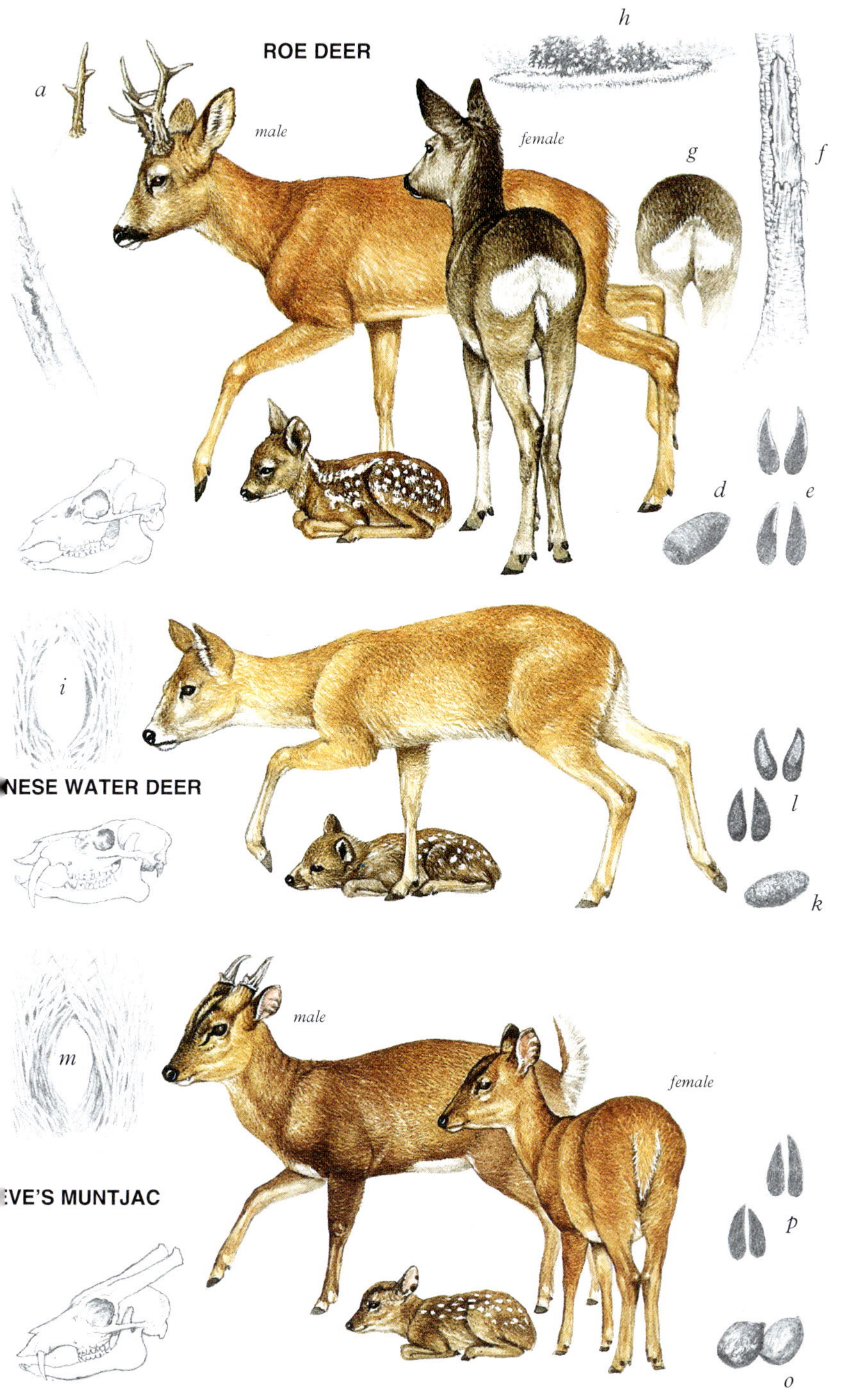
ROE DEER
a
male
female
h
g
f
d
e
i
NESE WATER DEER
l
k
m
male
female
EVE'S MUNTJAC
p
o

Plate 40

LARGE, UNSPOTTED DEER

RED DEER *Cervus elaphus* **p. 200**

Key features: Second largest European deer. Coat short and reddish in summer coat, dark brown in winter. Creamy coloured rump patch, not clearly outlined with black as in Sika and Fallow Deer. Mature stag (7 yrs) and two year old (*a*) from same population. Young calf is generally spotted and lies low. **Illustrated signs:** *b*, Female skull. *c*, Droppings, 1.5cm diameter, black, acorn-shaped, deposited in groups, but appear as a 'string' if animal was moving. *d*, Track relatively broad and with the outer edges of each half of the hoof curve symmetrically towards the tip. Front hoof particulary curved. Adult stag forefoot track is 8–9cm long, 6–7cm broad; hindfoot smaller, 6–7cm long, 4–5cm broad. *e*, Bark gnawed from trees; chisel marks of front teeth 16mm diameter. *f*, Saplings stripped of bark and frayed. *g*, Mud wallows.

WHITE-TAILED DEER *Odocoileus virginianus* **p. 209**

Key features: Size intermediate between Red and Fallow Deer. Red-brown summer coat, greyish in winter. White rump and underside. Broad tail, dark above, white below – largely conceals white rump when down. Tail raised in alarm. Main beam of antlers curved. Fawns lose spotted coat at 3–5 months. **Illustrated signs:** *h*, Male skull. *i*, Track about 7cm long. *j*, Bark gnawed from trunks. *k*, Saplings stripped of bark and frayed.

Left, Development of a
in a single stag from a
ling to 8 years old. Th
of growth varies depe
on the availability of fo

a
g
f
male, summer
female, winter
ED DEER
e
d
c
male
k
female
j
VHITE-TAILED DEER
i

Plate 41

SPOTTED DEER

SIKA DEER *Cervus nippon* **p. 203**
Key features: The summer coat is chestnut-red with rows of whitish spots; in winter males dark grey to black, females light brown or grey. Top of tail and the rump is white, with a dark brown outline. **Illustrated signs:** *a*, Mature male (*c.* 7 yrs) and antlers from a 2 year old stag from same population. *b*, Female skull. *f*, Tracks similar to Fallow Deer but slightly broader. *g*, Dropping a glossy black pellet with one end indented and the other pointed.

FALLOW DEER *Dama dama* **p. 204**
Key features: Older bucks have palmate antlers. Coat varies from fawn to black. In general, summer coat chestnut-brown, winter more uniform grey with less distinct spots. Tail relatively long. Stag illustrated is mature; antlers (*h*) are from a young stag from same population.
Illustrated signs: *i*, Female skull. *j*, Droppings generally clearly separate pellets (16 by 11mm) but, especially in summer, may congeal in to so-called fewmets (*k*). *l*, Track narrower and more elongated than that of Red Deer, often very pointed, with outside edges at the rear almost parallel; size varies with sex and age (adult male *c.* 6.5cm long).

AXIS DEER *Cervus axis* **p. 206**
Key features: The white spots are present throughout the year. White patches on throat. No black on tail or rump. Three points on each antler. *m*, Male skull.

Common features: *c*, Browsed saplings sprout bushily from the base. *d*), Gnawed bark. *e*, Frayed bark on saplings. *n*, Browse line where deer have fed from lower branches.

Below, A Red Deer calf. All spotted and European unspotted deer have spotted calves.

SIKA DEER
male, summer
female, winter
FALLOW DEER
male, summer
female,
winter
AXIS DEER
male
female
g
f
e
d
c
l
k
j
n

Plate 42

GOATS

ALPINE IBEX *Capra ibex* **p. 217**
Key features: Horns in both sexes: in the male curved backwards and ribbed with transverse ridges; much smaller in female. Coat brownish-grey above, belly pale. Male has small beard. Male illustrated is in the low-stretch courtship posture.

SPANISH IBEX *Capra pyrenaica* **p. 218**
Key features: Male horn shape varies with locality, but in general twisted in slight spiral upwards and outwards. Lacks the transverse parallel ridges of Alpine Ibex; female horns much smaller. Beard present in males. Pale coat with black stripe along the back and dark legs, flanks, chest and forehead. Extent of dark areas varies between localities: Pyrenean male illustrated.

CRETAN WILD GOAT *Capra aegagrus* **p. 219**
Key features: Similar to Alpine Ibex but the horns have sharper leading edges and the transverse ridges irregularly spaced, female horns much smaller. Male has pronounced beard.

Common features: Females very much alike. *a*, Skull of young male, horns thicker than a female's. *b*, Droppings are slightly cylindrical *c.* 1 cm in diameter. *c*, Track. All three species are very difficult to separate from feral domestic goats.

Left, Feral goats occur in many localities, and their appearance varies according to original domestic stock. This male is characteristic of Wales.

ALPINE IBEX
male
female
male
SPANISH IBEX
CRETAN
WILD GOAT
male
b
c

Plate 43

CHAMOIS, SHEEP & PIG

CHAMOIS *Rupicapra rupicapra* **p. 214**

Key features: Summer coat pale brown, dark brown legs, dark stripe along back. Whole body darker in winter, when coat longer. Dark stripe from muzzle, over eye to ear, contrasting with paler head. Long horns bent backwards at tip in both sexes. **Illustrated signs:** *a*, Male skull. *b*, Individual droppings almost spherical *c.* 1.5 cm in diameter. *c*, Track 6cm long by 3.5cm broad, hooves well separated but usually parallel.

MOUFLON *Ovis orientalis* **p. 220**

Key features: Smaller than Chamois. Short coat. Dark tail contrasting with white buttocks. Mature ram has white saddle-patch, 'socks' and muzzle, and dark neck, shoulder and upper legs. Female hornless, red-brown above, white below. Ram's horn curved in almost one full revolution. Horns in young male curved slightly down (*d*). **Illustrated signs:** *e*, Droppings consist of spheres of 1cm in diameter (ie smaller than hare's) which are deposited in compressed spheres or clumps. *f*, Track registers the two halves of the slender hoof, which do not join completely. The hoof tips are almost always splayed and the track is very angular posteriorly. In fully grown ram track is *c.* 5.5cm long and 4.4cm broad.

WILD BOAR *Sus scrofa* **p. 196**

Key features: Heavily built body, laterally flattened appearance, large head, short stocky legs. Dark, thick winter coat illustrated. In males upper canine tusks (*h*) point upwards. Piglets horizontally striped. **Illustrated signs:** *g*, Wallows. *i*, Dropping cylindrical but variable according to diet, *c.* 7 cm long. *j*, Whole track trapezoid shape as dew claws almost always leave clear impression. In adults the breadth of the main hoof is about 6–7cm. *k*, Ground much churned and rooted where Wild Boars forage.

Below, Wild Boars give birth in nests constructed by the female.

male, summer
female, winter
CHAMOIS
c
b
male
female
OUFLON
f
e
k
female
LD BOAR
j
i

Plate 44

ARCTIC GRAZERS

REINDEER *Rangifer tarandus* **p. 208**

Key features: Coat is dark grey-brown in summer, lighter in winter. Irregular branched asymmetric antlers in both sexes, length 52–130cm in males and 23–50cm in females.

Illustrated signs: *a*, Skull. *b*, Droppings in winter are firm pellets, and very dark; *c*, those in summer are yellow-brown and soft or semi-liquid. *d*, Track very characteristic half-moon shaped hooves which leave an almost circular impression with sharply marked edges. In a fully grown bull the forefoot is 8.5cm long and 10cm broad, and the hindfoot about 8.5cm by 9.5cm. *e*, Patches of closely grazed lichen.

MUSK OX *Ovibos moschatus* **p. 216**

Key features: Heavily built; long dark coat with guard hairs covering tail and ears. Winter coat hangs almost to hooves. Both sexes with horns that grow from broad bases in centre of crown. *f*, Male has bigger, broader horns than female.

Below, When threatened, for example by Wolves, Musk Oxen form a defensive ring, each individual presenting its horns to the perimeter, with the young protected in the core of the ring.

REINDEER
male, winter
female, summer
e
a
b
c
d
female
male
f
MUSK OX
g

Plate 45

VERY LARGE UNGULATES

BISON *Bison bonasus* **p. 214**

Key features: Unmistakable massive cattle-like animal. Calf chestnut brown.

Illustrated signs: *a*, Skull. Droppings (*b*) and tracks (*c*) cattle-like. *d*, Mud wallow.

ELK *Alces alces* **p. 207**

Key features: Very large (largest European deer) with roman nose. Male has spreading, palmate antlers (occasionally cervine, *e*) and beard. Calf reddish-brown.

Illustrated signs: *f*, Female skull without antlers. *g*, Droppings differ in appearance according to season. *h*, Track can be confused with domestic cattle only but has long, pointed hooves which leave an almost rectangular track, while cattle tracks are more rounded; track varies considerably with age and sex. Adult bull's forefoot track is 13–15cm long and 11–13cm broad, and the hindfoot is 11–15.5cm long and 10.5–11cm wide. *i*, Saplings frayed and bark stripped. *j*, Mud wallows.

Below, Elk in velvet. The antlers of deer re-grow each year in time for the annual rut, and are shed at the end of the breeding season. The growing antlers are covered in richly vascular skin which protects them and nourishes the growing bone. This skin is called velvet, and it dies and is shed once the antlers have fully developed.

BISON
male
female
d
c
b
e
j
male
i
ELK
h
g
female

Plate 46

TREE SQUIRRELS

GREY SQUIRREL *Sciurus carolinensis* **p. 224**
Key features: Larger and heavier build than the Red Squirrel, with small or absent ear tufts. Grey back, sides and limbs often tinged with brown in summer coat; winter coat less brown. The skull similar to Red Squirrel's (*a*) but longer and shallower with relatively longer nasal bones. **Illustrated signs:** *g*, Sometimes dens in tree hole. *h*, Drey situated away from tree trunk.

PERSIAN SQUIRREL *Sciurus anomalus* **p. 226**
Key features: The fur is reddish-brown over the back and yellow on the belly. In general similar to Red Squirrel but lacking ear tufts.

RED SQUIRREL *Sciurus vulgaris* **p. 226**
Key features: Summer coat reddish-brown, in winter changes to dark brown. The fur on the upperside is more uniformly coloured than that of the Grey Squirrel. Small ear tufts become longer in winter. **Illustrated signs:** Ball-shaped drey generally situated near to tree trunk (j). May den in tree-hole (i).

Common features: *b*, Stripped spruce cone with seed scales removed whole, always lying out in the open. *c*, Hazelnuts gnawed at the top and cracked open leaving clean edges in the two halves. *d*, Bark also gnawed off trees. *e*, Droppings are pellet shaped and 8mm in diameter. *f*, Tracks with five toes on front feet and four toes on hind foot .

FLYING SQUIRREL *Pteromys volans* **p. 232**
Key features: 'Gliding wing' formed by membrane of skin extending from wrist to ankle. Eyes very large. **Illustrated signs:** *k*, Skull. Tracks with four toes on front feet and five toes on rear print, with spur on wrist visible on front footprint (*l*). *m*, Dens in hollow tree.

Below, Uncertain food supplies during the winter are a problem for mammals. A Red Fox might cache a Red Squirrel for future use.

GREY SQUIRREL
h
g
a
b
c
d
e
f
PERSIAN SQUIRREL
j
i
winter
RED SQUIRREL
summer
m
FLYING SQUIRREL
k
l

Plate 47

GROUND SQUIRRELS

ALPINE MARMOT *Marmota marmota* **p. 228**

Key features: Large ground squirrel with a robust body, large head and short legs.

Illustrated signs: *a*, Skull. *b*, Tracks with four toes on forefoot, five on hindfoot; prints 5cm long and 4cm wide. *c*, Occupies burrows on hillside.

SPOTTED SOUSLIK *Spermophilus suslicus* **p. 230**

Key features: The creamy white spots on the back distinguish this species from the European Souslik. Tail shorter and thinner.

EUROPEAN SOUSLIK *Spermophilus citellus* **p. 231**

Key features: The fur on the back can be mottled but no defined spots are present. Tail is bushy.

Features common to European sousliks: *d*, Skull. *e*, Tracks with four toes on forepaws, five on hindpaws. *f*, Conspicuous warren system.

Below, An Alpine Marmot giving a shrill alarm whistle. These whistles are often the most obvious clue to the presence of Marmots.

c
a
ALPINE MARMOT
b
f
SPOTTED SOUSLIK
EUROPEAN
SOUSLIK
d
e

Plate 48

CHIPMUNK, MOLE RATS AND MOUSE-TAILED DORMOUSE

SIBERIAN CHIPMUNK *Tamias sibiricus* **p. 231**

Key features: Easily recognizable by the 5 dark stripes along the back. **Illustrated signs:** *a*, Skull. *b*, Droppings. *c*, Tracks with four toes on fore foot and five on hindfoot. *d*, May take refuge in hollow log, or den in burrow.

GREATER MOLE-RAT *Spalax microphthalmus* **p. 239**

Key features: Larger than the Lesser Mole-rat. Fur different shades of grey; tail-less; a thin membrane covers the eyes; well developed incisors adapted for digging. *g*, Skull similar to Lesser Mole-rat but back of skull (*e*) is wide and lacks perforations beside the posterior opening.

LESSER MOLE-RAT *Nanospalax leucodon* **p. 240**

Key features: Smaller size but very similar to the Greater Mole-rat. *g*, Skull similar to Greater Mole-rat, but recognizable by the presence of two small perforations at the rear (*f*).

Features common to European mole rats: *h*, Abundant molehills.

MOUSE-TAILED DORMOUSE *Myomimus roachi* **p. 274**

Key features: The tail has short hairs that distinguish this species from the other dormice. The coat is pale grey-brown and the ears are short. **Illustrated signs:** *i*, Skull.

Below, The nipple positions of European squirrels.

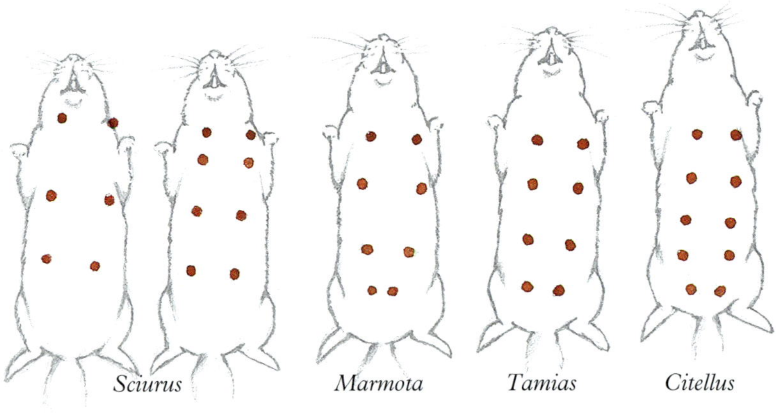

BERIAN
HIPMUNK
a
b
c
d
GREATER MOLE-RAT
e
f
g
h
LESSER MOLE-RAT
MOUSE-TAILED DORMOUSE
i

Plate 49

DORMICE

EDIBLE DORMOUSE *Glis glis* **p. 275**

Key features: Largest dormouse. Grey with brown tinge; hint of dark stripe along the back. Tail bushy. May nest in buildings, nest-boxes, hollow trees or ground burrows.

FOREST DORMOUSE *Dryomys nitedula* **p. 276**

Key features: Fur light grey to reddish-brown. Distinguishable from the Garden Dormouse by smaller size, smaller and more rounded ears and less evident mask around the eyes. The bushier tail also lacks the black and white tuft present in the Garden Dormouse.

GARDEN DORMOUSE *Eliomys quercinus* **p. 276**

Key features: Fur reddish-brown above with white underside and feet. The long tail is dark on top, progressively lighter towards tip, ending in a white bushy tuft. Distinctive black mask around the eyes and under the large ears.

Features common to Forest and Garden Dormice: *a*, Skulls similar. *c*, Tracks with four toes on forefoot and five on hindfoot. *d*, Spherical nest of grasses and leaves in fork of branch. May also nest in tree holes, nest boxes, crevices in walls or ground burrows.

Below, The Edible Dormouse bears a slight resemblence to a small Grey Squirrel. However, even a glimpse of the dormouse's face reveals the dark eye rings which, with its much smaller size, clearly differentiates it from a Grey Squirrel.

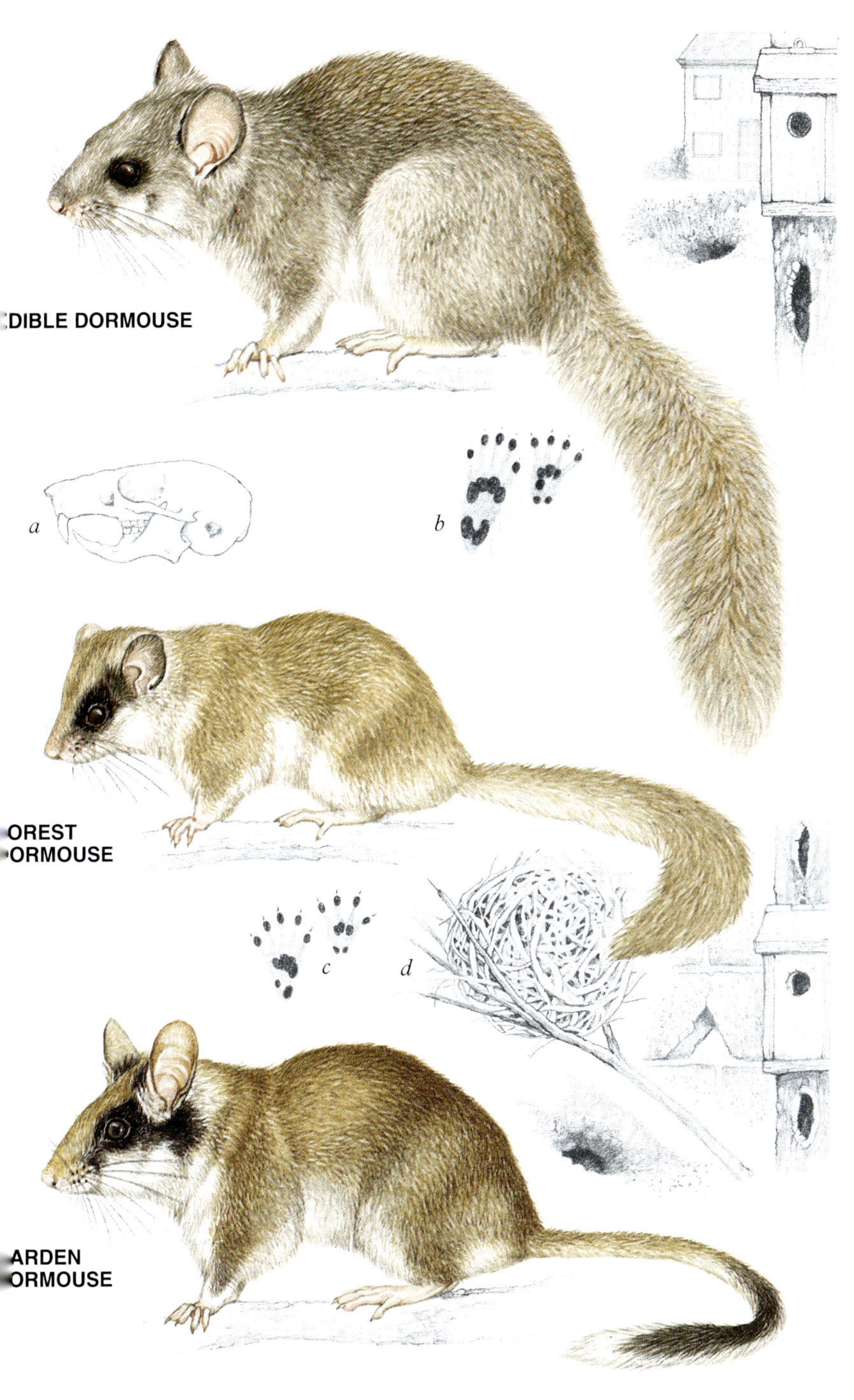

:DIBLE DORMOUSE
a
b
OREST
ORMOUSE
c
d
ARDEN
ORMOUSE

Plate 50

DORMICE AND MICE

COMMON DORMOUSE *Muscardinus avellanarius* **p. 277**

Key features: Small. Chestnut colour above. Long whiskers on the short muzzle. The tail is long and thickly furred with long hairs, but not bushy.
Illustrated signs: *a*, Skull. *b*, Chewed hazelnuts which are opened by a distinctive round hole with oblique marks around the edge. *c*, Droppings less uniform than those of mice; their surface is rough, usually black, plaited. *d*, Tracks with tail drag, four toes on front foot, five on hindfoot. *e*, Spherical nest of grass and leaves up to 15cm in diameter, sometimes in nest-boxes.

HARVEST MOUSE *Micromys minutus* **p. 257**

Key features: Smallest European rodent. The fur on the upperside is reddish, white on the underside, with a sharp division between the two. Rounded muzzle. Prehensile tail.
Illustrated signs: *f*, Skull. *g*, Small droppings of about 2mm length, usually cylindrical. *h*, Tracks with four toes on wide forefoot, five on the hindfoot. Tail drag may register. *i*, Nest of grass leaves woven into spherical ball in tall grasses or cereals.

SPINY MOUSE *Acomys minous* **p. 274**

Key features: The dark buffy fur above contrasts cleanly with the white underside. Soft bristles overlie fur. White feet and fragile tail.
Illustrated signs: *j*, Skull. *k*, Droppings. *l*, Tracks reveal four toes on forefoot and five on hindfoot. Dens in buildings, rocky crevices and ground burrows.

Left, Dormice get their name from their sleepy habits. The Common Dormouse may hibernate for seven months of the year. The tightly curled position helps to conserve heat.

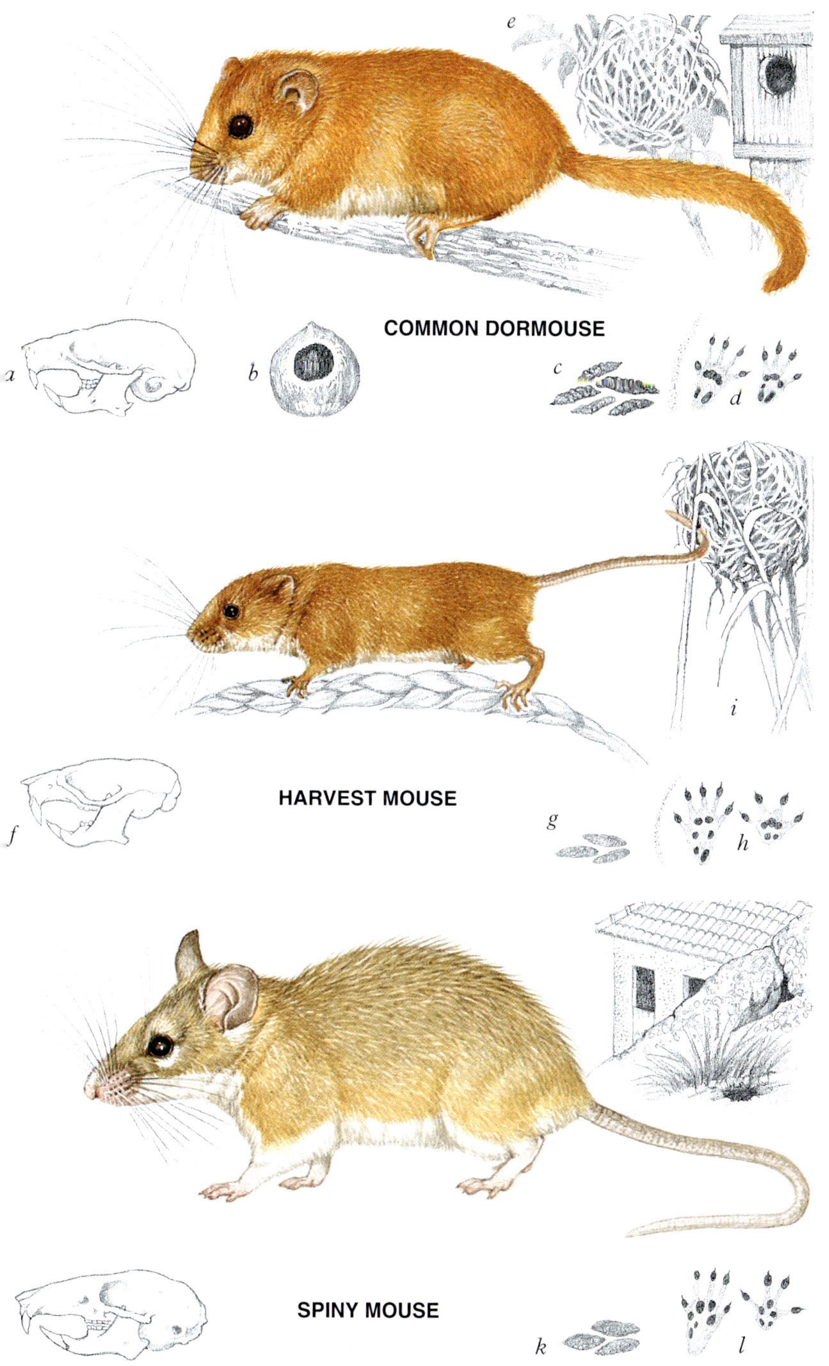
e
COMMON DORMOUSE
a
b
c
d
i
HARVEST MOUSE
f
g
h
SPINY MOUSE
k
l

Plate 51

HOUSE MOUSE AND ALLIES

STEPPE MOUSE *Mus spicilegus* **p. 269**

Key features: Similar to the Algerian Mouse. The coloration of upper and lower parts contrast sharply (cf House Mouse). The length of the tail slightly shorter than the head-body length. *a*, Upper incisors notched. **Illustrated signs:** *b*, Storage mounds; burrows in ground (*d*), as does Algerian Mouse. *f*, Skull similar to House Mouse.

ALGERIAN MOUSE *Mus spretus* **p. 270**

Key features: Small mouse with short tail. Similar to Steppe Mouse. The fur is yellowish brown above and white or pale grey below, separated by a yellowish line. Upper incisors not notched (*c*). *f*, Skull similar to House Mouse.

HOUSE MOUSE *Mus musculus* **p. 270**

Key features: Fur uniform greyish-brown above, slightly lighter below. Body and tail roughly same length. Sharp muzzle. Pale or Eastern form shown in the upright scanning position. Upper incisors notched (*e*). **Illustrated signs:** *f*, Skull. *g*, Field-signs include gnawed (kibbled) grain, soiled with droppings. *h*, Droppings are 6mm long and 2.5mm wide often found in clusters . *i*, The track of the forefoot has four toes and measures 10 by 13mm, the hindfoot has five toes and leaves a track measuring 18 by 18mm. Tail can also leave a mark. *j*, Dens in burrows and in crevices in buildings.

Left, Dead House Mice and Wood Mice are often brought in by Domestic Cats. Here a juvenile Wood Mouse (bottom) shows the longer tail and larger eyes and ears of this species.

STEPPE MOUSE
ALGERIAN MOUSE
HOUSE MOUSE
a
b
c
d
e
f
g
h
i
j

Plate 52

WOOD MOUSE AND ALLIES

ROCK MOUSE *Apodemus mystacinus* **p. 258**
Key features: Largest *Apodemus* species with greyish fur and long whiskers (*c.* 47mm long). Big ears. **Illustrated signs:** *a*, Dens in rocky crevices.

PYGMY FIELD MOUSE *Apodemus microps* **p. 259**
Key features: Smaller and greyer than the Wood Mouse, eyes and ears relatively smaller. Chest occasionally with small spots.

WOOD MOUSE *Apodemus sylvaticus* **p. 259**
Key features: Fur is dark brown above and greyish-white below with yellow-brown flanks, but not well delineated. Small or absent yellow collar on neck which, when present, is longer than it is broad (cf Yellow-necked Mouse). Larger ears, eyes and hindfeet distinguish this species from the House Mouse.

YELLOW-NECKED MOUSE *Apodemus flavicollis* **p. 262**
Key features: Brown above and pure white below with a sharper delineation between the two than in the Wood Mouse. A larger yellow patch on throat, which is broader than long, also distinguishes this species from the similar Wood Mouse.

Common features: Pygmy Field, Wood and Yellow-necked Mouse all burrow in ground (*b*) and generally travel on surface, but may develop runs through leaf litter (*c*). *d*, Skull similar in these four species. *e*, Hazelnut gnawed at one end or on the side leaving a hole with a row of incisor marks below the hole's edge. *f*, Spruce cone with its base stripped by mice is rounded and smooth, scales tidily gnawed off. Remains of cones usually found in sheltered feeding places. *g*, Elliptical droppings usually of 3–5mm in length and deposited at random. *h*, Tracks with four toes on forefeet and five on hindfeet.

Below, The size of the chest spot varies in Wood Mice (three animals on left) and in Yellow-necked Mice (three on right), but is always smaller in Wood Mice.

a
ɔCK MOUSE
b
ʼGMY FIELD MOUSE
c
OOD MOUSE
LLOW-NECKED
OUSE
d
e
f
g
h

Plate 53

STRIPED MICE

STRIPED FIELD MOUSE *Apodemus agrarius* **p. 263**
Key features: Distinguishable from the Wood Mouse by a distinctive bold dark stripe along the length of the back, from nape to rump. The ears are smaller than the Wood Mouse's and the tail shorter than the head-body length. **Illustrated signs:** *a*, Skull similar to Wood Mouse's. *Droppings (b)* similar to Wood Mouse's. *c*, Tracks with four toes on forefoot and five on hind foot. Den in ground burrow (*d*).

NORTHERN BIRCH MOUSE *Sicista betulina* **p. 278**
Key features: A long stripe running from nape to rump along the back differentiates this species from the other mice, except from the Southern Birch Mouse, which has a shorter tail and pale borders to its dorsal stripe. The Striped Field Mouse is larger. **Illustrated signs:** *f*, Skull similar to Southern Birch Mouse. *e*, Spherical nest of grass and leaves, or dens in hollow logs and ground burrows (*h*).

SOUTHERN BIRCH MOUSE *Sicista subtilis* **p. 279**
Key features: Semi-prehensile tail. Shorter than in the Northern Birch Mouse. Also with a central dark stripe along the back but in this case it is bordered on either side by a slightly paler band. **Illustrated signs:** *f*, Skull. Dens in hollow logs, ground burrows (*h*) or rock crevices (*g*).

Above, Birch mice will travel on the ground in characteristic bounds.

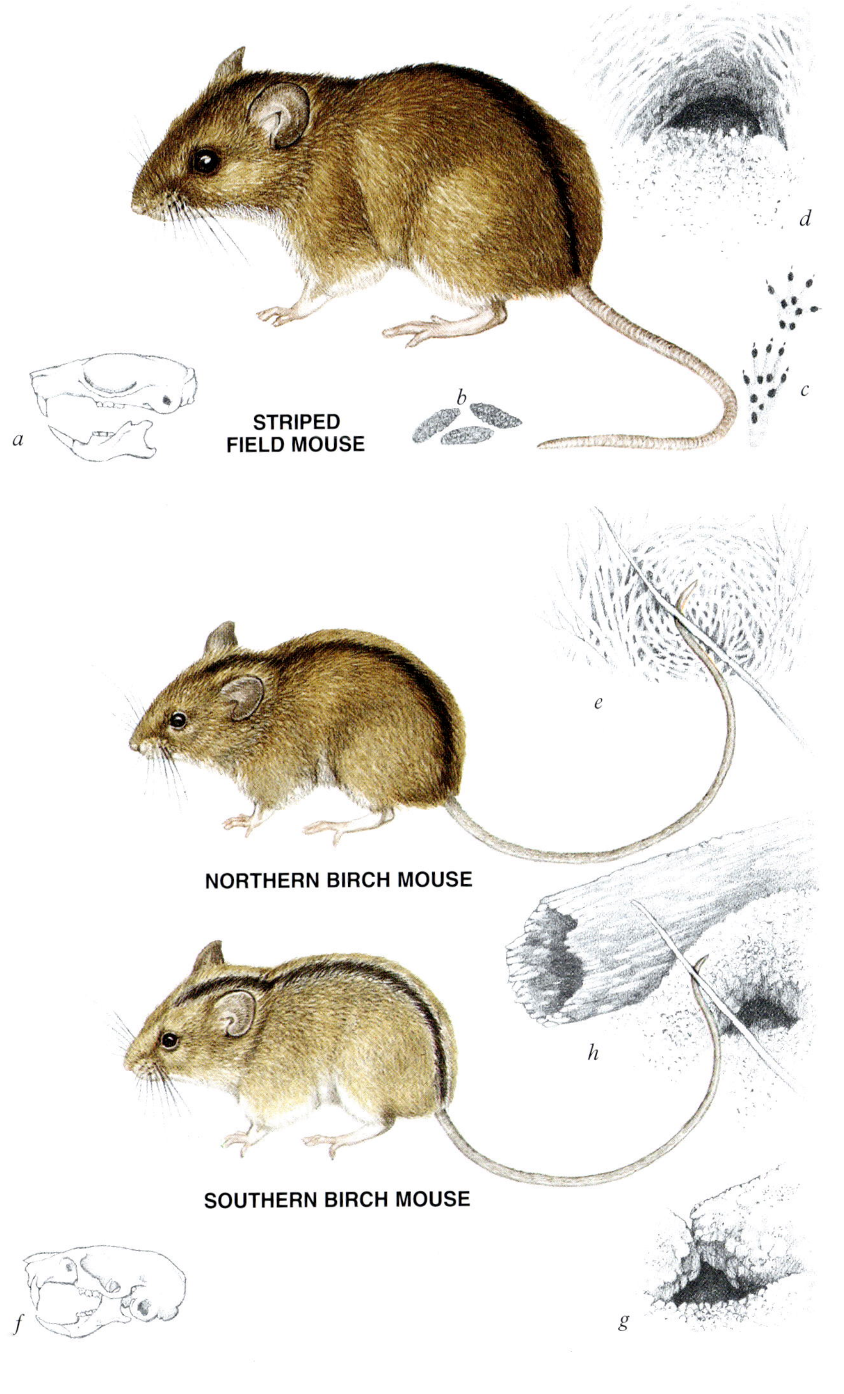
STRIPED
FIELD MOUSE
a
b
c
d
NORTHERN BIRCH MOUSE
e
h
SOUTHERN BIRCH MOUSE
f
g

Plate 54

RATS

BROWN RAT *Rattus norvegicus* **p. 263**

Key features: The fur varies from brown to black above, paler grey below. Ears are shorter and more hairy than in the Black Rat and the tail is shorter and thicker, and is dark on top and paler below. The skull (*b*) is similar to the Black Rat's but is more angular with parallel ridges along the top of the braincase (*a*).

Illustrated signs: *c*, Gnawed (ie. kibbled) grain soiled with droppings. *d*, Trails in buildings are dirty due to fatty smears and urine trails, and smear around joist forms a loop (cf. Black Rat). *e*, Forefoot, 1.8cm long by 2.5cm wide, leaves four toe marks; hindfoot, *c*. 3.3 by 2.8cm, leaves five toe marks. *f*, droppings are cylindrical, flat at one end and often pointed at the other. *g*, May den in ground burrow, typified by well worn track, or in buildings.

BLACK RAT *Rattus rattus* **p. 267**

Key features: Coat varies from brown to black. Longer, hair-less ears, larger eyes and a longer thinner tail distinguish this species from the Brown Rat. Shaggy appearance due to the longer hairs of the back. The skull (*b*) similar to Brown Rat's except shape more rounded with curved ridges flanking the braincase (*h*).

Illustrated signs: Signs similar to Brown Rat, except that greasy smears around joists (*i*) form a twin loop. *e*, Forefoot tracks, 1.5cm long by 1.7cm wide, with 4 toes; hindfoot, 2.1cm by 2cm, have five toes; claw marks visible in both. *j*, Generally confined to buildings, or ships. *k*, Droppings are shorter and narrower than Brown Rat's.

Below, Brown Rats swim readily and well, sometimes leading to confusion with Water Voles.

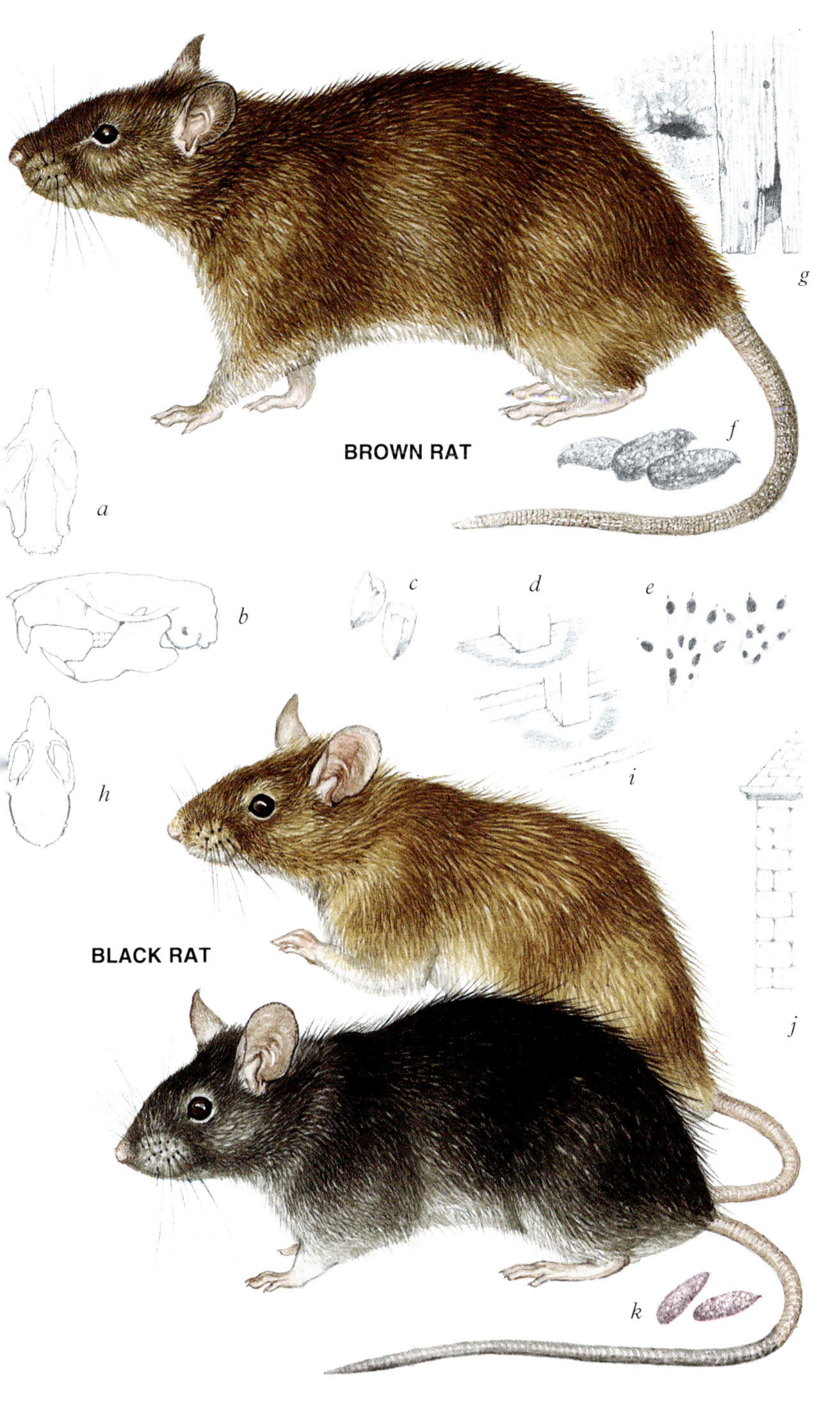

BROWN RAT
a
b
c
d
e
f
g
h
i
j
k
BLACK RAT

Plate 55

AQUATIC RODENTS

SOUTHERN WATER VOLE *Arvicola sapidus* **p. 246**
Key features: Fur dark brown above, yellowish below. Ears hidden by fur. This species is darker, larger and has a slightly longer tail than the Northern species (except in France and the Pyrenees where the two species coexist and the southern form of the Northern Water Vole is smaller than the northern form of the Southern Water Vole).

NORTHERN WATER VOLE *Arvicola terrestris* **p. 246**
Key features: Dark brown, but lighter than Southern Water Vole. Slightly paler below. Shorter tail than the Southern species, and smaller body (especially in France and Pyrenees where the two species coexist).

Common features: *a*, When swimming both species float relatively high in the water (cf. Muskrat, *j*). *b*, Skulls similar. Food signs include gnawed hazel nut (*c*) with irregular serrated appearance, and cut stems of grass and reeds (*d*). *e*, Small and cylindrical droppings of 5–8mm in length, with rounded edges, usually found in latrine sites. *f*, Tracks. *h*, Gnawed tree bark, with discarded chips of wood. Dens in ground burrow, often characterised by heavily grazed 'lawn' of short grass nearby (*g*) and, away from water, 'molehills' of spoil material near entrance (burrows of Brown Rat distinguished by spoil heap positioned away from entrance).

MUSKRAT *Ondatra zibethicus* **p. 248**
Key features: Hindfoot trimmed with bristles. The tail is compressed laterally.
Illustrated signs: *i*, Channels cut through aquatic vegetation. *j*, When swimming floats low in water (cf Water Voles, *d*). *k*, Skull. *l*, Tail drag very obvious in tracks. Most conspicuous sign is lodge (*m*) made of cut vegetation. Burrows in river bank have ventilation shaft (*n*).

Below, American Mink and Muskrat were both introduced to Europe by accidental escapes from fur farms in the 1950s. Mink now predate Muskrat, as they do in their native North America.

SOUTHERN WATER VOLE
h
g
a
NORTHERN WATER VOLE
d
c
e
f
i
n
m
j
MUSKRAT
l

Plate 56

AQUATIC RODENTS AND PORCUPINE

EUROPEAN BEAVER *Castor fiber* **p. 233**

Key features: Large. Fur yellowish to reddish or almost black. Tail distinctive, wide and horizontally flattened. Small eyes and ears, rounded head and short legs. Hindfoot webbed. Skull (*d*) has longer nasal bones (*e*) than Canadian Beaver (*f*). **Illustrated signs:** Lodge (*a*) in lake dammed with sticks and cut vegetation (*b*). *c*, Outline when swimming is relatively low in water. *g*, Droppings distinctive but seldom seen on land, *c.* 2.5cm long and almost as thick, made of coarse material like sawdust and containing undigested fragments of bark. *h*, Forefoot tracks show claws, hindfoot has conspicuous webbing, but tail drag so wide that it normally obliterates tracks. *i*, Gnawed trees and discarded chips of wood. Channels worn through aquatic vegetation (*j*).

COYPU *Myocastor coypus* **p. 281**

Key features: Large. Brown to yellowish-brown guard hairs and grey underfur. Cylindrical tail tapering only towards the end. Small ears and eyes. Incisors orange and protruding. **Illustrated signs:** *k*, Parting through bankside vegetation where Coypu slides into water . *l*, Outline in water involves both head and rump breaking the surface, with shoulders submerged. *m*, Skull. *n*, Droppings elongated. *o*, Tracks show five clawed toes on hindfoot, which is *c.* 12cm long by 7cm broad, with visible web between inner four toes; forefoot, *c.* 6cm by 6cm, also with five toes. Mark from tail usually apparent in tracks. *p*, Aquatic vegetation cut to form rest bed. *q*, Burrows in bank beside water.

CRESTED PORCUPINE *Hystrix cristata* **p. 279**

Key features: Very large. Unmistakable long black and white quills dispersed among more numerous shorter and thicker ones. Quills on the tail (*v*) and the lower back (*w*) detach easily when embedded in an adversary. **Illustrated signs:** *r*, Skull. Droppings about 2.4cm long, sometimes linked in a chain (*s*) or cluster (*t*). *u*, Tracks show five toes on forefoot, five on hindfoot. *x*, Dens in ground burrow.

Right, Young Coypus suckling from their mot laterally positioned teats

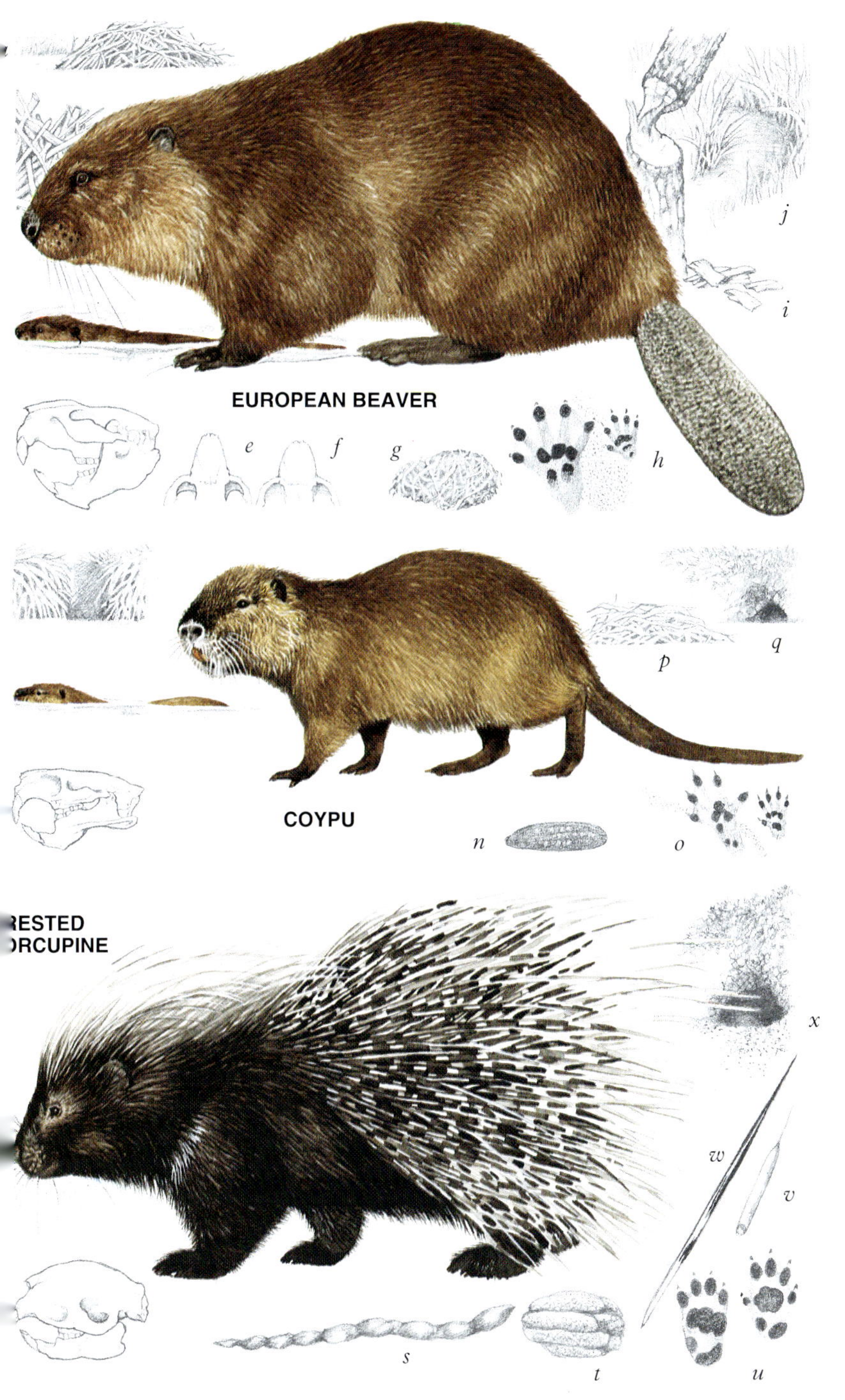
EUROPEAN BEAVER
e
f
g
h
i
j
COYPU
n
o
p
q
RESTED
ORCUPINE
s
t
u
v
w
x

Plate 57

RED-BACKED VOLES

Ears and eyes larger and tail longer than those of *Microtus* and *Pitimys* species.

RUDDY VOLE *Clethrionomys rutilus* **p. 242**
Key features: Reddish-brown above and creamy below; lighter than Bank and Grey-sided Voles. Very short tail ending in a tuft of hairs.
Illustrated signs: *a*, Last molar in upper tooth row always similar to the complex form of Bank Vole teeth, ie. with four outer ridges.

GREY-SIDED VOLE *Clethrionomys rufocanus* **p. 242**
Key features: Largest of red-backed voles. Pure grey flanks with a red band along the back.
Illustrated signs: *b*, Last upper molar in tooth row similar to the simple form of Bank Vole teeth, ie. with three outer ridges; in general appearance cheek teeth are more angular than those of Bank Vole.

BANK VOLE *Clethrionomys glareolus* **p. 243**
Key features: Reddish fur; flanks often greyish but red descends further than on Grey-sided Vole. Large variation in size of island subspecies e.g. the Skomer Vole is one third larger than mainland form.
Illustrated signs: *c*, Cheek teeth less angular than *Microtus* and *Pitimys* and show some degree of variation.

Common features: *d*, Skulls similar. *e*, Hazel nut gnawed in a large neat circular hole leaving no teeth marks. *f*, Spruce cone stripped by voles with characteristic smooth, round shape. *g*, Dark brown to black droppings circular in cross-section, with length four times the width. *h*, Tracks similar, hindfoot track is 16 by 17mm. *i*, Runways through leaf litter and tall grasses. *j*, Bark stripped and trees gnawed several metres above the ground.

Below, Baby Bank Voles, at seven days old, have a coppery sheen to their emerging coats. Later they acquire the dark grey coat typical of juvenile voles.

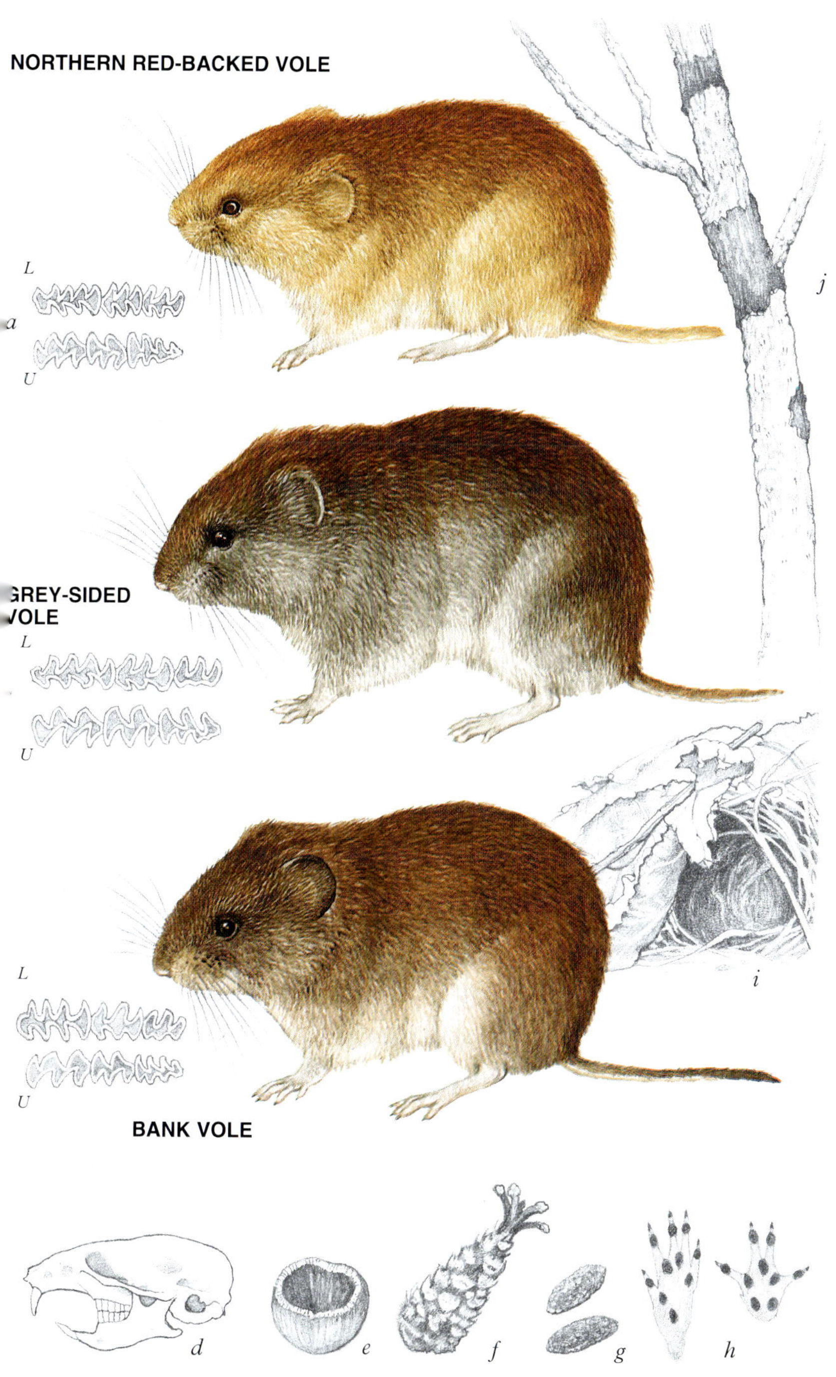

NORTHERN RED-BACKED VOLE
L
a
U
j
GREY-SIDED VOLE
L
U
L
i
U
BANK VOLE
d
e
f
g
h

Plate 58

GRASS VOLES

Small ears and eyes and blunt snout.

ROOT VOLE *Microtus oeconomus* **p. 251**
Key features: Slightly larger, darker, inner ear less hairy and tail longer than Field Vole. Tail bi-coloured. *a*, First lower cheek tooth distinctive, with a simple structure shared only by the Snow Vole among grass voles.

FIELD VOLE *Microtus agrestis* **p. 252**
Key features: Darker than Root Vole. Brownish-grey, slightly shaggy dorsal fur with grey underside, throat and feet. Longish fur partly covers ears (cf. Bank Vole). Tail darker above. Shorter tail than in the Common Vole.

Common features: *c*, Skulls similar. *d*, Piles of chewed grass stems. *e*, Green oval droppings usually found in a pile. *f*, Tracks with four toes on forefoot, five on hindfoot; hindfoot 17mm by 18mm. *g*, Runways through grasses. *h*, Spherical nest in tussocks of grass or in fork of low branches. *i*, Bark stripped from trees at ground level.

SNOW VOLE *Microtus nivalis* **p. 254**
Key features: Distinguished by its prominent whiskers, and dense coat of pale grey tinged with brown. The tail is shorter than that of the Balkan Snow Vole. First lower and last upper cheek teeth distinctive.
Illustrated signs: *k*, Skull. *l*, Track typical of grass voles but tail drag reputedly prominent. *m*, Dens in crevices in rocks or in ground burrow.

Below, Juvenile water voles may resemble field voles but can always be distinguished by their longer tails and hindfeet.

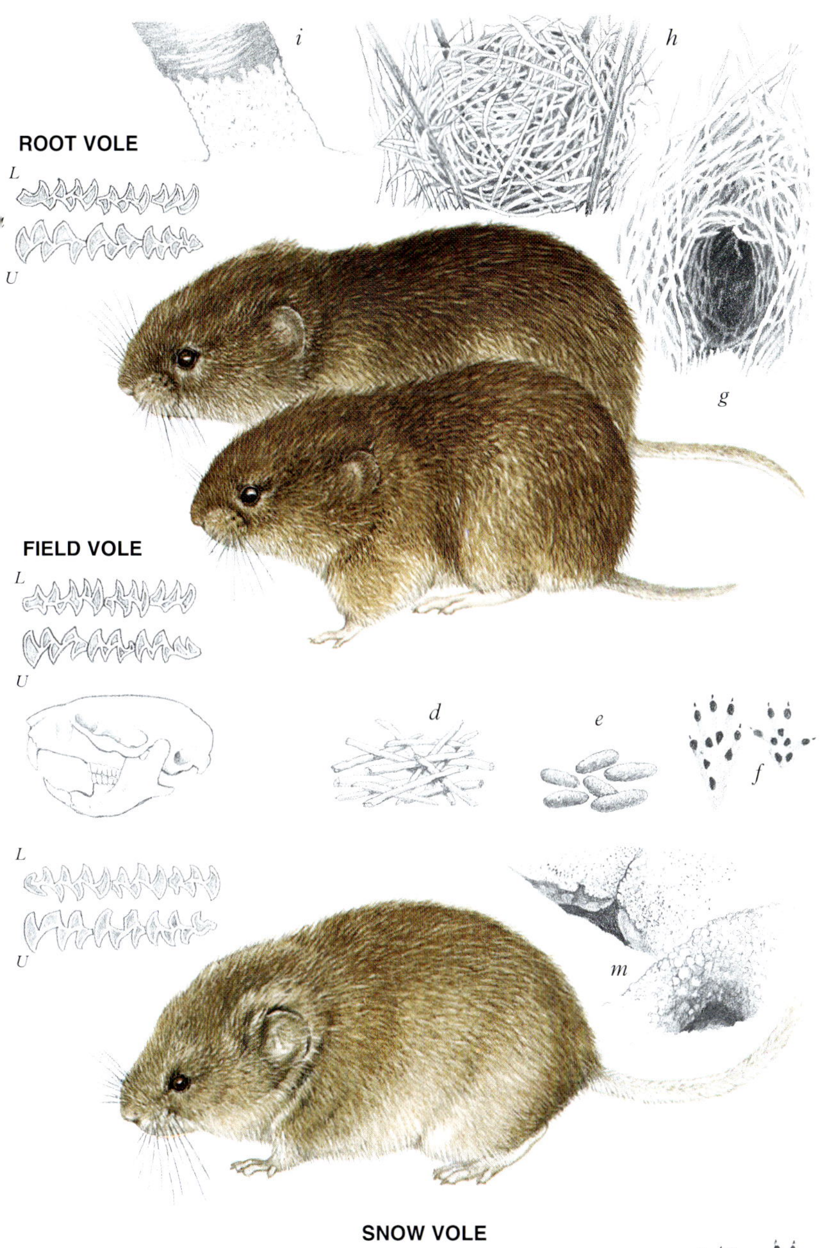

i
h
ROOT VOLE
L
U
g
FIELD VOLE
L
U
d
e
f
L
U
m
SNOW VOLE
k
l

Plate 59

GRASS VOLES

CABRERA'S VOLE *Microtus cabrerae* **p. 254**
Key features: Dorsal fur darker than in the Common Vole; grey underside more suffused with buff. Long guard hairs on rump. *a*, In profile the skull has a convex shape.

GUNTHER'S VOLE *Microtus socialis* **p. 255**
Key features: Similar to Cabrera's Vole, but lacks pronounced guard hairs on rump, and the tail and hind feet are very pale. *e*, Well separated triangular cusps on the grinding surfaces of the molar teeth distinguish this species from the Common Vole.

COMMON VOLE *Microtus arvalis* **p. 256**
Key features: Compared to Field Vole, the coat is smoother and lighter, the hairs shorter. Tail slightly darker on top than below. *f*, Extra loop on second upper molar absent (cf. Field Vole); first lower molar has four cusps on outer surface (cf. Root Vole). Skull (*d*) flatter than Cabrera's Vole (*a*).

SIBLING VOLE *Microtus rossiaemeridionalis* (Not Illustrated) **p. 257**
Key features: Only distinguishable from the Common Vole by chromosome count.

Common features: Skulls similar, except as indicated. *e*, Small piles of cut grass stems. *f*, Cylindrical black or green droppings, *c.* 3–4mm in length. *g*, Footprint extremely small and difficult to find. *h*, runways through tall grass. *i*, burrows with small spoil heap of soil outside. *j*, Gnawed bark on trees at ground level.

Below, Black individuals of the Common Vole are common on Orkney, where the so-called Orkney Vole generally has a darker, thicker coat than mainland forms.

j
i
h
RERA'S
E
THER'S
E
MMON
E
e
f
g

Plate 60

PINE VOLES

Smallest eyes of all the European voles.

ALPINE PINE VOLE *Pitymys multiplex* **p. 249**
Key features: Similar to Common Pine Vole, but larger and somewhat more yellow.

MEDITERRANEAN PINE VOLE *Pitymys duodecimcostatus* **p. 250**
Key features: Dense fur with velvety appearance, yellowish brown to reddish upperside and silvery to dark reddish-grey underside. *a*, Three loops on the upper third molar (cf. Common Pine Vole) and outside ridges of different sizes.

SAVI'S PINE VOLE *Pitymys savii* **p. 250**
Key features: Coat slightly lighter and tail shorter than the Common Pine Vole. *b*, Only three loops on inside upper third molar and outside ridges equal in size (cf. Mediterranean Pine Vole).

COMMON PINE VOLE *Pitymys subterraneus* **p. 251**
Key features: Fur dark grey. *c*, Four loops on inner upper third molar.

Common features: *d*, Skulls similar. *e*, Footprints tiny, seldom found. *f*, Dens in ground burrow. *g*, Spoil heap from the burrow construction.

Below, Nipple positions of field voles (left), pine voles (centre) and the Muskrat (right).

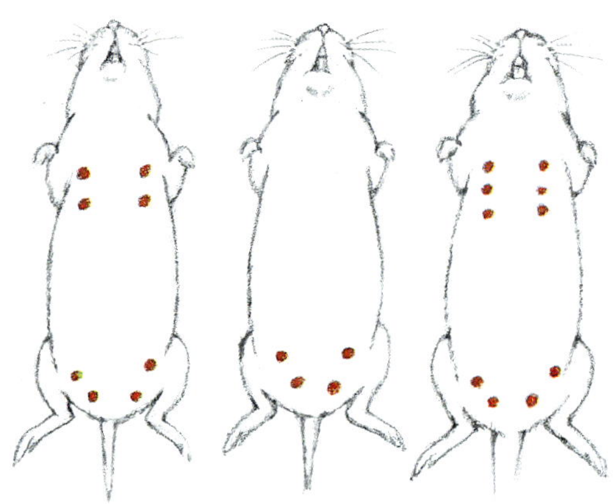

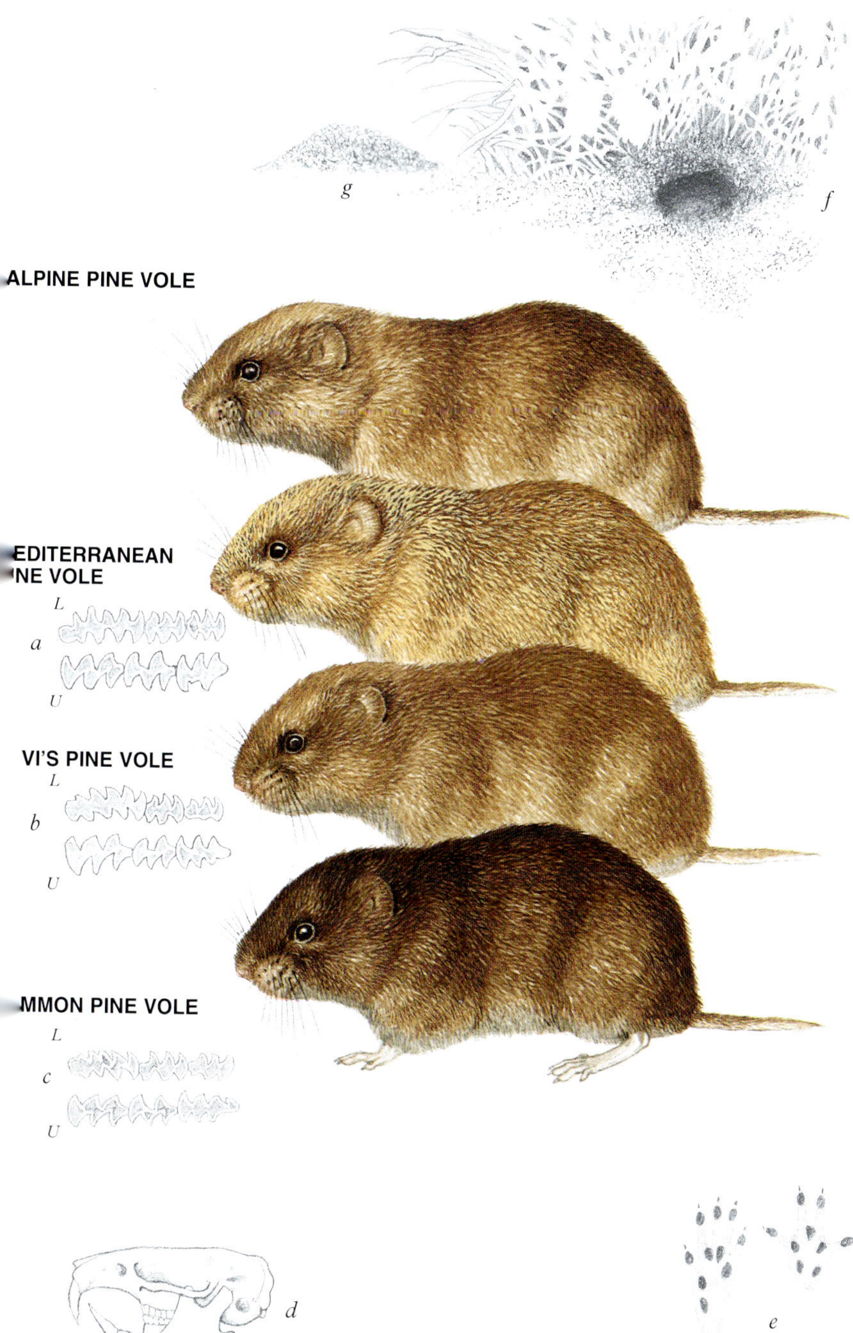
g
f
ALPINE PINE VOLE
EDITERRANEAN
NE VOLE
L
a
U
VI'S PINE VOLE
L
b
U
MMON PINE VOLE
L
c
U
d
e

Plate 61

BALKAN SNOW VOLE AND LEMMINGS

BALKAN SNOW VOLE *Dinaromys bogdanovi* **p. 245**
Key features: Long-tailed with dense fur: grey-blue above, greyish-white below. Even less brown in coat than Snow Vole. Feet white. Hindfeet large (over 22mm). Ears large, tail dark above and thinly haired. Cheek-teeth (*a*) similar to *Microtus* (Plates 31 and 32).
Illustrated signs: *b*, Skull. *c*, Dens in ground burrows and rock crevices.

WOOD LEMMING *Myopus schisticolor* **p. 240**
Key features: Unmistakable dark slate grey coat. Inconspicuous brown rump patch on adults. Tail short, ears very short.
Illustrated signs: *j*, Runways under moss.

NORWAY LEMMING *Lemmus lemmus* **p. 241**
Key features: Distinguished by a bold yellow and black pattern that varies in intensity. Very short tail.
Illustrated signs: *e*, Tunnels of dead grass appear after thaw. *f*, Dens in ground burrow, with trail leading to entrance. *g*, Spherical winter nest of grass and moss in crevice visible after snow melts.

Features common to lemmings: *d*, Skulls similar. *h*, Droppings resembling those of Field Vole, in piles. Trail very distinctive because tend to trot rather than bound. *i*, Tracks with very long claws, especially in winter. Clawmarks on Wood Lemming track reputedly less distinct than those of Norway Lemming.

Below, Norway Lemmings are notorious for their mass migrations.

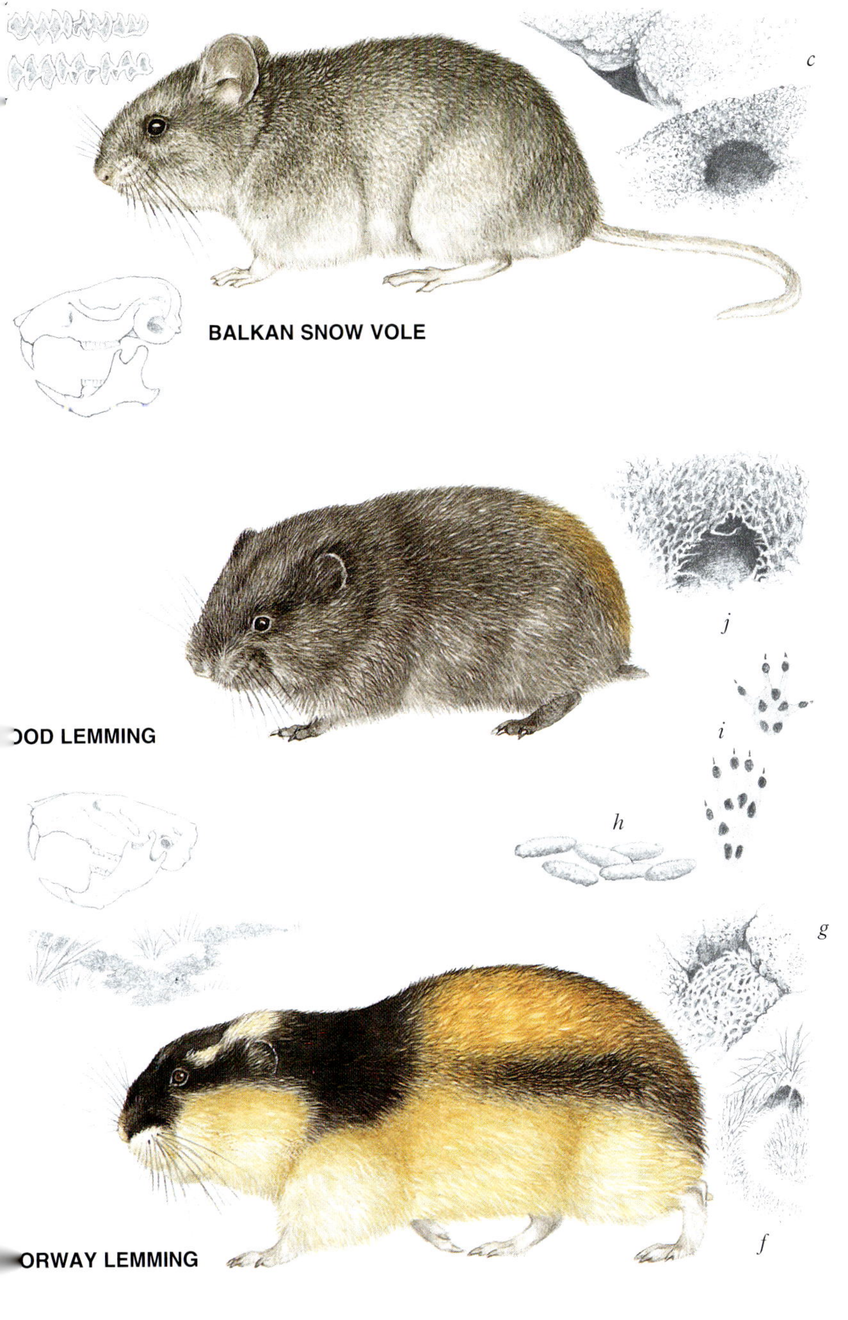
c
BALKAN SNOW VOLE
j
OOD LEMMING
i
h
g
ORWAY LEMMING
f

Plate 62

HAMSTERS

GREY HAMSTER *Cricetulus migratorius* **p. 237**
Key features: Small size, grey coat. Compared to voles, the ears and eyes are larger and the tail shorter.

ROMANIAN HAMSTER *Mesocricetus newtoni* **p. 238**
Key features: The top of the head has a dark area that extends backwards to the nape and a dark cheek stripe extends back to shoulder.

COMMON HAMSTER *Cricetus cricetus* **p. 236**
Key features: Brown dorsal fur with white patches, mainly black below. Large size, robust body with short, furred tail. *c*, Tracks, forefoot *c.* 1.5cm long and 1cm broad, hindfoot *c.* 2.0cm long and 1cm broad.

Common features: *a*, Skulls similar but for size. *b*, Droppings. *c*, Tracks similar, again but for size. *d*, Den in ground burrows with a single entrance, used by only one individual.

Below, Hamsters, like this Romanian Hamster, stuff food into their cheek pouches and carry it back to their burrows for storage or for consumption in safety.

GREY HAMSTER

d

ROMANIAN HAMSTER

COMMON HAMSTER

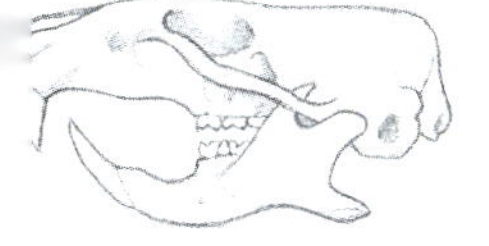

b

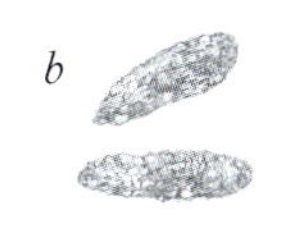

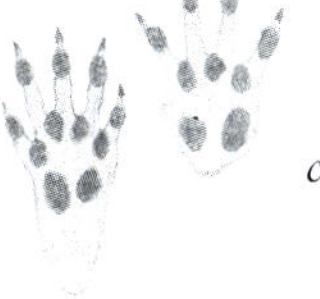

c

Plate 63

HARES AND RABBIT

BROWN HARE *Lepus europaeus* **p.284**

Key features: Ears black-tipped and longer than Rabbit's. The Brown Hare has longer ears, dark upperside of tail and more yellowish fur than the Mountain Hare. Tail carried away from body when running. *a*, Skull is distinguishable from the Rabbit's (*m*) by the shorter bony palate and wider nasal passage and lacks suture (*b*) of Rabbit's skull (*n*).

Illustrated signs: *c*, Well-trodden paths , similar to, but narrower than, Badger's, and with much less heavily registered paw prints. *d*, Grassy couch (called form) in long grass.

MOUNTAIN HARE *Lepus timidus* **p.287**

Key features: Smaller and more rounded profile than the Brown Hare, with pale tail. Colour varies with season, being grey-brown in summer and whitish in winter. The underfur is dark blue. Black-tipped ears.

Illustrated signs: *g*, Well-trodden paths, often running up and down hillside. *h*, Couch or form in heather.

RABBIT *Oryctolagus cuniculus* **p.289**

Key features: Smaller than hares, tips of ears brown, top of tail brownish-black, white below. Tail carried upright against body when running. Back of the neck rufous, belly white to grey. Skull (*m*) is distinguishable from the hare's (*a*) by the longer bony palate and narrower nasal passage and can be differentiated from hare's (*b*) by the suture delimiting the interparietal bone (*n*).

Illustrated signs: Communal warren of ground burrows (*l*) with accumulation of droppings at latrines on bare ground (*o*).

Common features: *i*, Skulls similar. *j*, Droppings vary in colour and size (7–12mm). *k*, All tracks similar but hares slightly larger than Rabbit.

Below, 'Mad March' Hares are actually unreceptive females attempting to repel the sexual advances of males.

a
b
c
d
BROWN HARE
e
summer
g
f
winter
MOUNTAIN HARE
h
j
k
l
m
n
o
RABBIT

Plate 64 PRIMATES

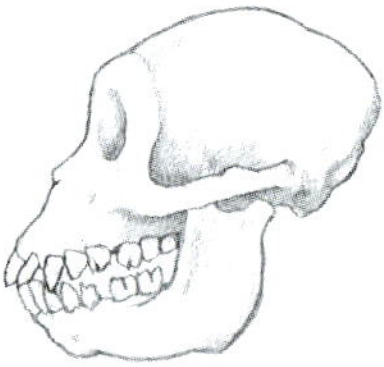

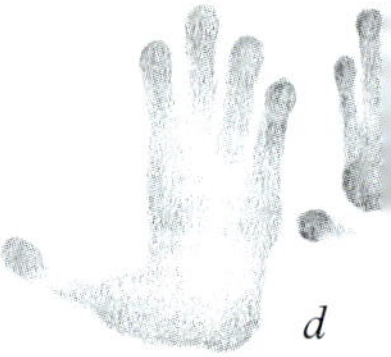

BARBARY APE *Macaca sylvanus* **p. 89**

Key features: Only non-human primate in Europe. Tail-less. Short muzzle, large eyes orientated forewards, short ears and long legs with prehensile hands and feet. Thick shaggy fur. Young (*a*) born black.
Illustrated signs: *b*, Skull. *c*, Dropping. *d*, Tracks.

Left, Barbary Apes carry their infants on their backs. Parental care is highly developed in the intricate societies of many primates, whose young are born helpless and with much to learn.

generally have small, low triangular dorsal fins. Also, they have many spade-shaped teeth. Monodontidae are the white whales and narwhals. 2Narwhals and white whales have many differences, but neither has a dorsal fin and in both groups the flippers turn upwards at the tips as they grow older.

The Physeteridae are sperm whales, including the Sperm Whale itself which is the biggest toothed whale. All have barrel shaped heads and a very narrow underslung lower jaw which lack functional teeth.

The Ziphiidae are the beaked whales and have a distinct beak; very few teeth – none in the upper jaw and only emerging from the gums in adult males, where they are relatively large and may project from the mouth as small tusks.

ROUGH-TOOTHED DOLPHIN *Steno bredanensis* Pl. 30

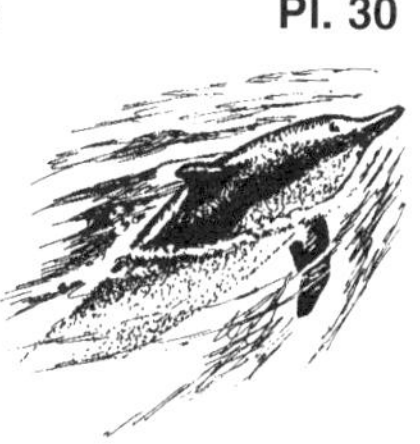

Name. *Stenos* = narrow (Gk.); *bredanensis* = of Breda (name of artist who painted first specimen).
Recognition. Long conical beak, continuous with forehead. Dark grey tinged with purple on back and flanks; white underside with pink to buff-white blotches. Often marked with white streaks. Beak, with white tip and sides, not distinct from forehead. Sickle-shaped dorsal fin in mid-back. Keel above and below stock of tail. Similar species include *Stenella* (junction of beak and head delineated by white line, and has numerous small regular spots); *Tursiops* (stubby beak, also delineated from forehead by line, but no spots). Surfaces every 7–10 minutes.
Habitat. Pelagic. Tropical and subtropical waters and occasionally Mediterranean Sea. Very rarely further north.
Habits. SOCIAL: Uncertain, probably in groups of up to 50.
Measurements. Head-body length: 2–2.6m. Weight: male 140kg, female 120kg. Dental formula: 20–37 teeth in each ramus of jaw (fine ridges give rough surface).
General. Although widely distributed, it is apparently nowhere abundant. Some morphological variation between Atlantic and Indo-Pacific populations. Attracted to fast-moving vessels, and can swim at 30km per hr. Sometimes schools with tuna.

< *Rough-toothed Dolphin.* English Channel, south along Iberian coastline.

> *Striped Dolphin.* Atlantic waters south of Scotland and including the Mediterranean Sea.

STRIPED DOLPHIN *Stenella coeruleoalba* **Pl. 31**

Name. Euphrosyne Dolphin, Blue-white Dolphin. *Stenos* = narrow (Gk.), referring to shape of beak. *Caeruleus* = sky blue; *albus* = white (Lat.).

Recognition. Similar to Common Dolphin, but distinguished by pale grey V-shaped 'shoulder blaze' originating above and behind the eye and narrowing to a point below and behind the dorsal fin. In N. Atlantic, may be confused with Spotted and Bottle-Nosed Dolphins due to shoulder blazes. Adult Spotted Dolphin likely to have spotting on shoulder and blaze usually muted; any similar blaze on the Bottle-Nosed Dolphin usually muted and less clearly defined (Atlantic form of Spotted Dolphin much less spotted than Pacific or Indian Ocean forms). Other features of the Blue-White Dolphin are an elegant pattern of white and grey on the flanks together with black lateral stripes running from eye to flipper and eye to anus.

Habitat. Pelagic, in temperate and warm seas. Rare in British waters, common offshore in the Atlantic and Mediterranean Sea, at depths more than 200m.

Habits. FOOD: Fish, also squid and decapod crustaceans. Hunts (and navigates) using echolocation. SOCIAL: Generally schools of 5–300 (occasionally up to thousands), sometimes in association with Common Dolphin. Some evidence for segregation of sexes outside breeding season and also segregation by age: large schools containing subadults may have no adult males and only few adult females.

Breeding. Breeding in spring and autumn in Atlantic, autumn only in Mediterranean. Sexual maturity: Atlantic males 205cm, females 200cm, between 6–10 years old; Mediterranean males about 190cm, females about 180cm. Gestation *c.* 10–12 months. Litter size 1. Calving interval 1 to 4 years. Mammae 2. YOUNG: Weaning age 12–20 months.

Lifespan. Max. *c.* 35 years, maybe more.

Measurements. Head-body length: 2.1–2.5m, occasionally up to 2.7m; males slightly larger than females. Weight: up to 120kg (at 2.0m = 90kg). Dental formula: 43–55 (usually 45–50) teeth (3mm diameter) in each ramus of jaw, sharp and slightly incurved.

General. Widely distributed across all temperate, subtropical and tropical seas. Thought to be most common cetacean in Mediterranean. Often come to play around boats.

ATLANTIC SPOTTED DOLPHIN *Stenella frontalis (attenuata)* **Pl. 31**

Name. Bridled Dolphin. *Attenuata* = reduced (Lat.).

Recognition. Back black with white spots, light underneath. Black circle around eye forms 'bridle' which extends from back of beak to eye and dark line from beak down to pectoral fin. Dark spots on lower flanks and belly absent at birth but enlarging with age, spotting also decreases away

<*Atlantic Spotted Dolphin.* In European waters only sighted S of entrance to Mediterranean.

>*Common Dolphin.* Atlantic waters S of Iceland and Mediterranean and Black Seas.

from both Pacific and Atlantic coasts of North America. Body more robust in coastal populations; slender beak longer in females with upper and lower lips lighter colour. Sickle-shaped dorsal fin in mid-back. Keel below, and sometimes above stock of tail.

Habitat. Pelagic. Found once stranded on French coast, but quite common off the Azores. Generally found in N. Atlantic, mainly in Caribbean and off W. Africa.

Habits. SOCIAL: Schools of 5–500, but sometimes over a thousand. FOOD: Fish and squid.

Breeding. Births in spring and autumn. Sexual maturity: male 12 years (at *c.* 194cm); female 9 years (at *c.* 181cm) (NW Pacific, 1 year younger in tropical Pacific). Gestation 9–12 months. Calving interval: 2.5–3.9 years. YOUNG: Weaning age 13–27 months.

Measurements. Head-body length: 1.9–2.3m. In Pacific, males and coastal types slightly larger than females and offshore individuals. (Birth: NW Pacific *c.* 89cm; tropical Pacific *c.* 82.5cm). Weight: *c.* 110kg. Dental formula: 29–34 pairs of small sharp-pointed teeth in upper jaw, 33–36 pairs in lower jaw.

General. Population unknown but one of the commonest of all cetaceans, with *c.* 3.5 million estimated in eastern tropical Pacific. Declines due to tuna purse-seine fishery recorded in tropical Pacific. Four species of Spotted dolphin (*Stenella attenuata*, *S. frontalis* (G. Cuvier, 1829), *S. plagiodon* (Cope, 1866) and *S. dubia* (G. Cuvier, 1829)), with different distributions and appearances now recognised as probably representing two species (*S. attenuata* and *S. frontalis*).

COMMON DOLPHIN *Delphinus delphis* Pl. 31

Name. Fraser's Dolphin. *Delphinus* = dolphin (Lat.); *delphis* = dolphin (Gk.).

Recognition. V-shaped black or dark grey saddle with downward-orientated apex on sides directly below dorsal fin. Conspicuous white thoracic patch. 'Hourglass' effect on side is yellowish tan anterior to the posterior edge of the dorsal fin, and partially created by the light grey coloration

of the flank sweeping over the dorsal aspect of the tail stock. Slender well-defined black beak, often tipped with white, 11–12cm long. Many pigmentation features, particularly the intensity of the striping, differ between regions, as do such characteristics as length of beak and body size. Most easily confused with the Striped Dolphin *Stenella coeruleoalba*, but the black pointed V-shaped 'saddle' below the dorsal fin of the Common Dolphin, and the complex pattern of dark stripes (eye to anus and eye to flipper) of the slightly larger Striped Dolphin are the primary clues. Can bowride for 20 mins or more, keeping up with fast ships.

Habitat. Pelagic; prefers temperate seas but widespread (excluding polar seas).

Habits. FOOD: Blue Whiting, Pilchard, Whiting, Pollack and Lantern Fish (cophids) and any small fish (depending on what is available). Sometimes cephalopods. Can dive to depths of at least 280m and for as long as 8 minutes. HOME RANGE: Well-defined migrations unknown. SOCIAL: Mixed sex groups with equal sex ratio. Schools of a few tens to a few thousands. Often mixed schools with Striped Dolphin at productive feeding sites but generally not while travelling.

Breeding. Mating in July–October. (North Atlantic), births in June–September. Sexual maturity: male 5–9 years (c. 200cm); female 6–9 years (c. 190cm). Gestation 10–11 months. Litter size 1. Calving interval 1–3 years. Mammae 2. YOUNG: Weaning age 19 months or more. Members of school attentive to young (and wounded).

Lifespan. Max. probably *c.* 30 years, maybe more.

Measurements. Head-body length: males bigger than females, with lengths ranging to 2.5m, but most adults 1.7–2.4m (newborn *c.* 80cm long). Weight: at 2m = 100kg, but seldom weigh more than 75–85kg. Dental formula: 33–58 (usually 40–55) (*c.* 3mm diameter) in each ramus of jaw.

General. Very widely distributed, occurring in all waters to limits of tropical and warm temperate waters. Several distinct forms that probably deserve racial or subspecific status. Often accompany ships or play around boats. Also seen to ride bow-waves of large whales (eg fin whales in N. Atlantic). Hunting has reduced numbers in the past century. Pollution may be a problem nowadays, for example in Mediterranean, although effects as yet not determined. Tuna fisheries' impact on this species thought to be less severe than on other species. Throughout the world, are killed directly by harpoon or drive fisheries, and inadvertently in fixed or active fishing gear. Previously, 120,000 per year killed in Soviet direct fishery in the Black Sea. Reports of 'sportsmen' shooting at Common Dolphins for fun in the Mediterranean.

BOTTLE-NOSED DOLPHIN *Tursiops truncatus* Pl. 30

Name. *Tursio* = porpoise (Lat.); *ops, opos* = face (Gk.); *truncatus* = truncated i.e. foreshortened = reference to bottle-nose.
Recognition. Short beak (bottle nose) with lower jaw protruding beyond upper, clearly separated from the melon. Robust yet streamlined. Dark grey on back, light grey underside.
Habitat. Predominantly coastal, often around estuaries, but groups can be found offshore, sometimes in association with other species (eg *Globicephala*).
Habits. FOOD: Largely fish (the species varying with season and availability, although mullet are a typical prey). Cuttlefish seasonally important in some areas. HOME RANGE: Thought to be usually only tens of kilometres for coastal populations, with forays of several 100km at a time. Coastal groups occupy linear home ranges of 200–300km. SOCIAL: Schools varying from 2–10 up to a thousand. Hierarchical social relationships. Some groups contain adults of both sexes, calves and subadult females, other groups composed of adult females with calves and immatures of both sexes; or subadult males only. COMMUNICATION: Whistles (2–20kHz) and a large variety of burst-pulsed sounds. Some whistles are specific 'signatures' of individuals. Echolocation using clicks (for navigation and prey detection). Excellent eyesight, exceptional auditory sense, good taste and touch.
Breeding. Mating and births occur throughout the year but with apparent peaks between April and September. Sexual maturity: males *c.* 11 years (245–260cm); females 12 years (220–235cm). Gestation 12–13 months. Litter size 1. Calving interval 1.3–3 years. Mammae 2 – one on each side of genital slit. YOUNG: Weaning age 19 months. Females nurse calves for about a year and a half, the suckling calves gaining assistance as they swim by placing their flipper on the flanks of the mother who will also protect them from the approaches of other animals. May leave the calf with another adult ('baby-sitter') while making long dives. The juvenile may stay with the mother even if she has another new calf. Males seem to avoid the company of calves.
Lifespan. Max. at least 35 years, possibly up to 40.
Measurements. Head-body length: 2.5–4.1m (Birth: 98–130cm). Appears to be a larger and a smaller form of this species, although only the larger form exists in western European waters. Weight: average 150–200kg (at 3.4m = 394kg), can attain or even exceed 400kg. Dental formula: 20–26 teeth in each ramus, each *c.* 12mm in diameter.
General. Common, but potentially vulnerable where inshore populations come into contact with human activities. Species most commonly exhibited in 'dolphinaria'.

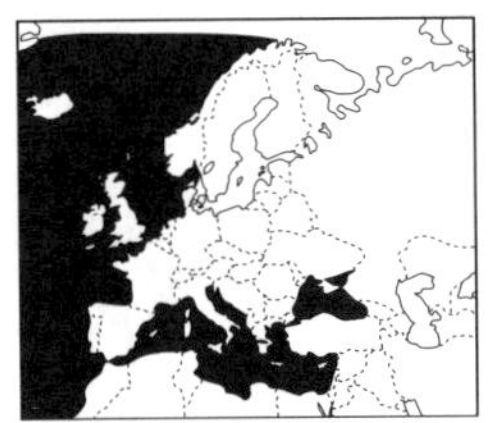

< *Bottle-nosed Dolphin.* Very widespread throughout Atlantic from Arctic Ocean southwards, excluding Baltic but including Mediterranean.

> *White-beaked Dolphin.* Atlantic N to Arctic ocean.

WHITE-BEAKED DOLPHIN *Lagenorhynchus albirostris* Pl. 32

Name. Squid Hound. *Labenos* = flask (Gk.); *rhynchos* = snout (Gk.); *albus* = white (Lat.); *rostrum* = prow (Lat.).

Recognition. Individuals from NW European waters invariably have white beak (those with dark beaks occur in NW Atlantic, but so do many Bottle-nosed Dolphins, so not a diagnostic character). White patch immediately behind fin, on back. Individual variation in black and white patterning particularly related to age (young animals are less patterned and may lack distinct white patch over back behind pectoral fin). Dorsal fin large and sickle-shaped.

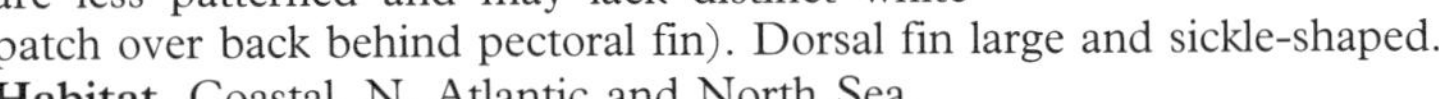

Habitat. Coastal, N. Atlantic and North Sea.

Habits. FOOD: Fish (mackerel, herring, cod, whiting), squid and benthic crustaceans. HOME RANGE: Annual migration between temperate and subpolar waters; otherwise seasonal movements mainly onshore vs. offshore. SOCIAL: Schools of 2–20 individuals but aggregations may number 1000 or more.

Breeding. Mating in July–October, births in May–August. Gestation *c.* 10 months.

Measurements. Head-body length: 2.3–2.8m. Weight: at 2.3m = 180kg; male 10% heavier than female. Dental formula: 22–28 teeth (6mm diameter) in each ramus of jaw.

General. Common in cool temperate and subarctic seas of N. Atlantic.

ATLANTIC WHITE-SIDED DOLPHIN *Lagenorhynchus acutus* Pl. 32

Name. *Acutus* = sharp (Lat.).

Recognition. Thick, short (5cm) black beak delineated from head by a deep crease. Distinct ridge between dorsal fin and tail. Black or dark grey with long white anal blaze from below dorsal fin to area above anus, and elongated yellow-ochre band extending backwards from upper edge of white blaze towards tail.

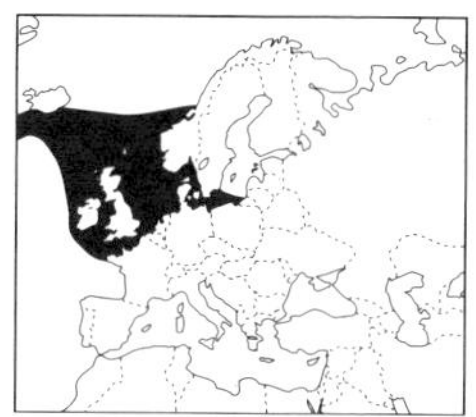

<*Atlantic White-sided Dolphin.* Around British Isles, north to Iceland.

>*False Killer Whale.* W of British Isles and S, also into W Mediterannean, but several strandings on Orkney Island and E coast of Scotland.

Habitat. Usually farther offshore than White-beaked Dolphin, in deep waters particularly along continental slope.

Habits. FOOD: Fish (eg blue whiting, mackerel, herring, cod, whiting), squid and gammarid crustaceans. HOME RANGE: Seasonal migration between Arctic and temperate waters is marked in the Western Atlantic stock, but not in the European stock. European population is confined to cool temperate waters. SOCIAL: Schools of 10–100 individuals; aggregations may number 1000 or more.

Breeding. Calving in May–July. Sexual maturity: male 4–6 years (230–240cm); female 5–8 years (201–222cm). Gestation *c.* 10–12 months. Litter size 1. Calving interval possibly 2–3 years. Mammae 2. YOUNG: Weaning age 18 months.

Lifespan. Max. 27 years or more.

Measurements. Head-body length: 1.8–3.0m (mainly 2.25–2.5 m); N. Atlantic: male *c.* 250cm; female *c.* 224cm (birth: 110cm). Weight: male *c.* 215kg; female 165kg. Dental formula: 30–40 teeth (5mm diameter) in each ramus of jaw.

General. Common in cool temperate and subarctic seas of N. Atlantic.

FALSE KILLER WHALE *Pseudorca crassidens* Pl. 33

Name. *Crassus* = thick, *dens* = tooth (Lat.); *pseudos* = false (Gk.).

Recognition. Third largest of the Delphinidae. Dorsal fin is sickle-shaped, and situated just behind mid-point of back. Slender, tapered head. Pectoral fins are pointed but fairly short (broad ridge-like shape to front margin). General colour black or slate grey on body, slightly paler on throat or chest, and dorsal fin, flippers and caudal flukes deep black. The lips are white or cream coloured internally.

Habitat. Worldwide oceanic: pelagic.

Habits. FOOD: Cuttlefish, cod, tuna. HOME RANGE: Follows ocean currents over huge distances. SOCIAL: Large schools of up to several thousand but usually smaller (occasionally stranded en masse eg 150 in Scotland in 1927).

Breeding. Sexual maturity 8–14 years (males 396–475cm, females 366–

427cm). Gestation 11–15.5 months. Litter size 1. Calving interval perhaps 3 years. Mammae 2. YOUNG: Weaning age 18 months.
Lifespan. Max. recorded *c.* 40 years.
Measurements. Head-body length: 3.7–5.5m (male *c.* 541cm; female *c.* 452cm) (birth 193cm). Weight: 1.2–2.0 tonnes (at 5.3m = 1.7 tonnes). Dental formula: 8–12 (usually 11) strong teeth in each ramus of the jaw (approx. 25mm diameter); 9th and 11th teeth bigger than those of Long-finned Pilot Whale.
General. Cosmopolitan, although mainly in warm temperate and tropical waters. Very fast swimmer (up to 30 knots). Attracted to vessels.

KILLER WHALE *Orcinus orca* Pl. 33

Name. Orca. *Orca* (Lat.) from *oryx, orygos* (Gk.) = pickaxe.

Recognition. Largest dolphin. Grey 'saddle patch', white lateral flanks, white patch behind eyes. High black dorsal fin, falcate in females and triangular in males, may be up to 1.8m high in the latter. Snout blunt and rounded. Noisy, explosive low blow, visible in good light. Normal travel pattern is diving for 2–5 minutes followed by 3–5 blows at the surface.
Habitat. Nearshore, coastal and pelagic.
Habits. FOOD: Large fish, squid, seals, other species of whale, and penguins (also carrion). In Europe, particularly schooling fish such as herring and mackerel. HOME RANGE: often quoted as 320–480km² for a pod (based on pods in Puget Sound, Washington), but recent photo-identification studies report ranges of 700km², and in the north-eastern Pacific up to about 1650km². Travels hundreds of kilometres to follow prey movements. Killer Whale pods show one of the lowest dispersal rates of individuals among mammals. In the NE Pacific, individuals of both sexes have been found to stay in their maternal subgroups into adulthood, probably for life. No whale has been seen to leave its group for another on a permanent basis, and maternal groups have thus had a stable composition (except for births and deaths) for more than 15 years of study. As pods grow large, splitting seems to take place between subgroups. Insufficient studies in the northeastern Atlantic for scientists to be sure these habits occur there too. SOCIAL: Matrilineal kinship groups, called 'pods', of 2–40 individuals, but sometimes smaller groups and solitary, as well as temporary schooling of up to hundreds. COMMUNICATION: Short (0.8–25 msec) broad band pulses for echolocation (energy content up to at least 80kHz, peak energy usually lower); pulsed calls (pulses produced with high repetition rate, up to 5000 a second; energy to at least 30kHz, with most energy concentrated usually at 1–6kHz) and narrowband whistles (frequency range about 1.5–18kHz) for communication. Pods have been found to produce pod repertoires of 7–17 stereotyped pulsed calls, which differ to varying degrees between different, sympatric pods

<*Killer Whale*. Very widespread throughout Arctic and Atlantic Oceans.

>*Risso's Dolphin*. North Sea southwards, Baltic Sea and Mediterranean.

(i.e. pod-specific dialects). As yet unknown as to whether European Killer Whales also have such dialects. Highly developed acoustical and hearing senses, echolocation, as well as high visual acuity.

Breeding. Mating in October–December, births in October–January. Sexual maturity: males 15–16 years (at *c.* 579cm); females 8–10 years (457–488cm). Gestation 12–16 months. Litter size 1. Calving interval 3–8 years. YOUNG: Weaning age 12 months. Young accompany mother for several years.

Lifespan. Max. *c.* 50–80 years. Mortality very low, 1–3% in NE Pacific.

Measurements. Head-body length: 3.8–9.5m, male 6.4–9.5m; female 3.8–6.5m (birth: 208–220cm). Weight: male 4.5–5.5 tonnes; female 2.5–3.0 tonnes. Dental formula: 10–14 (often 12) teeth in each ramus of jaws (each 25–30mm diam. and slightly oval in cross-section).

General. Cosmopolitan, but locally populations are small except perhaps in Antarctica, where populations are thought to total in the low hundreds of thousands. In the Northern Hemisphere they are most abundant in the northeastern Pacific Ocean (Alaska and British Columbia) and the northeastern Atlantic. European populations number 4–10,000 and around Iceland and Norway, the regions where they have been studied most, they are thought to number at least 1500. Largest dolphin. Swimming speeds can reach 60km per hr. Often curious towards boats, particularly younger individuals. Individual-specific natural markings (dorsal fin shape, nicks and tears in it, saddle pigmentation patterns and body scarring) are widespread and make it possible to identify and study individuals over several years.

RISSO'S DOLPHIN *Grampus griseus* Pl. 30

Name. *Griseus* = grey (adapted from French).

Recognition. Blunt head bisected by deep crease, no beak, prominent eye. Large anchor-shaped light grey patch on chest and chin. Pectoral fins: long and narrow. Dorsal fin: large and pointed. Often parallel scars along sides (probably tooth wounds from aggression within species but also squid sucker marks). Coloration: dark grey-brown becoming paler with age, especially around head. Oldest individuals may be pale grey with white heads.

Habitat. Pelagic in tropical and temperate seas (rarely further north than British Isles).
Habits. FOOD: Squid and cuttlefish. SOCIAL: Schools of 5–25 (up to 300). Thought to be polygynous.
Breeding. In N. Atlantic, births from April–September. Sexual maturity reached at 3m length or more.
Measurements. Head-body length: 3.3–3.8m (newborn: *c.* 1.5m). Weight: male *c.* 400kg; female *c.* 350kg (at 3.4m = 343kg). Dental formula: 2–7 peglike teeth in each ramus of the jaw. Generally no maxillary teeth.
General. Common in warm temperate and tropical seas. Less acrobatic than many dolphins. Usually rides vessel bow-waves only very briefly.

LONG-FINNED PILOT WHALE *Globicephala melaena* Pl. 33

Name. *Globus* = sphere (Lat.); *cephalus* = head (Gk.); *melaena* = black (Gk.).

Recognition. Second largest delphinid, after the Killer Whale. Bulbous head. Pectoral fins: 15–20% total length. Short beak. Curved dorsal fin positioned well forward. Resembles Short-finned Pilot Whale and False Killer Whale, but former has fewer teeth and shorter flippers, latter has more tapered head and a higher, falcate dorsal fin. Around European coasts, generally seen in schools of up to 75 in number. Can be seen floating at surface, with a synchronous blow every 3–10 minutes.
Habitat. Pelagic, with one variety in N. Atlantic and another in the Southern Hemisphere. Cold temperate and polar waters. Replaced by *G. macrorhynchus* in warm temperate and tropical regions.
Habits. Activity pattern uncertain, but probably related to vertical migration of squid. FOOD: Cuttlefish, squid and fish. HOME RANGE: Unknown, but probably wide ranging. Well marked migrations unlikely in that these whales probably follow available food, which will be spatially variable from year to year. SOCIAL: Schools 10–200 (aggregations of up to *c.* 3000). Schools probably matrilineal. More than 1 large male (and up to 15) in each group. Tight social bonding implied by mass strandings (see vignette). COMMUNICATION: Clicks *c.* 2.8–4.7kHz, whistles up to 35kHz.
Breeding. In N. Atlantic mating peaks between April–July, with births in July–September. However, births recorded throughout year and little synchrony in breeding. Sexual maturity: males 10–12 years (at *c.* 490cm) possibly up to 20 years, females 6–10 years (at 365–400cm). Gestation 14.5–15.5 months. Litter size 1. Calving interval 3–4 years. Mammae 2. YOUNG: Light grey at birth and have, like many, white vertical 'crease' marks for several months. Colour darkens and creases disappear during first 12 months of life. Females probably remain within natal school for life, males staying up to seven years before generally moving to other schools. Weaning age 22 months.

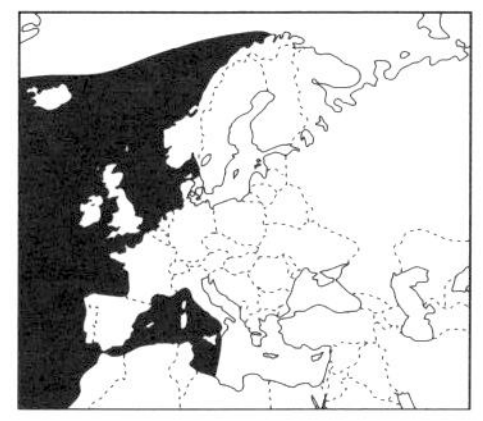

<*Long-finned Pilot Whale.* Widespread in Atlantic waters, N to Norwegian Sea and E in Mediterranean to Italy.

>*Harbour Porpoise.* Atlantic coastal waters, NW to Iceland.

Lifespan. Max. *c.* 60+ years. Higher adult mortality of males reflected in the adult sex ratio of *c.* 1 male : 2 females.
Measurements. Head-body length: 3.8–6.3m (male *c.* 5.5m; female 4.6m) (birth 177cm). Condylo-basal length: averages 630mm (618–655mm range). Weight: 1–3.5 tonnes (max.: male 3.5 tonnes; female 2.5 tonnes). Dental formula: 8–12 (generally 10) teeth in front of jaws on each side (less than 13mm in diameter).
General. The species most commonly involved in mass live strandings in Europe. Range of Long-finned Pilot Whale overlaps with that of the Short-finned Pilot Whale *G. macrorhynchus.* Seabirds, such as Great Shearwaters, often follow Long-finned Pilot Whales, either hunting squid or eating the whales' faeces.

HARBOUR PORPOISE *Phocoena phocoena* Pl. 32

Name. Common Porpoise. *Phocaena* = porpoise (Gk.) of *phoke* = seal.
Status. Vulnerable.

Recognition. Dorsal fin: low, triangular (concave) trailing edge. Blunt snout, small, spade-shaped teeth, small rotund body. Generally found in small groups in shallow water or river estuaries. Seldom swim or dive in formation, rather cutting hither and thither across each others' paths. Rises for a single breath between dives of up to 3 minutes. Often four short dives followed by one longer one.
Habitat. Coastal, bays, estuaries and large rivers, inhabiting cold water. found at depths between 3–100m, but mainly less than 50m.
Habits. Rests by floating on the surface, but appears not to sleep. Rarely breaches clear of water and does not bow-ride. Rests interspersed by foraging pattern which involves about 3–4 short dives (up to 10sec) followed by a maximum dive of 3 min. Activity pattern of Harbour Porpoise influenced by activities of herring. Herring follow vertical diurnal migrations of copepods; herring hunt by night and require optimal light conditions, so they adjust their depth for this reason, swimming deeper in daylight, but coming to the surface at dawn and dusk to feed which is when the porpoise hunt them. FOOD: Fish (eg herring, whiting, cod, mackerel, capelin), crustacea, cuttlefish, 50–80% of diet is Atlantic Herring. HOME RANGE: Unknown, some apparently resident in coastal waters

throughout year. Migratory. Onshore and offshore migrations in spring and autumn/winter respectively. Similar migrations into/out of the Baltic have now ceased. SOCIAL: Solitary or in small groups of 2–4. For feeding or migration, loose aggregations of up to several hundred animals may form.

Breeding. Mating in June–August, births in May–July. Sexual maturity 3–4 years for both sexes in North Atlantic; in North sea: males 5 years; females 6 years (males at *c.* 135cm; females at *c.* 145cm). Gestation *c.* 11 months. Litter size 1. Calving interval 1–2 years, depending perhaps on nutritional levels. Mammae 2. YOUNG: Sexual dimorphism present from birth, with female being slightly larger. May stay with mother after weaned until up to a month after the birth of the next young. Average length at birth 67–80cm, weight 5–8kg. Weaning age *c.* 8 months. Tight mother-calf bonds.

Lifespan. Max. recorded 23 years, 16–17 years probably more common. Accidental catches in fishing gear may be important cause of death.

Measurements. Head-body length: 1.3–1.9m. Condylo-basal length: 20–30cm. Weight: 35–90kg (mainly 54–65kg). Dental formula: 22–28 flattened spade-like teeth (partly hidden in gums) in each ramus of jaw.

General. Three major subpopulations exist but the species has a worldwide coastal distribution in northern temperate seas. Of the North Atlantic group, a sub population in the Baltic Sea appears to be in critical condition, and that in the Mediterranean (probably never very large) appears virtually extinct. European populations have shown marked declines, particularly in southernmost N. Sea and the English Channel since 1960. Probably has well-developed echo-locating ability. Short life span with 3–4 calves produced on average by each female renders species highly vulnerable to excessive additional mortality through exploitation. Vulnerable to entanglement in gill nets (eg for salmon) and to any other type of set-net or drift net as it pursues schooling fish. The modern synthetic monofilament twines are probably almost visually and acoustically invisible to these animals, especially in turbid coastal waters, and the struggling fish in the nets may attract Harbour Porpoises to them; fishing nets mainly catch young and or their mothers. Catches of Harbour Porpoises, which are probably heavy relative to their population size, occur in the North Sea subpopulations; this may be an important cause of mortality but changes in fish stocks (eg herring) may have been the major factor in their decline. Easiest to watch August–October (eg around Shetland Islands). Often seen in Straits or places where surface slicks indicate sites where current convergencies and upwellings concentrate food.

BELUGA *Delphinapterus leucas* Pl. 34

Name. White Whale. *delphis* = dolphin, *a* = without, *pteron* = wing (back fin); *leucos* = white (Gk.).

Recognition. Adults milky white (juveniles grey). Short, broad beak, often overhung by a distinct melon which develops with age (vignette

<*Beluga*. Sightings in Baerents Sea.

>*Narwal*. Arctic Ocean near Greenland and Baerents Sea.

shows shape of yearling melon). Short, rounded flippers. No dorsal fin. Free cervical vertebrae allow it to nod and turn its head as few other whales can. Transverse slit of blowhole located just in front of neck crease.
Habitat. Coastal, arctic waters, occasionally south inshore N. Sea. Sometimes mouths of rivers.

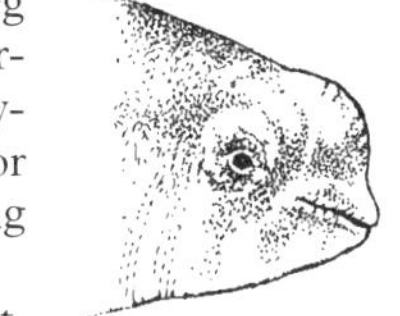

Habits. FOOD: Mainly squid, but also schooling fish (eg cod, herring and salmon), crustaceans and other invertebrates. HOME RANGE: Migrant. SOCIAL: Pairs or polygynous groups of 5–20 (some aggregations of 1000 or more). Sexes form separate schools outside breeding season.
Breeding. Mating in April–May, births in July–August. Sexual maturity: males 8–9 years (at *c.* 360cm); females 4–7 years (at *c.* 300cm). Gestation 14–15 months. Litter size 1. Calving interval 3 years. YOUNG: Weaning age 20–24 months.
Lifespan. Max. 40 years (average 20–25 years).
Measurements. Head-body length: 3–5m (Arctic: male *c.* 420cm; female *c.* 360cm) (birth: 150–160cm). Weight: 500–1500kg. Dental formula: 8–11 teeth in each upper and lower jaw.
General. Circumpolar, mainly in arctic, but extending to subarctic. Total population size at least 40,000–55,000 (mainly in Baffin Bay, Davis Strait, Barents, Kara and Lapteu Seas).

NARWHAL *Monodon monoceros* Pl. 34

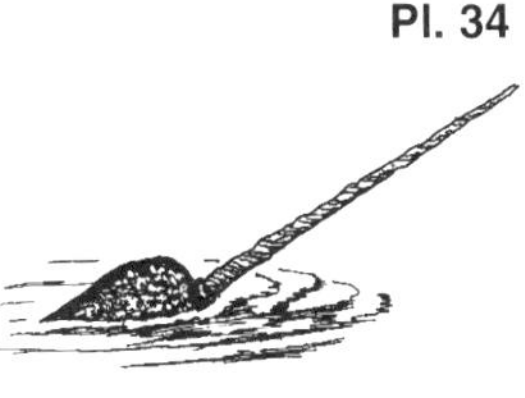

Name. *Monos* = single, *odous, odontos* = tooth; *keras* = horn (Gk.).
Recognition. Readily recognised by male's tusk. Swollen forehead. No dorsal fin. Trailing edges of flukes strongly convex in adults: mottled coloration on back of adults. Old animals nearly all white. Newborn young blotchy slate grey or bluish grey, juveniles completely bluish-black or black. As they mature, white patches appear around genital slit, anus and navel, spreading with age. Slight, insignificant beak. Crescent-shaped blowhole to left of centre on top of head. Noticeably upturned tips of flippers becoming more exaggerated with age. Sometimes 'stand' vertically in water with head (and tusk if male) exposed. Lunging as they surface or rolling on surface, may also expose tusk.

Habitat. Circumpolar distribution in Arctic littoral zone, mainly above 65°N. Generally not found far from loose pack ice. Inshore movement in summer dramatic and predictable, coinciding with break up of ice cover.

Habits. FOOD: Cuttlefish, fish (eg polar cod), crustacea, squid. HOME RANGE: Vagrants occasionally travel to North Sea. Annual migrations appear to be responsive to ice formation and drift. SOCIAL: Probably polygynous; schools of 6–20 (occasional aggregations of 500–1000). Males joust with tusk in breeding season (tusk's primary function probably aggressive), but these subdivided into smaller, more cohesive units: groups of juveniles, 'nursery' groups of females and calves, and aggregations of bulls seen, but mixed grouping not unusual. COMMUNICATION: Shrill whistle (mainly 0.3–10kHz) when surfacing. Female moans to young. Echolocating clicks mainly 0.5–5.0, 12–24kHz.

Breeding. Mating in April, births in July–August. Sexual maturity: male 11–13 years; female 5–8 years (length at sexual maturity: male 390cm; female 340cm). Gestation 14–15 months. Calving interval 3 years. YOUNG: Weaning age 20 months.

Lifespan. Max. 50 years. Trapped in fast-forming ice in autumn, may be a significant cause of death. Predators include Killer Whales, rarely Polar Bears, possibly Walruses, Greenland Sharks (weak or badly wounded whales only).

Measurements. Head-body length: 3.95–5.5m excluding tusk (male's tusks 1.5–3.0m long and can weigh up to 10kg) male 410cm; female 350cm (birth 150–170cm). Weight: 800–1600kg (maximum). Males larger than females. Newborn weigh *c.* 80kg. Dental formula: two teeth in upper jaw, hidden in gums of female; male's left tooth grows into spiral tusk (and, exceptionally in females too). Occasionally both teeth protrude, but then the right one is usually imperfectly developed.

General. Locally abundant in high Arctic, especially Western Hemisphere. Large concentrations also in Davis Strait, Baffin Bay and adjacent waters, Greenland sea. Several thousand in Soviet Arctic. Almost extinct in Asiatic waters due to whaling, but population estimated at 10,000 individuals in Canadian waters. Total population estimate 25–50,000. Females and calves more accessible in some areas since large males seem to remain further from shore. Sometimes seen in close company of Belugas.

PYGMY SPERM WHALE *Kogia breviceps*

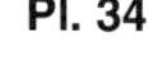

Name. Lesser Cachalot. *Brevis* = short; *ceps* = headed (Lat.). *Kogia* either Latinised form of English 'codger' (old fellow) or name of a Turk called Cogia Effendi who observed whales in the Mediterranean.

Recognition. Very similar to the Dwarf Sperm Whale, but somewhat larger, with more numerous teeth, a smaller, less erect dorsal fin, and a rounder head; also similar to the Sperm Whale, but distinguished by small size, slightly larger dorsal fin, relatively smaller head, only 1/6 length of body – that of Sperm Whale = 1/3. Body dark blue-grey (back, outer margin of flippers, upper surface of tail flukes), lightening to pale grey on flanks, and dull white belly (sometimes with pinkish tinge). Head shark-like with underslung lower jaw, no beak. 12–16 pairs of teeth (lower jaw only) – thin, incurved and sharp. Sexes similar in size and appearance. Snout – shape changes with age, becoming blunter and more rectangular. False gill on side of head. Skull resembles that of miniature Sperm Whale with marked asymmetry of bony nostrils and numerous teeth confined to lower jaw. Low and inconspicuous blow during slow sluggish roll. May bask on surface with head and back exposed, reportedly floats higher in water (more head and back exposed) than Dwarf Sperm Whale. Skin may have wrinkled appearance. When stranded, often mistaken for a shark. Dark bluish grey dorsally, fading to lighter grey laterally and dull white/pink on belly. Outer margin of flipper and upper fluke surface steel grey.

Habitat. Poorly known but apparently temperate and tropical latitudes, deep water.

Habits. FOOD: Predominantly cephalopods off continental slope, outer continental shelf, or oceanic, but also deep water fish and crabs. HOME RANGE: Non-migratory apparently in S. Africa, and stranding records for every month of the year in NE America. SOCIAL: Occur in small groups. Semi-mass strandings have occurred on coasts of New Zealand and Japan.

Breeding. Unknown, but calving probably largely between autumn and spring, but present data scattered throughout the year. Sexual maturity: female 2.6–2.8m; male 2.7–3m, but age unknown. Gestation *c.* 11 months (based on S. African coastal strandings). Litter size 1. Annual reproduction suggested on evidence of stranded females (with calves) simultaneously pregnant and lactating. Mammae 2. YOUNG: Newborn 120cm length. Weaning age unknown.

Lifespan. Assuming 1 dentine layer on teeth is an indication of 1 year of life, then some individuals at least 19 years.

Measurements. Head-body length: female 3.03m; male 3.07m, up to 3.5m (c. 1.2m at birth). Condylo-basal length: *c.* 470mm. Weight: 318–408kg.

General. Rare in Europe but second most commonly stranded cetacean in SE USA. Occasionally stranded on W. European coasts. Most information comes from beached animals. Taken opportunistically by hand harpoon off the coast of Southern Japan and in the Timor Sea from Indonesia. *K. breviceps* may be distinguished from *K. simus* (at all ages including late foetal stage) by position of blowhole: more than 10% of body length from snout in former, and generally less in latter.

< *Pygmy Sperm Whale.* Strandings reported from S Wales and SE Ireland, less certain sightings from NE England and NW Ireland.

> *Dwarf Sperm Whale.* Sightings in Atlantic near entrance to Mediterranean.

DWARF SPERM WHALE *Kogia simus*

Name. *Simus* = snub-nosed (Gk., Lat.).

Recognition. Similar to *K. breviceps*, but smaller with 7–12 pairs of short, slender teeth in lower jaw, and 1–3 pairs of rudimentary maxillary teeth (sometimes present), several irregular throat grooves (occasionally present), a larger, more erect dorsal fin positioned further forward on the back. Reported to have inconspicuous blow, and to surface slowly showing only the back when breathing.

Habitat. Widely distributed in temperate and tropical latitudes, perhaps centering along edge of continental shelf. Immatures probably live closer inshore than do adults.

Habits. FOOD: Primarily squid, although fish and crustaceans also eaten. Stomach contents indicate adults can dive to at least 300m depth. Shift in feeding habits from immatures to adults due to habitat differences. Cephalopod species typical of continental shelf comprise 45% of diet of calves (and adult females accompanying them), whereas oceanic cephalopods comprise 71% of adult diet. HOME RANGE: Seasonal migrations not documented. SOCIAL: Usually in groups of no more than 10 animals. 3 types of pod – females and calves, immatures, adults of both sexes without calves.

Breeding. Calving season prolonged. Occurrence of foetuses in stranded whales suggests matings in summer and births in early summer in the Southern Hemisphere. Sexual maturity at 2.1–2.2m length for both sexes. Gestation *c.* 9.5 months. YOUNG: Newborn *c.* 1m length.

Measurements. Head-body length: 2.1–2.7m. Condylo-basal length: *c.* 320mm. Weight: 136–272kg. Dental formula: 7–12 or 13 pairs of short, slender teeth in lower jaw, but may have additionally up to 3 pairs of small teeth in upper jaw.

General. Only recently recognised as distinct species, hence many records lumped together with Pygmy Sperm Whale.

SPERM WHALE *Physeter macrocephalus* (*P. catadon*) Pl. 38

Name. Catchalot. *Physeter* = whale (Gk. and Lat., literally 'blower'); *macro* = long (Gk.); *cephalus* = head (GK.); *kat* = below (Gk.); *odous, odontos* = tooth (Gk.).

Recognition. Largest species of toothed whale. Single blow hole at the

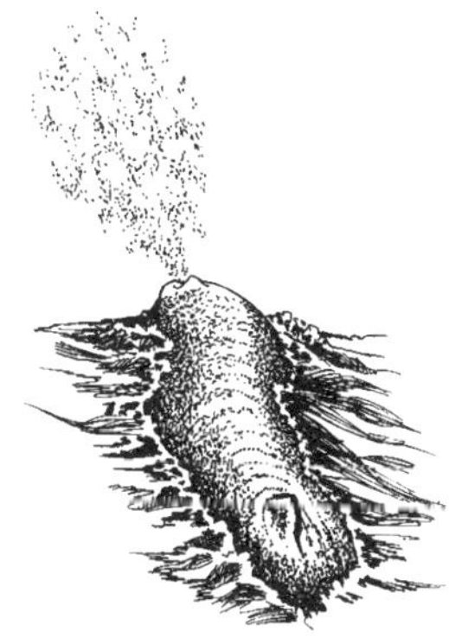

front and to the left of a huge (25–35% of total length) blunt-ended barrel-shaped head (other whales have blow holes situated well back on the head). Blow directed forward at approximately 45° and slightly to the left (see vignette). Dorsal fin is small with spinal ridge from fin to tail. Characteristic corrugations of skin give a shrivelled appearance. Typical behaviour while feeding is to make dives of 20–30 mins (max. 90 mins) followed by periods of 10–15 mins on the surface recovering. Sperm whales may dive to depths of over 2000m. They raise their flukes above the surface when initiating deep dives often revealing the ventral keel on the tail stock. During surface periods they typically blow 30–50 times. Blow intervals: female 10sec; large male 15sec. If they are moving they may submerge between blows. Surface swimming speeds are usually slow (less than 3 knots). At other times Sperm Whales spend longer periods on the surface, sometimes resting quietly and at other times interacting with other individuals in large pods. Mature males often have scars on their bodies, particularly on their heads (probably inflicted by other males during mating conflicts). Sperm Whales slough their skin continuously; there is no evidence that this is seasonal. Most readily found by listening with hydrophones for their distinctive loud clicks.

Habitat. Open ocean. Rarely found in waters less than 500m deep. In some areas found in association with the break in the continental shelf and with sea mountains. But also found in very deep, topographically featureless waters. Found throughout N. Atlantic and in Mediterranean.

Habits. Active both night and day. FOOD: A large range of items have been recovered from Sperm Whale stomachs, including stones and marine debris. Squid are their main food in most regions, with fish being of varying importance in different areas. HOME RANGE: Individuals may stay within a discrete area and return to it in successive years. Males disperse into colder waters as they become sexually mature – there is a tendency for larger males to be found nearer the pole. Females mainly remain in warmer waters (more than 20°C, between 40°N and 40°S, approx.). Males swim south to join female groups and breed, but not all mature males do this every year. Both male and female population may show seasonal N.–S. movements. SOCIAL: Females and young males are members of 'mixed schools' of 20–25 individuals which are fairly discrete and stable over time. They are typically seen in smaller subgroupings, 'pods' (up to 10 individuals). Older males also live in 'bachelor' groups which become smaller as males get older. Oldest males often found alone. Very large associations (100+ individuals) may occur (probably temporarily) during migrations and on feeding grounds. Groups of mature males, or singles, are believed to associate briefly with mixed schools to breed. Fierce fights reported among males. Recent evidence suggests that males do not guard harems for long periods as had previously been supposed. They may travel between several groups searching for receptive females. COMMUNICATION: Certain stereotyped patterns of clicks called 'codas' thought to be adapted for communication. There is potential for long range communication (certainly over tens of miles). Certain surface displays such as breaches and lobtails may have a function in communication both visually and acoustically. SENSES: No sense of 'smell' in water. Chemo-sensory cells at the base of tongue may be used to sense chemicals in the water (possibly including secretions from other sperm whales) as well as tasting food. Vision may be of limited importance especially away from the surface, where light is reduced or completely absent. The eyes and optic nerves are large in absolute terms, but small in relative terms, i.e. in proportion to the size of the body. Sense of hearing acute, probably used both in echolocation and communication as well as passive listening.

Breeding. Sexual maturity 7–19 years (male 18–19yrs, length 11.9m; female 7–12 yrs, length 8.8m). Gestation 16–17 months. Litter size 1. Calving interval 5.2–6.5yrs. YOUNG: Males associate only briefly with mixed schools and are not present when offspring are born. Within mixed schools a number of individuals, including mature males, may help in calf rearing. On occasions, helpers seem to look after calves while their mothers are on deep dives. There are some indications that communal suckling may also occur.

Lifespan. Max. *c.* 65–70 years. Annual adult mortality estimated at 6%.

Measurements. Head-body length (max.): male 18.5m; female 20m). Weight (max.): male 70 tonnes, female 20 tonnes. Dental formula: 18–25 conical teeth in each half of lower jaw. May grow as large as 27cm and fit into sockets in the upper jaw when the mouth is closed. Rudimentary teeth are found at the apex of these depressions.

General. Population greatly reduced (to about half) during commercial whaling: catches from Azores, Madeira, Iceland, Spain, Faroes, Greenland, Norway and Canada during the modern era. Largely female catch around Madeira seemed to lead to local collapse of population by 1970. Mid-1970s estimated (very approximately) N. Atlantic population of 38,000. Quota for whales in N. Atlantic set by International Whaling Commission in 1981: 130 males. Valuable products include blubber, meat for fertilizer, ambergris, and (most valuable) spermaceti wax from the head. Male's head is much bigger than female's and contains a larger spermaceti organ with a larger reservoir of oil. Nowadays also teeth which are carved or engraved and sold to tourists.

< *Sperm Whale.* Widespread in Atlantic waters, excluding North Sea and English Channel.

> *Gervais's Beaked Whale.* Sightings in English Channel.

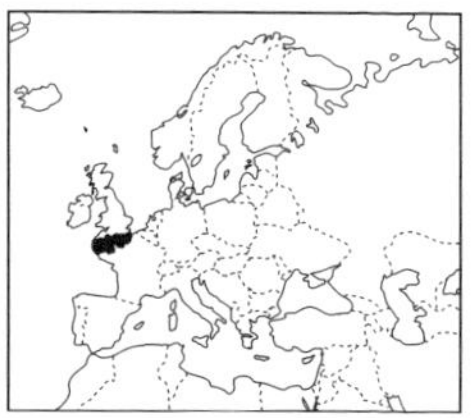

GERVAIS' BEAKED WHALE *Mesoplodon europaeus* **Pl. 36**

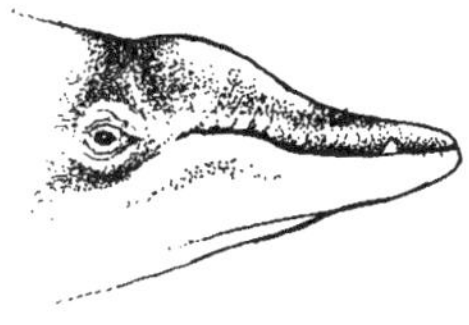

Name. Antillean Beaked Whale. *Mesos* = middle (Gk.); *hoplon* = weapon (Gk.); *odous, odontos* = tooth (Gk.); *europaeus* = european (Lat.).

Recognition. The largest, and seemingly most laterally compressed, species of *Mesoplodon*. Flippers positioned particularly low down on the side of the body. Small dorsal fin, varying in shape from shark-like to blunt and triangular. No notch in the tail fluke. Uniformly dark grey marine blue or black coloration with a slight lightening on the undersides. In some animals, there is an irregular white blotch around the anal region. Scarring common (presumably from fighting). No recorded sighting in the wild. Might be confused with Cuvier's Beaked Whale, but has a sharper, narrower beak and a shorter fin.

Lifespan. Max. recorded *c.* 27 years.

Habitat. Apparently confined to deep waters of warm temperate and subtropical Atlantic (mainly north of Equator). Range from New York and English Channel south to Gulf of Mexico, Trinidad and Caribbean Sea across to Ascension Island and West Africa. May be associated with Gulf Stream.

Habits. FOOD: Squid.

Measurements. Head-body length: 4.5–5m, newborn *c.* 210cm. Weight: *c.* 5.6 tonnes. Dental formula: 1 pair of teeth in lower jaw, protruding about ⅔ from tip of beak in adult males, fitting into grooves in skin of outer upper jaw.

General. Only one specimen has ever been found in Europe; more than a dozen have been found stranded in N. America and the Caribbean.

BLAINVILLE'S BEAKED WHALE *Mesoplodon densirostris* **Pl. 36**

Name. Dense-Beaked Whale

Recognition. General body form and shape are typical of other *Mesoplodon* species. In male, entire middle third of each mandible is dominated by a huge tooth, positioned 30cm from tip of jaw, that rises up above top of beak and is enveloped and supported by a massive bony protuberance. This gives mouth a high arching contour which sweeps up over rostrum, only dipping down again in front of eye, making it virtually impossible for whale to see forwards at all. Short flippers, inconspicuous throat grooves. Dorsal fin relatively large and curved backwards to a point. No notch in tail flukes, which sometimes extend outwards and backwards in centre. Coloration black or dark grey, marked with paler (grey-white or pink) blotches on flanks, with slightly lighter grey throat and chest, and often white anal region. Scarring, especially on head, particularly in males (presumably from fighting). Sometimes seen moving together in small groups (up to 10 individuals) at the surface.

< *Blainville's Beaked Whale.* Sightings off Portugese coast and W Mediterranean.

> *True's Beaked Whale.* Sightings in coastal waters W of British Isles and W France.

Surface head- first so it is clear of the water before blowing. In this position, may be possible to make out high contour of lower jaw of any large males in group. Smaller animals also show a curve at back of mouth, though less distinct. Blow indistinct, but on a calm day can be seen to shoot forwards at a sharp angle. These Beaked Whales spend several minutes at or near the surface, breathing at 15- or 20-second intervals, before diving together. Not seen to raise flukes before diving. Fin prominent and clearly visible at the end of a long curve of back with every breath.

Habitat. Distribution worldwide: the only ziphiid apart from Cuvier's Beaked Whale which is found on the equator.

Habits. FOOD: Analysis of stomach contents of one stranded animal contained only squid. SOCIAL: Appears to live in small family units of 3–6 animals. COMMUNICATION: Some stranded animals reported to produce 'roars', 'lowing sounds' and 'sobbing groans'. One stranded male, of which a recording was made, produced a pulsed sound as well as audible chirps and whistles.

Measurements. Head-body length: 4.7–5.2m. Weight: *c.* 3.6 tonnes. Dental formula: one pair of large teeth in middle of lower jaw in adult males.

TRUE'S BEAKED WHALE *Mesoplodon mirus* — Pl. 36

Name. *Mirus* = wonderful (Lat.).

Recognition. Coloration dark grey to grey-black on back, lighter slate grey on sides, grey on belly, with light spots usually present, particularly in anal and genital regions. Triangular or slightly sickle-shaped dorsal fin ⅔ along back. Un-notched tail flukes have trailing edge slightly concave. Blow barely visible, not seen to raise tail flukes before diving.

Habitat. North Atlantic (stranded on Irish Coast and once on Outer Hebrides). Sightings off South Africa indicate range wider than previously thought. Deep waters of temperate Atlantic extending to SW Indian Ocean, with records from N. America, NW Europe and S. Africa.

Habits. FOOD: Presumably deep water squid, but no data.

Breeding. YOUNG: 233cm long reported minimum.

Measurements. Head-body length: 4.9–5.5m. Weight: 3.2 tonnes. Dental formula: single pair of teeth of male laterally compressed and situated at tip of lower jaw.

< *Sowerby's Beaked Whale.* Atlantic Ocean and North Sea south of 60°N including sightings in the Mediterranean off western coast of Italy.

> *Gray's Beaked Whale.* Sightings in the English Channel.

SOWERBY'S BEAKED WHALE *Mesoplodon bidens* Pl. 36

Name. North Sea Beaked Whale. *Bidens* = having 2 teeth (Lat.).

Recognition. Rarely seen at sea, where it would be difficult to distinguish from other members of the genus *Mesoplodon.* May have prominent bulge on forehead, with moderately long slender beak. Position of teeth a diagnostic feature. Teeth project backwards then slightly forwards. Dark grey coloration may grade into sandy colour on head, paler grey on belly and light spots scattered over back and flanks. Relatively smaller flippers, often tucked within 'flipper pockets'. Dorsal fin triangular or sickle-shaped and positioned almost ⅔ along back. Un-notched tail flukes have slightly concave trailing edges. Fast swimmer, often swimming at surface.

Habitat. Distribution appears centred on northern North Sea. Clearly mainly pelagic.

Habits. FOOD: Takes squid, but otherwise poorly known. Oral musculature indicates a sucking action when feeding. SOCIAL: All records are of individuals or 'pairs'. COMMUNICATION: Echolocating sound pulses have been recorded from a young animal kept in a dolphinarium for a few hours. Stranded animals reported as lowing like a cow.

Breeding. Mating and births both thought to occur in late winter and spring. YOUNG: Newborn *c.* 2.4m long. Young have a lighter belly and fewer spots.

Measurements. Head-body length: *c.* 5m. Weight: *c.* 3.4 tonnes. Dental formula: 1 pair of triangular teeth at midpoint of lower jaw, but exposed above gum only in males, in which the teeth protrude outside of the mouth.

General. Thought to be relatively common, being the most commonly stranded *Mesoplodon* species. Rarely seen but many strandings, mainly on the N. Isles of Scotland and along coast of eastern Britain. Generally single animals, during all months, but particularly July–September. Nowhere observed frequently enough to be hunted.

GRAY'S BEAKED WHALE *Mesoplodon grayi* Pl. 34

Name. Scamperdown Whale.

Recognition. Coloration dark brownish- grey to black on back, grey (often mottled) on flanks, and light grey to white on belly; long, very

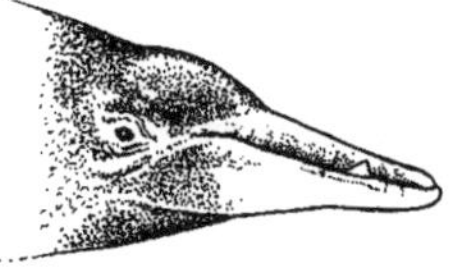

slender beak, often white or with white flecks (may extend to throat), may frequently be projected above water when surfacing. Noticeable throat grooves; conspicuous white markinngs around navel, genital region and anus; one pair of large teeth (about 10cm high and 8cm wide) about 20cm from the tip of the jaw. These have a serrated edge with one or more sharp points and lean slightly forwards. Also, up to 26 vestigial teeth, usually on the upper jaw behind the point of origin of the large teeth, breaking through the gum far enough to be functional. Emerge fast from the water at an angle of 30° until all but the tail is exposed, then fall back with a splash.

Habitat. Poorly known (*c.* 50–100 records) but apparently circumpolar in Southern Hemisphere. A single stranding on the Dutch coast suggests that it also exists in the Northern Hemisphere.

Habits. SOCIAL: Appears to be highly gregarious. Mass strandings and recent sightings suggest that it commonly occurs in groups of 6 or more.

Measurements. Head-body length: *c.* 5.5–6.0m. Weight: *c.* 4.8 tonnes. Dental formula: 1 pair of moderate-sized triangular-shaped teeth (upright, and erupting only in adult males) in lower jaw, towards back of mouth, and usually 17–22 (max. 26) small teeth on each side of upper jaw.

CUVIER'S BEAKED WHALE *Ziphius cavirostris* Pl. 35

Name. Goosebeaked Whale

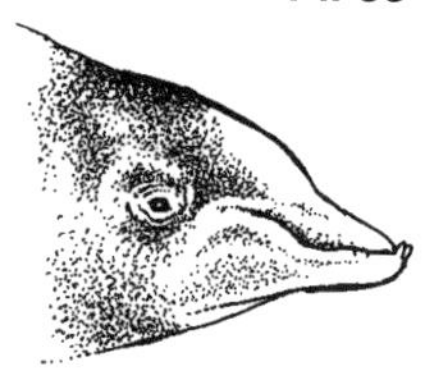

Recognition. Long stout body with small slightly concave head (likened to goose's beak). Beak ill-defined, particularly in older animals. Coloration grey or blue-grey, paler grey or white head (particularly in older males) with linear pale scars often on back and sides; pale blotches on sides and belly. Dorsal fin variable – small and triangular to relatively tall (up to 38cm) and sickle-shaped; about ⅔ along back. Somewhat concave tail flukes lacking distinct median notch. V-shaped pair of throat grooves. Low, inconspicuous blow directed forwards and obliquely to left.

Habitat. Mainly deep-water species.

Habits. FOOD: Mainly squid and deep-sea fish. HOME RANGE: Strandings occur in British waters every month, but particularly between January–

< *Cuvier's Beaked Whale.* Atlantic and North Sea south of about 60°N, and throughout Mediterranean waters.

> *Northern Botttlenosed Whale.* Widespread in Arctic Ocean and N Atlantic S to Iberian Peninsula.

March and June–July. SOCIAL: Rarely seen, but most sightings of single individuals or small groups (3–10, occasionally up to 25).
Lifespan. Max. at least 36 years.
Measurements. Head-body length: male *c.* 6.7cm; female *c.* 7.0cm; newborn are *c.* 270cm long. Weight: *c.* 5.6 tonnes (male); *c.* 6.5 tonnes (female). Dental formula: 1 tooth at tip of each ramus of lower jaw, usually erupting only in the males.
General. Small numbers taken in Japanese fishery and occasionally from Bequia, St Vincent (Lesser Antilles). Wary of vessels.

NORTHERN BOTTLE-NOSED WHALE *Hyperoodon ampullatus* Pl. 35

Name. *Hyperon* = pestle (Gk.); *odons, odontos* = tooth (Gk.); *ampulla* = bottle (Lat.).
Status. Vulnerable.
Recognition. Largest of British beaked whales, distinguished by large bulbous forehead and short dolphin-like beak (15–17.5cm long). Minute vestigial teeth in upper and lower jaws, otherwise older males (15–17 years) have single pair of pear-shaped teeth at tip of lower jaw – rarely appear through gum in females. Pair of V-shaped throat grooves. Coloration very variable: chocolate brown to greenish-brown above, often lighter on flanks and belly with irregular blotches. Lightens to buff/cream all over with age. Single crescent-shaped blowhole. Dorsal fin moderate height (30cm), often strongly hooked and ⅓ along back from tail. Broad un-notched tail flukes with deeply concave trailing edge. Single, low (2m) bushy blow, slightly forward-pointing.
Habitat. Temperate and arctic North Atlantic, particularly in deep waters.
Habits. Migrate to Arctic in spring and summer, and back to warmer areas in autumn and winter. FOOD: Mainly squid; also herring, and other pelagic fish. HOME RANGE: N. Atlantic deeper waters, usually 1000m+. Thought to migrate off the edge of the continental shelf, Iceland, Jan Mayen and Greenland from Cape Verde to Spitzbergen, passing northward west of Britain and Ireland in spring and returning south again in winter. Some onshore movement during summer (July–September). SOCIAL: Solitary or in small groups of up to 10, occasionally 30–40. Groups may separate by sex and age during migrations: females with calves may form basic social unit. Polygynous males thought to associate with these groups (but description of social system needs verification). Reputed to take care of injured companions. COMMUNICATION: High frequency echolocation clicks, but otherwise poorly known. Unlike most beaked whales, the Northern Bottle-nosed breaches and lobtails.
Breeding. Births in April–May. Sexual maturity: males 7–9 years (length

730–760cm); females *c.* 11 years (length 670–700cm). Gestation *c.* 12 months. One calf every second year. Mammae 2. YOUNG: Uniform chocolate brown all over, *c.* 300–360cm at birth. Born April–May. Weaning age *c.* 12 months.

Lifespan. Max. recorded *c.* 37 years.

Measurements. Head-body length: male 9.0–9.5m (average 9.2m); female 7.0–8.5m (average 7.5m). Weight: on average 7500kg (male), 5800kg (female). Dental formula: 1 tooth on each ramus of the jaw in older males.

General. Some indication of a decline in NE Atlantic in recent years, with number of strandings declining. Previously hunted from Norwegian shore stations (1882 to late 1920s and 1938–1972) for oil and animal food. Strong tendency to approach vessels.

BLUE WHALE *Balaenoptera musculus* Pl. 37

Name. *Musculus* = little mouse (Lat.); *Balaena* = whale (Lat.); from *phallaena* = whale (Gk.); *pteron* = wing/fin (Gk.).

Status. Endangered.

Recognition. Body huge; very narrow when viewed from side. Pectoral fins long (*c.* 15% of total body length), pointed and white underneath. Dorsal fin small. Baleen plates short, broad and all black. 70–120 throat grooves (average 90). Blow is narrow and high (up to 9 m). Blow shaped like inverse-cone for 4–5 secs. Stock and tail flukes lifted slightly before dive, but often no tail flukes actually visible. Bluish-grey coloration over most of the body, mottled with grey or greyish-white. Does not unfurl flukes on last dive before prolonged submersion.

Habitat. Pelagic, in all oceans.

Habits. FOOD: Almost exclusively planktonic crustacea. The throat of Blue Whales, and all other rorquals, distends massively while feeding (see vignette). HOME RANGE: Migratory. Summer: polar and cold temperate regions to feed. Winter: subtropics to breed. SOCIAL: Nowadays usually seen singly or in groups of 2–3, but formerly larger herds. COMMUNICATION: Vocal high and low pitched sounds. 'Moans' of 20–32 Hz; 12–231 kHz clicks (possibly echolocation).

Breeding. Southern Hemisphere mating in June–July, births in May. Northern Hemisphere *c.* six months out of synchrony, hence mating in

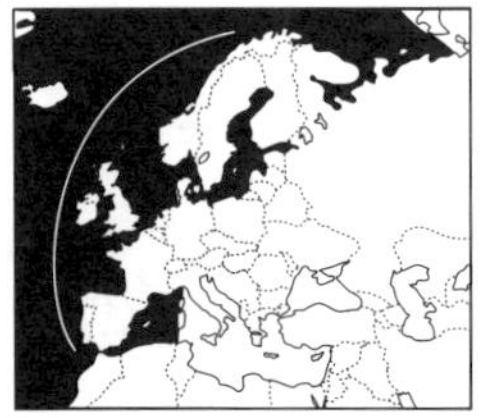

< *Blue Whale.* Previously widespread in all W waters, including W Mediterranean, recently not seen E of arc.

> *Fin Whale.* Throughout Arctic and Atlantic waters.

December–January, births in November. Sexual maturity *c.* 5 years (length at sexual maturity: male 22.6m, female 24m). Age at sexual maturity has declined with over-exploitation. Gestation 11–12 months. Litter size 1. Calving interval 2–3 years. Mammae 2. YOUNG: Weaning age 6–7 months (in Antarctic and South African waters).

Lifespan. Max. *c.* 80 years. Preyed upon by Killer Whales.

Measurements. Head-body length: 22–30.5m (eg males 25m; females 26.5m) (birth: 7m). Weight: at 27m = 120 tonnes (max. 150 tonnes). 260–400 baleen plates in each ramus of the upper jaw; deep black; 80–90cm long.

General. Numbers crashed due to whaling. No hunting allowed since 1967. Fast swimmer (30 knots). The existence of a pygmy race (*Balaenoptera musculus brevicaudata*) has been reported on the basis of various anatomical features, but requires further study. *Ultrasaurus*, weighing *c.* 150 tonnes and *c.* 35m in length, is the only vertebrate known to have been larger than the Blue Whale. Recently not seen east of arc marked on distribution map around European coastal waters

FIN WHALE *Balaenoptera physalus* Pl. 37

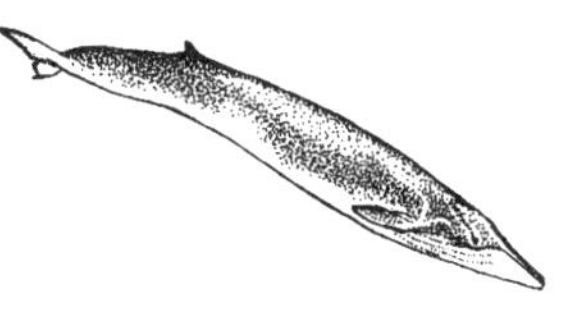

Name. Razorback. *Physis* = nature (Gk.); *alos* = sea (Gk.).

Status. Vulnerable.

Recognition. Dorsal fin *c.* 10% total body length. Dorsal fin small, but relatively longer than Blue Whale's. Narrower head (more V-shaped) than Blue Whale. 70–110 throat grooves (average 85). Lower jaw: right side is white, left side is dark and lacks mottling of Blue Whale. Blow shaped like inverted cone, for 2–4 secs up to 6m. Usually dives without showing tail flukes. Ridged back between dorsal fin and flukes, hence 'Razorback'. Occasionally breaches, unlike Blue and Sei Whales, falling back with resounding splash (unlike Minke Whale which enters water smoothely and head first, whereas Humpback and Right Whales breach-crash too).

Habitat. Pelagic – throughout oceans, up to pack ice, but excluding tropical waters.

Habits. FOOD: Krill (copepods and enphausids) and small fish such as capelin. HOME RANGE: North Atlantic: 20°N to almost 80°N. Some migrate. Winter: north to polar seas. Summer: south to breed in subtropical waters. However, at least part of the population does not migrate to subtropical waters in summer. This is the least migratory of the balaenopterids. SOCIAL: Usually single or in pairs. Rarely in larger loosely aggregated groups up to 30 individuals. COMMUNICATION: Low frequency 20 Hz 'moans', high frequency 'clicks' of 16–289 kHz.

Breeding. Mating in December–January, births in November–December. Sexual maturity *c.* 4–6 years (eg N. Atlantic length at sexual maturity: males 16.8–17.6m; females: 17.7–19.1 m). Before over-exploitation sexual maturity was 5–12 years. Gestation 11–12 months. Litter size 1. Calving interval 2–3 years. Mammae 2. YOUNG: Weaning age *c.* 6–7 months. Parental care by female only, ends at weaning.

Lifespan. Max. *c.* 90 years. May be preyed upon by Killer Whales.

Measurements. Head-body length: 18–23m maximum in North Atlantic (eg N. Atlantic: male 18.5m, female 20m) (birth: 6.4 m). Weight: at 21m *c.* 53 tonnes (max. 80 tonnes). Baleen plates yellow-creamy at front on right side of body, otherwise banded yellow and grey; fringe yellow and white. Up to 80cm long.

General. The Fin Whale's asymmetrical colouring is a puzzle. Perhaps it functions as 'counter-shading' camouflage against prey, such as fish, with good vision. One suggestion is that Fin Whales advance fast on fish and roll on to their right sides with mouths open and throat distended; this manoeuvre would keep the dark side upward and the pale side downward. In contrast, krill-eating rorquals, such as Sei and Blue Whales, tend to be spotted like murky water. Regular sightings from Faroe Islands and SW Norway, with occasional sightings from N. Scotland, Outer Hebrides, Shetlands and the west coast of Ireland.

SEI WHALE *Balaenoptera borealis* Pl. 37

Name. *Borealis* = belongs to North, (Lat.).

Recognition. Similar to Fin Whale, but smaller. Beak sharply pointed and slightly convex in profile and from front view. Pectoral fin small (*c.* 9% total body length). Dorsal fin relatively large, about 5% body length (i.e. much larger than fin and blue whales and positioned less than two-thirds from whale's front). 32–80 (average 50) ventral grooves ending well short of umbilicus. Tip of snout slightly turned downward in lateral profile (cf Fin Whale) with single rostral ridge. Baleen fringe is very fine. Blows up to 3m for 1–2 secs (shows less body and head when blows than Fin Whale). Dorsal fin clearly visible when inhales 1–3 breaths at surface followed by dive of 5–10m. Mainly feeds by skimming surface, so head surfaces at shallow angle (see vignette). Fin is relatively longer, and more deeply curved, than that of Fin Whale.

Habitat. Pelagic, in all oceans, but particularly subtropical to polar seas.

Habits. FOOD: Krill and small crustaceans, especially the copepod *Calanus finmarchicus* in N. Atlantic. Feeds in more northern waters and returns to the southern North Atlantic for mating and calving. Migration route includes West coast of UK, but does not include shelf edge region on the eastern side. SOCIAL: Seen in small groups, pairs and solitarily. COMMUNICATION: 3 kHz pulses have been recorded, but function unknown.

Breeding. Mating in November–February, births in November–Decem-

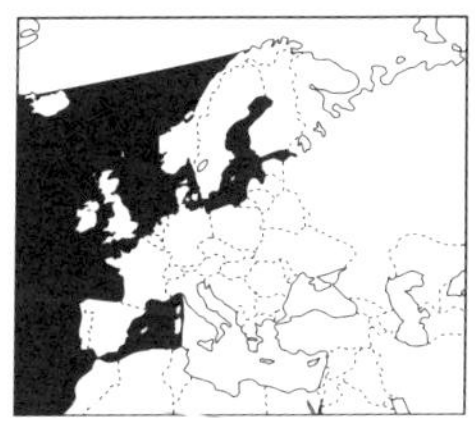

<*Sei Whale.* Widespread in Atlantic Ocean south of Iceland.

>*Minke.* Very widespread in Arctic and Atlantic Oceans and in W Mediterranean.

ber. Sexual maturity: in N. Atlantic, average length at sexual maturity for male 12.7–12.8m (at 7 years); female 13.1–13.4m (at 6–8 years). Gestation 10–11 months (N. Atlantic); litter size 1. Calving interval 2–3 years. Mammae 2. YOUNG: Weaning age *c.* 6–7 months.

Lifespan. Max. *c.* 65 years. Adult mortality estimated at 6–7% annually.

Measurements. Head-body length: 12–16m (eg average maximum lengths in N. Atlantic: male 14.0m; female 14.5m) (Birth: *c.* 4.5m). Weight: *c.* 20 tonnes. *c.* 330 baleen plates in each ramus of upper jaw, maximum length *c.* 77cm.

General. Can swim fast (30 knots), although only in short bursts and is generally slower. Often swims and feeds near to the surface. Difficult to establish whether the Icelandic Sei whaling has had a major effect, since the status of the North Atlantic Sei is far from clear. Hunted relatively recently from Iceland and Canada, and at present a few are taken each year as part of an Icelandic research programme.

MINKE WHALE *Balaenoptera acutorostrata* Pl. 35

Name. Lesser Rorqual, Piked Whale. *Acutus* = sharp (Lat.); *rostrum* = beak, snout (Lat.).

Status. North east Atlantic stock classified by IWC as 'Protected Stock', meaning it is believed to be below 54% of initial (i.e. pre-exploitation) stock level. No agreement in IWC as to classification of the central stock.

Recognition. Pointed snout, with top of head flattened with mid rostral ridge (as for other rorquals). Pectoral fins have diagonal white band above, white below, 10–12% of total length. 50–70 ventral throat grooves. Blow: small (2m) and hazy, often not visible. Blow usually seen at same time as relatively tall sickle-shaped dorsal fin. Typical breathing sequence: 5–8 blows at intervals less than 1 minute. Usually 3–8 shallow dives before each deep dive of 3–5 mins. Re-enters water cleanly after breaching. Tip of snout often breaks water at start of surfacing. Dorsal fin fairly prominent and set farther back on body than that of the Pilot Whale. Pale grey patches on sides visible in favourable lighting. Does not lift flukes when sounding. Smallest of the baleen whales in the Atlantic. Most reliable fieldsign is white flipper band and yellowish baleen plates. Also the sharply V-shaped pointed rostrum together with small size (cf other baleen whales which possess a dorsal fin). Most likely to be confused with Northern Bottle-nosed Whale. Can be confused with Sei and perhaps Fin Whales, but rostrum much more pointed than that of Sei, and lacks white right lower lip of the Fin Whale.

Habitat. Widely distributed in temperate waters, especially around the lower temperatures of the continental shelf; more coastal than other rorquals.

Habits. FOOD: Various small fish (eg Capelin, Sand Eel, Herring, Cod), krill, small squid. HOME RANGE: Individuals may return year after year to same location for feeding in summer. Migratory: winter northern feeding grounds; summer southern breeding grounds (reversed in Southern Hemisphere). Little evidence of substantial latitudinal migrations (as with larger rorquals). Winter range poorly known: no winter aggregations observed (cf Humpback, Right and Grey Whales). SOCIAL: Solitary, or in families of 2–3 (occasional larger aggregation of unknown composition) when feeding. COMMUNICATION: Vocalisations include grunts (80–140Hz frequency), trains of thumps (mainly 100–200Hz frequency) and very short ratchet-like pulses at 850Hz frequency, and high frequency (mainly 4–7.5 kHz) clicks. Appears to vocalise less often than other baleen whales.

Breeding. Mating in January–May, births in December–January. Sexual maturity *c.* 7 yrs (average length at sex maturity in N. Atlantic : males 6.9m; female 7.3–7.45m. Gestation *c.* 10 months (female often simultaneously lactating and pregnant). Litter size 1. Calving interval 1–2 years. Mammae 2. YOUNG: Weaning age 6 months – the earliest known weaning for baleen whales. Parental care unknown.

Lifespan. Max. *c.* 40–50 years. Adult mortality 9–10% annually. Many reports of attacks by Killer Whales. Also strandings and massive infections of parasitic nematodes. Frequently trapped in ice around Greenland, and probably other N. Atlantic areas.

Measurements. Head-body length: 8–10.5m (N. Atlantic: males 8m, females 8.5m) (birth *c.* 2.6 m). Weight: *c.* 10 tonnes. 260–325 white or yellowish baleen plates in each ramus of upper jaw, up to *c.* 30cm long.

General. Occasionally stranded but never en masse. Hunted around Iceland and Norway: 1981 quota 2554, gradually reduced to 877 for Norway and Iceland in 1985, when moratorium for commercial whaling came into effect. Norway now takes small numbers for 'scientific' reasons and 12 animals annually from central stock by Inuits in E. Greenland. This is the species most likely to be observed from land, particularly headlands along coasts of Iceland, W. Norway and the Northern Isles of Scotland, W. Scotland and W. Ireland.

HUMPBACK WHALE *Megaptera novaeangliae* Pl. 37

Name. *Mega* = large (Gk.), *pteron* = wing/fin (Gk.); *novus* = new (Lat.), *anglia* = England (Lat.).

Status. Vulnerable.

Recognition. Pectoral fins very long (up to 33% total length) and predominantly white and lumpy (as is head). Underside of tail often white.

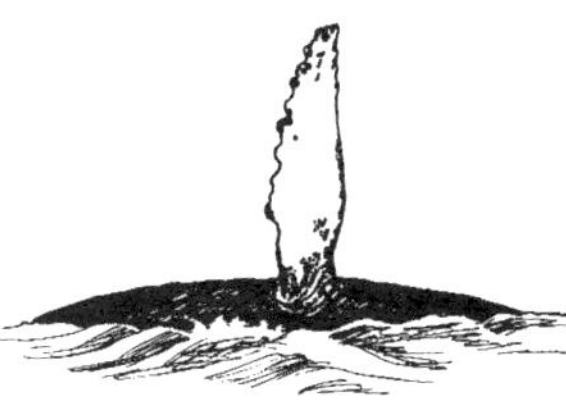

14–35 throat grooves. Strongly arched, broad back when diving. Dorsal fin usually small and indistinct. Tail often comes clear of water when dives. Frequently leaps clear of water and individuals can be recognised for many years from photographs of the black and white pattern of the underside of the tail in combination with dorsal fin shape. Heavily infested with barnacles and whale lice, though most will appear parasite-free when viewed from boats. Blow up to 3m, spreading at top, for 2–3 secs, squatter and bushier than that of Fin Whale. Great variety of acrobatics: breaching (throwing most of body out of water, then crashing down), lobtailing (smacking water with tail), spyhopping (poking head out of water), or waving long white flippers (see vignette). At close range, any one of the following may be seen: dark grey body, head and snout bearing knoblike swellings, each of which houses a stiff hair (may help to detect prey or water movements). Dorsal fin, seemingly perched atop a large hump, usually smaller than Fin Whale's, more variably shaped and often scarred or notched. Series of small bumps (the tops of the caudal vertebrae), which appear on back, aft of fin, in some individuals.

Habitat. Coastal, or less often pelagic in all oceans. Have distinct habitat requirements, thus in some areas very abundant (reaching 1.0 per km^2) and absent in adjacent areas. Preferred summer feeding grounds may shift from week to week or year to year, depending where prey present. Some relatively discrete breeding areas in tropical waters.

Habits. FOOD: Planktonic crustaceans, small schooling fish (eg lancets, capelin, herring), schooling squid, krill eaten when abundant. Wide variety of prey. May return to same feeding site year after year. Feeding methods depend on prey type, depth, abundance of prey, and local conditions – different methods used to concentrate prey before single mouthful of prey and water is gulped. Fish herded or stunned by lobtailing, flipper slapping at surface or underwater displays. Several whales may work a school of fish together. Individuals have been seen to swim in an upwardly spiralling circle beneath a group of krill, releasing a stream of bubbles as they go. Bubbles 'corral' crustaceans as the whale swims upwards with its mouth open. May lunge partway out of water with jaws open when food (particularly shoaling fish) is abundant. HOME RANGE: Migratory routes, directly between polar and temperate seas in summer and autumn, eg Norway, Iceland, Greenland, Newfoundland and Labrador (and in the Gulf of Maine in summer) where they feed, and subtropical and tropical coasts in winter and spring, where they mate and give birth (i.e. on shallow banks in West Indies, and possibly Cape Verde islands in winter). These routes are predictable and well known to whalers. SOCIAL: Few long-term bonds except mother and calf under one year of age. Larger temporary aggregations (up to 20 animals) when feeding or mating – during breeding up to 6 males cluster around female

< *Humpback Whale.* Widespread in Atlantic waters, excluding North Sea and English Channel.

> *Bowhead Whale.* NW Arctic Ocean.

or male-female pair and jostle for access to female; this competition may produce some of the scars on the flanks, dorsal fins or backs. Calves may learn particular migration routes when follow mothers north for the first time. COMMUNICATION: Great variety of calls. Males having long (6–35 mins.), complex stereotyped 'songs'. The most complex non-human vocalisation yet known. Heard mainly during winter breeding season, sometimes at over 30km, when songs repeated throughout day and night.

Breeding. Mating in February (NW Pacific and Caribbean); births in April–February (NW Pacific and Caribbean). Sexual maturity 4–5 years (length at sexual maturity: males 11.5m; females 12m). Age at first breeding, however, not yet known. Gestation 10–12 months. Litter size 1. Calving interval 2–3 years normally, although interval can be as short as 1 year. Mammae 2. YOUNG: Calves grow at rate of 45cm per month. Weaning age *c.* 6–12 months, when *c.* 7.6–8.5m long. Mothers do not abandon calves even in extreme danger.

Lifespan. Max. 60 years. Data from known living animals suggest natural mortality rates of 3–7% per year. Humpbacks were a species severely affected by historical whaling (Right Whales, Bowheads and Sperm Whales were even more affected), nowadays most serious threat is commercial fishing and pollution (depletes prey stock or alters prey behaviour – either way, less available to whales) and entanglement in fishing gear.

Measurements. Head-body length: 11–18m (eg male 13.5 m; female 15m) (Birth: *c.* 4.2 m). Weight: at 12.4m = 31 tonnes. 270–400 grey-black baleen plates per ramus of upper jaw, up to 64cm long.

General. Both stocks (Western North Atlantic and Eastern North Pacific) classified as Threatened. Evidence suggests that the NW Atlantic population has recovered a substantial proportion of its pre-whaling numbers, but latter still vulnerable and should retain original threatened status. Total world population probably 5,000–10,000 animals in 10 stocks. NE Pacific and NW Atlantic are largest populations. Since protection, there has probably been a very slow increase in numbers in the east of the N. Atlantic. The International Whaling Commission has not approved quotas since 1955 in the N. Atlantic, and from 1965 in the N. Pacific. Boat and aeroplane traffic and over-enthusiastic whale watchers may affect these whales adversely, eg around Hawaiian islands. In Newfoundland frequently become entangled in cod fishing nets, eg when fish stocks fall and they move inshore to feed. Several observations of attack by

Killer Whales (which may also attack Fin Whale). Slow swimmer (1–3.5 knots, max *c.* 15 knots). Flippers may be used to manoeuvre, herd fish, guide calves or slap the water for signalling position, aggression or to stun fish. Waving flippers in air may also help cool the whale, which may be necessary in the warm tropical/subtropical waters of its breeding ground.

BOWHEAD WHALE *Balaena mysticetus* **Pl. 38**

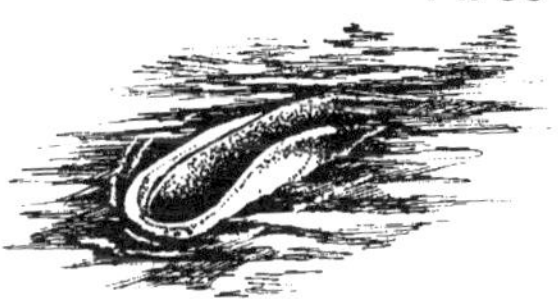

Name. Greenland Right Whale. *Mystax, mystakos* = moustache (Gk.).
Status. Vulnerable.
Recognition. Pectoral fins large and broad, tail flukes very broad. Head 33–40% total length, white on chin (see vignette). No dorsal fin or ridge. Blow holes widely separated so V-shaped blow, directed forwards and sideways 3–5m. Tail often in view when dives. Slow swimmer (6km per hr). Longest baleen of any whale (more than 2.8 m).
Habitat. Pelagic, in Arctic ice-drift regions, generally in vicinity of margin of sea-ice.
Habits. FOOD: Crustacea (copepods, euphausiids, amphipods, mysids, isopods and pteropods). HOME RANGE: Range in Arctic, with major populations near ice in Bering, Chukchi and Beaufort Seas. SOCIAL: Alone or female and calf or mixed herds of 2–10 (occasionally up to 50). COMMUNICATION: Complicated sets of low moans. Eyesight and hearing probably very good. Taste and touch may also be especially well developed.
Breeding. Mating in March–May, births in April–May. Sexual maturity 4–6 years (length at sexual maturity: males 11–12m; females 12.2–14m). Gestation 12–13 months. Litter size 1. Calving interval: probably 2–3 years. Mammae 2. YOUNG: Weaning age: 5–6 months. Mother tends young for slightly less than one year.
Measurements. Head-body length: 15–18.5m (birth: 4.5 m). Weight: 60–80 tonnes. 230–360 long (up to 450cm), black, narrow baleen plates in each ramus of upper jaw.
General. Dramatic decline following whaling in 18–19th centuries; protected since 1935 but only slow increase. Estimated population 7,000 individuals. Prolonged dives (usually 15–20 mins, up to 1 hour) during feeding. Aboriginal whaling continues throughout much of Alaskan range on a quota system.

NORTHERN RIGHT WHALE *Eubalaena (Balaena) glacialis* **Pl. 38**

Name. North Atlantic Right Whale, Black Right Whale, Biscayan Whale. *Glacialis* = of the ice (Lat.) from *glacies* = ice (Lat.).
Status. Endangered.
Recognition. Yellowish-white, irregularly shaped horn-like bonnet on top of head (variation allows recognition of individuals). Often infested with parasites (eg whale lice). No dorsal fin or ridge. Broad tail flukes deeply

notched with concave trailing edge, often lifted before dive. V-shaped blow directed slightly forward (8m up) for 3–5 seconds; typically blows once per min during surface cruising, for 5–10mins.

Habitat. Pelagic in cold temperate waters of N. Hemisphere (between 30–60°N) and in Gulf Stream.

Habits. FOOD: Plankton, especially 3–4mm long, swarming crustacea (mainly Copepods). HOME RANGE: Migratory. Summer: north to Arctic and cold temperate seas to feed on plankton. Autumn and winter: south to warm temperate and subtropical seas. In NE Atlantic used to breed off coast of North Africa. SOCIAL: Solitary or in pairs or small family groups (previously herds of up to 100). COMMUNICATION: Many underwater sounds.

Breeding. Sexual maturity 4–6 years. Gestation 10–12 months. Litter size 1. Calving interval 2–4 years. YOUNG: Weaning age 12 months (milk less than 50% water). Strong maternal ties.

Measurements. Head-body length: 15–18m (birth: *c.* 4.4–4.8m), female larger than male. Weight: 50–56 tonnes. 220–260 black baleen plates (*c.* 2–4m long) in each ramus of upper jaw.

General. Formerly migrated along European coast between Spitzbergen and Iberian waters. Almost extinct in E. Atlantic (protected since 1935) and small numbers in W. Atlantic and possibly occurring in N. Sea and English Channel. Estimated population in early 1980s, 200–500 in N. Pacific, 200–500 in N. Atlantic. No increase since protection. Blubber (36–45% of total weight (up to 62cm thick). Dives generally less than 50 fathoms, but up to 250 fathoms if frightened. Known to be attacked and killed by Killer Whales.

<*Northern Right Whale.* Atlantic waters, generally not NW of Iceland.

Order Artiodactyla – Deer, Goats and Sheep

Origins: The origins of the even-toed ungulates or Artiodactyls are lost in the Palaeocene, more than 55 million years ago. Pigs are the most primitive surviving Artiodactyls, their omnivorous habits reflected in low-crowned molars with many small cusps (which are similar to the teeth of Humans and bears). Artiodactyls arose from the condylarthrans (as did the Cetacea), a group which filled a wide variety of niches after the demise of the dinosaurs. The condylarthrans had, in turn, descended from insectivore-like ancestors. There was an early radiation of Artiodactyls in the Eocene from rabbit-sized four-toed creatures (eg *Diacodexis*) with a rolling astragalus bone (see below); by the end of the epoch they had diversified to twenty families spanning all the northern continents. Today the only direct descendants of these early successes are the camels. The amazing success of the modern ruminant Artiodactyls came in the Miocene and coincided with the evolution of grasslands and savannah. Twenty million years ago the bovids (today's cattle and antelope) were small, horned ruminants of woodland savannah which, only 10 million years later, had given rise to a radiation of 70 or more new genera of fast running (cursorial) creatures of the open plains, equipped with high-crowned teeth for grinding abrasive grasses. At the peak of their success, in the Pleistocene, they comprised some 100 genera, but today that number has fallen to nearer 50. A famous, recently extinct, Artiodactyl is *Magalocerus*, a giant fallow deer (known as the Irish Elk although it was neither an elk nor confined to Ireland). *Megalocerus* flourished from Ireland through Siberia to China in a warm inter-glacial period which ended about 12,000 years ago. A few survived around the Black Sea and in Austria until 500 BC. At 45kg their antlers amounted to almost one seventh of their body weight and spanned 3.7m. Similarly, we have only just missed seeing Aurochs *Bos primigenius*, the ancestor of domestic cattle which was domesticated in *c.* 4000 BC and survived until the 17th century, and the Steppe Wisent *Bos priscus*, represented in the cave art of the Upper Palaeolithic of France, which died out in very recent times.

Main features of the Artiodactyla: The ungulates or hoofed mammals is divided into the odd-toed horses and rhinos (Perissodactyla) of which there are now none in Western Europe (excepting the feral horses of the Camargue), and the even-toed Artiodactyla which include about 90% of the hoofed genera living today. This situation used to be reversed: 65 million years ago in the early Tertiary, more than 60% of ungulate genera were odd-toed. The Perissodactyla have the distinction of including the largest ever land mammal, *Indricotherium*, a rhino standing 5.4m at the shoulder, able to reach browse at 8m and, at 30 tonnes, 4.5 times heavier than the heaviest living elephant. Modern artiodactyls vary in size from Mouse Deer (2kg) to Hippo (3200kg). They vary greatly in form too, and are split into three suborders: Suina (pigs, peccaries and hippos), Tylopoda (camels and llamas) and Ruminantia (giraffes, deer, cattle, antelopes etc.). There are two hallmarks of Artiodactyl success. First, their so-called rolling astragalus: an elegantly engineered double-pulley system on their ankle bones, which brings springiness and flexibility to their gate, and makes them very good at running away. Second is their ability to ruminate, and chew the cud, which enables them to thrive on unpromising vegetarian diets. In brief, the design of their stomachs distinguishes the three modern suborders: the pigs have a two-chambered stomach but do not ruminate (they also have upper incisor teeth), the camels have a three-chambered stomach and ruminate, whereas the cattle and antelope have four chambered stomachs and ruminate (they also have no upper incisors, but a horny pad in their place). All artiodactyls are predominantly terrestrial and cursorial, with long slender legs (even the Hippo, when on land to feed, can move quite fast). Their weight is carried mainly on the 3rd and 4th digits and toes; the 2nd and 5th ones are reduced, and usually bear no weight (and are altogether absent in camels and giraffes). All artiodactyls lack the 1st digit and toe.

Of the four stomachs of Ruminantia, the first two, the rumen and reticulum, are where the cellulose fibres are fermented; the cud then results, is chewed and then digested in the 3rd and 4th chambers.

A particularly interesting feature of Artiodactyls is their antlers found on the males of most species of deer (exceptions being Chinese Water Deer and Musk Deer). Only in Reindeer do both sexes have antlers. In most deer antlers grow for the first time in the second year of life, but in Elk, Roe and White-tailed Deer, mini-antlers appear in the fawn's first autumn, and Reindeer start sprouting them a few weeks after birth. Antlers are cast and regrown each year, a process which is under hormonal control. The young male deer's skull first differs from that of the female with the development of pedicles: special bony growths from the frontal bones. These are platforms which will support the external antler. As antlers grow they are covered with a furry external skin, rich in blood vessels and nerves, called the velvet, which protects and nourishes the developing antler in conjunction with blood supplies within. Once the antler is grown the blood supplies dry up, the velvet dies and is rubbed off, leaving the polished bone exposed. After 8 months (by which time it is spring), the antlers are shed, shearing at the junction with the pedicle. This happens very suddenly; the deer may be using the antlers in fights only a few days before they are shed. The cycle of re-growth begins almost immediately to produce a new antler for the next rut. However, the White-tailed Deer, the Elk and the Reindeer shed their antlers in the autumn/winter, and there may be a hiatus of several months before regrowth. Presumably this reflects the poverty of their habitat, where they use all the available energy to keep warm overwinter. Deer can grow antlers in 12–16 weeks; for a bull Elk this is a remarkably swift production of up to 30kg of new bone, and can mean effectively doubling the mineral content of the body. Deer increase their mineral intake by nibbling their old antlers, and by eating mineral-rich soil.

In all the European deer except the Muntjac, each successive antler set is bigger and more complex, more fully branched, with a wider beam, until the individual reaches old age, when antlers become less complex. Antlers have increased in size and complexity in evolution, in parallel with a trend towards polygyny. While primitive dagger-like antlers are used for stabbing and thrusting, more ornate antlers provide, instead, a 'hold' for pushing and wrestling. In general, antler weight is proportional to body weight and thus is a fair indicator of social dominance. But why do deer shed antlers each year (when bovids do not)? There is no definite answer, but one might speculate that it may be advantageous to save energy by avoiding carrying antlers during seasons when food is short. Having cut off the blood supply, and thus capacity for growth, damaged antlers can only be replaced by growing a new set. Clearly shedding provides the maturing male with an opportunity for repair and to build bigger and better antlers each year, but bovids achieve this by continual growth. The question thus becomes why do cervid antlers lose their blood supply whereas bovid horns retain theirs, and is there something in bovid life that makes them unable to be intermittently without antlers? Currently these questions lack conclusive answers.

Artiodactyla in Western Europe: There are at least 478 extinct genera and 76 modern genera of artiodactyls, with 187 modern species. Of these, three families and 19 species are represented in Europe.

The Suidae or pigs are characterised by fairly complete dentition, with well developed, upturned canines, molars with rounded cusps and a pre-nasal bone which supports the nose. Males are larger than females with more pronounced tusks and warts. In the genera *Sus* and *Potamochoerus* (there are 5 genera in all) the coat is striped at birth. Pigs usually forage in family parties, communicating in grunts, squeaks and chirrups. The only European species, the Wild Pig, is a generalist herbivore reaching sexual maturity at about 4 years old. Courtship involves the boar giving lateral displays and repeatedly attempting to rest his chin on the sow's rump which will cause her to stand if fully receptive; also, boars of many species produce lip-gland pheromones at this time. Litter size is up to 12 in *Sus* but only 1 or two in the Babirusa, a S.E. Asian suid. Each piglet has its own teat; though weaned at about 3 months they stay with their mother till she is ready to farrow again. After farrowing in isolation, young sows may rejoin their mothers, thus leading to larger matriarchal herds or sounders which may include several generations. When these sounders reach a certain size, they fragment, leading to kinship units or clans comprising a number of related sounders. These have overlapping home ranges, share feeding grounds, water holes, wallows, resting sites and sleeping dens. The other social units are solitary boars and bachelor groups.

The Cervidae or deer vary in size between the Southern Pudu at 8kg, to the Elk at 800kg. Cervids have graceful elongated bodies, slender legs and necks, short tails, angular heads and large round eyes which are placed well to the side of the head; their triangular/oval ears are set high on the head. A distinguishing feature is the males' antlers, which are shed and regrown each year, and these function as weapons and in display. Water Deer are, in evolutionary terms, the oldest cervid genus in Europe, and only have tusks. One suggestion is that tusks evolved first and that antlers originated as a form of defence against them. Some species have both tusks and antlers: fighting Muntjac interlock antlers and try to manoeuvre for a slash with their tusks. In many deer the young have light coloured spots which improve their camouflage – the adults of some species have this pattern too. A light coloured patch around the anus is often displayed by alarmed deer, which raise their tails and rump hair as they flee. Like other ruminants, deer bite off their food between the lower incisors and a callous pad on the upper gum. The dental formula in deer is always 0/3, 0/1 or 1/1, 3/3, 3/3. Premolars and molars grind the cud. Scent is very important to cervids, both in detecting predators and in communication; most species have interdigital glands and all have facial glands. Vocal communication is also quite important, both between mother and offspring and between competing males. Deer are typically woodland animals, but are found in almost all habitats, and they are generally grazers and browsers of plant matter. Almost all species are hunted for their meat, many for their antlers; they are also persecuted because of the damage they wreak upon crops and commercial forests.

The Bovidae include wild cattle, sheep, goats, Musk Ox, duikers and grazing antelope and are the most successful and diverse ruminants (and the largest ungulate family). Their most distinctive feature is their horns, which consist of permanent horny sheaths on bony cores developed from the frontal bones of the skull. Horns may be present in both sexes or only in males. Bovids are primarily an Old World group, reaching their peak of diversity in Africa.

WILD BOAR *Sus scrofa* Pl. 43

Name. Wild pig. *Sus* = pig (Lat.) from *hus* (Gk.) = pig; *scrofa* = sow (Lat.).

Recognition. Powerful, pig-like appearance, appears laterally flattened with large head, short stocky legs. Dark, bristly pelage (strong smell). Upwardly pointing upper canine tusks in male only. Summer coat of bristly guard hairs, still relatively long and dense (see vignette); winter coat dense, longer hair, with thick underfur (NB not found in Domestic Pigs). Fieldsigns include strong smell and disturbance of ground due to rooting – in pasture, much removal of grass. Splayed tracks (showing dew claw). Droppings *c.* 7cm long, often sausage-shaped but variable, depending on dry:wet ratio in food. When eating solely wheat, get sausage-shaped compact droppings composed solely of compacted wheat husks. Other fieldsigns include wallows and rubbing trees.

Habitat. Deciduous woodlands of Europe and East; thrives on cultivated land if access to good cover (eg reedbeds, woodland), but widely extinct in agricultural regions (eg extinct in UK 17th century, now only *c.* 300 left in Southern Sweden). Den in shallow pit, often in sunny location, or a 2m wide pile of dry leaves and small sticks (1m high) with hollow inside.

Habits. Mainly crepuscular and nocturnal. FOOD: Omnivore, with emphasis on vegetation (eg acorns, beech mast, chestnuts, green plants, potatoes). Animal food includes carrion, invertebrates (eg earthworms, insect larvae) and vertebrates (eg rodents). HOME RANGE: Female groups (travel with young) 200–2000 ha. Male 2000 ha. Home range size very variable however. Seasonal movements from forests onto cultivated land (eg Poland). Otherwise, if environment stable, more or less sedentary. May move up to 20–30km if subjected to serious hunting pressure. SOCIAL: Males mostly solitary except during rut; females in exclusive small herds with young and one to two year olds, although male subadults frequently exist in more or less open groups. COMMUNICATION: Grunts and snuffles while foraging. Alarm: gruff grunt and 'bark'. Boar chatters teeth when aggressive. Scent marking in males, including sex hormone on breath that prompts females to stand for copulation. Hearing and sense of smell well developed, sight mediocre. Sense of touch (especially on tip of snout) well developed, as is taste – can distinguish between varieties of potato.

Breeding. Extended breeding season with mating in September–March and births in February–June. Peak of births in March, occasionally a second peak in July (June–August births). Females and males sexually mature at 1 year, although males usually excluded from reproductive activity until older by dominant older males. Gestation 115 days. Litter size 3–10 (older females have larger litters), generally 4–7. Litters per year 1, but second litter if first is lost within a few days. Mammae 10. YOUNG: Horizontal stripes, fading within 3–5 months. Weaning age 3–4 months. Female builds nest for protection of newborn. Mother and young remain at birth place for *c.* 1 week, before piglets accompany their mother, and join the original female group until her next litter is born.

Lifespan. Max. recorded 20 years (captivity); 8–10 years (wild). Hunting accounts for large percentage of adult and yearling pigs: 2–60% during first year of life.

Measurements. Head-body length: female *c.* 100–146cm; male *c.* 105–167cm, though large males of up to 185cm have been known. Tail-length: female 16–28cm; male 17–30cm. Hind-foot length: *c.* 25–35cm. Shoulder-height: female *c.* 59–89cm; male *c.* 64–109cm. Condylo-basal length: 305–336mm. Weight: male *c.* 33–148kg (average 130kg); female *c.* 30–80kg (lighter weights in southwest; while Boars in the northeast up to 185kg or even larger though huge variation often the result of inter-

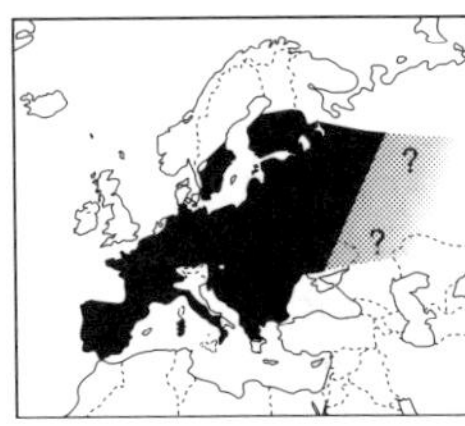

< *Wild Boar.* Widespread throughout C&W Europe.

> *Chinese Water Deer.* One isolated population in East Anglia, England.

breeding with feral Pigs). 200kg specimens recorded in Eastern Germany. Great seasonal variation in weight; can lose half body weight under bad conditions, and can double weight in a few weeks under good conditions. Record for Central Europe is 230kg. Birth weight *c.* 1.1kg. Weight increases with age. Dental formula: 3/3, 1/1, 4/4, 3/3 = 44.

General. In Poland, a large part of the population moves out of the forest in spring into cultivated fields (particularly maize) remaining till late autumn. Considerable economic damage is done. Pig-Boar hybridisation occurs. Remains widely distributed, and abundant in some areas, but range contracting in some areas and extinct in others. Male has thick layer of cartilage over thorax (4cm thick) acting as a shield in fights, protecting shoulders, heart, lungs etc.

CHINESE WATER DEER *Hydropotes inermis* Pl. 39

Name. *inermis* = unarmed (no antlers); *hydro* = water (Gk.); *poto* = I drink (Lat.).

Status. Introduced. Established in England in Norfolk Broads, Cambridgeshire, Bedfordshire/Hertfordshire – apparently unable to increase range in GB.

Recognition. No antlers. Upper canines greatly elongated in male (*c.* 8cm) to form tusks. Ears: large and sandy coloured. Tail: same colour as winter pelage, which is grey brown or pale fawn colour and very thick (unlike summer, when it is reddish and sleek); no prominent rump patch. Black nose and dark eye very conspicuous. Winter moult changes to summer coat *c.* April, looks very scruffy during transition. Most have winter coat by November. Faecal pellets more elongated than Muntjac's (10–15mm long × 5–10mm wide). Chunks of hair scattered during moult.

Habitat. Reed beds, grassland and bushy woodland (associated with marshland in China).

Habits. Crepuscular, but seen grazing at any time where undisturbed. FOOD: Mainly grasses, sedges, rushes, also root and other vegetable crops. Browse bramble, willow etc. HOME RANGE: Males territorial, territory typically occupied by male and female (Woodwalton Fen, UK). Woodwalton Fen density *c.* 3 deer per 10 ha. Whipsnade *c.* 20 in 10 ha. SOCIAL: Usually solitary, but may aggregate on good feeding area when food scarce. Male defends territory during rut, using canine teeth during fights.

COMMUNICATION: In rut, male whistling call. Screams like hare when alarmed. Bark at each other (and at intruders). Sense of smell is acute.
Breeding. Rut in December. Births in late May to early July. Sexual maturity 6 months. Gestation up to 180–210 days (mean 176). Litter size 2–4 (6 recorded in Asia). Litters per year 1. Mammae 4. YOUNG: Fawns – light spots in parallel lines – fade by 2 months. 0.8kg at birth. Weaning age *c.* 2 months.
Lifespan. Max. recorded 11 years. Causes of death include fawns being taken by foxes (and reputedly even by Stoats). Adults die during severe (wet and cold) winters in England.
Measurements. Head-body length: *c.* 100cm. Tail-length: 6–7.5cm. Hind-foot length: *c.* 19cm. Shoulder-height: male up to 60cm; female up to 55cm. Condylo-basal length: *c.* 16cm. Weight: male 11.4–14.0kg; female 9–11.5kg. Dental formula: 0/3, 1/1, 3/3, 3/3 = 34.
General. Introduced to UK. Very little known, even in its native China. Sometimes runs like a hare, flinging hind legs high.

REEVES' MUNTJAC *Muntiacus reevesi* Pl. 39

Name. Barking Deer (UK; although originally this term seems to have been used for Indian muntjac), Chinese Muntjac. *reevesi* after J.R. Reeves who collected type specimen.
Status. Introduced. Widespread and distribution increasing in England.
Recognition. Often stands with rounded 'humped' back. Pelage chestnut coloured above, buff below, ginger tail. Individual hairs banded (dark tops). Antlers (male only) backward pointing single spikes (c. 6–8cm), though very small brow point may or may not be present, dark brown or black stripes along sides of elongated permanent pedicles. Upper canines of male form tusks (*c.* 2cm) projecting below lip. Antlers shed May–July. Male stockier than female. When alarmed run off with tail erect to show white underside. Conspicuous moult in April–May, but second moult in September–October is less obvious. Droppings are black and rounded or cylindrical, sometimes with point at one or both ends, 10–13mm × 5–11mm. Droppings usually scatter on impact with ground. Footprints: small, delicate slots, *c.* 2.5cm long; outer toe sometimes longer and tip just curving over tip of inner toe. Fieldsigns include fraying of saplings at low height; after fraying may strip bark with incisors leaving characteristics twist of bark at top. May tread well-worn courtship rings on the ground.

Habitat. Dense deciduous or mixed woodland. Prefers diverse and dense understorey.
Habits. Active by day and night, but often seen at dusk and dawn. FOOD: Browse (eg bramble, ivy) and fruits including berries, acorns, chestnuts, beech mast. Grasses and herbs major component of diet only in

< *Reeve's Muntjac.* Confined to S England where spreading fast.

> *Red Deer.* Widespread in C Europe and Scotland, with pockets in Iberia, Italy, Yugoslavia and Bulgaria, and on Corsica and Sardinia.

spring/early summer. HOME RANGE: 14ha in coppiced woodland. Buck's range larger than females. Ranges of several females overlap; buck's territory overlaps ranges of several females. Young females often stay near mother's range, young males wander to new areas. SOCIAL: Largely solitary. COMMUNICATION: Barking may be repeated every few seconds for up to 45 minutes (especially during rut). Click when alarmed, also squeak and scream when very frightened. Small shoots frayed with teeth and marked with preorbital scent glands. Frontal glands wiped on ground (see vignette). Also has interdigital gland on hind feet. Senses of smell and hearing acute. Apparently short-sighted but nonetheless quick to detect movement.

Breeding. No fixed breeding season in UK. Post-partum oestrus. Sexual maturity: usually less than 10 months for females, longer for males. Gestation 210 days. Litter size 1. Up to two litters in a calendar year, i.e. at 7 month intervals. Mammae 4. YOUNG: Fawn born with buff spots, which disappear at 8–12 weeks. Lie up in vegetation for most of first few weeks. Weaning age 12 weeks, but still suckled intermittently at 3–4 months. Fawn may remain with mother till next one dropped. In captivity father also grooms fawn.

Lifespan. Max. 19 years known in captivity. Red Foxes may take young fawns.

Measurements. Head-body length: 90–100cm. Tail-length: *c.* 15cm. Hind-foot length: 22cm. Shoulder-height: 45–52cm. Condylo-basal length: *c.* 15–17cm. Weight: male mean 15kg; female mean 12kg (birth: 1kg). Dental formula: 0/3, 1/1, 3/3, 3/3 = 34.

General. Smallest deer in Europe. Introduced (from China).

RED DEER *Cervus elaphus* — Pl. 40

Name. *Cervus* = deer (Lat.); *elaphos* = deer (Gk.).

Recognition. Second largest European deer. Antlers up to *c.* 71cm long (max. 91cm), weigh up to 3kg each but usually 1kg; branched rather than palmate (cf Fallow) or rugose (cf Roe). Often more than 10 points (cf Sika). Pelage: spots generally only on calf (very occasionally on adults); coat short and reddish in summer, dark brown in winter. Rump patch creamy coloured extending dorsally above short beige tail – not clearly marked with black as in Sika and Fallow Deer. Beige tail a diagnostic feature. Winter coat starts growth in September, usually complete by December. Summer coat starts in May (older animals and those in good condition starting first), over by July–August. Moulting begins at

head, legs and anterior of body. Two moults in calves – first is 2 months after birth (spots lost), second in autumn to give winter coat. Droppings: acorn-shaped, black, 1.5cm diameter. Faeces variable according to diet and age of deer. Deposited in groups, but appear as a 'string' or 'runner' if animal moving when defecating. Norway spruce stripped throughout life, Lodgepole and Scots Pine only when young (up to 20–30 years). Chisel marks of front teeth less than 16mm width (cf Elk more than 16mm). Tree saplings of 5–10cm diameter frayed at less than 150cm when stags rubbing velvet and during the rut. Wallows used by both sexes in summer when shedding coat, and by stags during rut. Hoof prints, called 'slots', 5cm wide in calves, 7cm in adults. Overlap in size/shape with other medium-large deer, and sheep. Size of prints larger on snow, soft ground or when running, because hooves splayed out.

Habitat. Great diversity (eg woodland, grassland, moor, scrub). Generally occupies woodland and feeds at edge or in grassland. In some areas (eg Scottish Highlands, Exmoor and Quantocks in SE England) on upland moors, generally below tree line except in summer (although the moors often maintained below natural tree line). In Alps and also Norway moves above the tree line in summer.

Habits. Normally active throughout 24-hour period, but with peaks at dawn and dusk, often because of human activity. In Scottish Highlands on open range, ascend hill in early morning to rest, descend to grazing area in evening. In woodland, emerge from cover to graze in fields at dusk, but feed throughout 24-hour period. FOOD: In woodland browses on shrub and tree shoots for more than 80% of diet (eg Carpathians: shoots of 36 species, bark of 27 species of trees and bushes, especially spruce and sallow). On moorland graze on grasses, sedges, rushes, heather, plus conifers and holly in winter, if available. Hind more than doubles food intake during early lactation. HOME RANGE: Depends on habitat, woodland deer having a smaller home range. Also varies with distance between feeding grounds and resting harbourage (eg Scotland: male 800 ha; female 400 ha). Also seasonal variation (eg Alps, winter 50–150 ha, summer 400 ha). (Population density varies between 5–45 per km^2 depending on habitat). In Alps and Norway migrate between summer and winter ranges (up to 6km in Norway). Exploratory behaviour starts within 6 months of birth, earlier in males than females. Strong fidelity to home area in females – ranges overlap that of dam. Most young stags leave natal area, dispersal age varying from one year (plantations) to 2 or more (open ground). SOCIAL: Woodlands: reputedly hinds tend to centre, stags to periphery. Highlands: breeding hinds on lower slopes, younger hinds above; stags highest in summer, lowest in winter. Matriarchal herds (larger on moorland, smaller family groups of females in woodland). Dominance contests among fighting stags in rut – males defend harem of females and young, or area around them, 'moving territory'. Adult males and females otherwise segregated for rest of year, females staying in more fertile areas in winter. Dominance feeding hierarchy throughout year. COMMUNICATION: Roaring rate of rutting stag (see vignette) indicative of strength. Females and calves bleat. Alarm bark by hinds. Visual communication important, eg rump patch flared acts as warning signal. Scent produced by hinds in season. Scent glands on metatarsals (hock), between hooves, ventral surface of tail and lachrymal (preorbital) in both sexes.

Breeding. Mating in late September–November. Births in late May–June. Sexual maturity male: 1–3 years; female: 1–2 years, both depending on habitat quality (but physical maturity *c.* 7 years). Gestation 225–245 days. Usually 1 calf, twins very rare. Litters per year 1 – proportion of yearly pregnancies reflect environmental conditions: 0–80%. Mammae two pairs (abdominal) – udder only conspicuous in late pregnancy and early lactation. YOUNG: Calves spotted for 2 months; rump patch yellowish-buff. Weaning age *c.* 6–10 months but non-pregnant females may continue to suckle yearlings. Parental care by female only. Initially, calf left lying alone except during feeds. Accompany mother after 7–14 days. Can form 'creches' within large hind groups, with frequent playing.

Lifespan. Max. recorded 25 years, but few over 13–15 years. Annual mortality *c.* 3% in adults, (depending on habitat and increasing rapidly after 9th year) – highest in calves (between 8–11 months).

Measurements. Note: for all measurements size influenced by (a) habitat and (b) E-W cline. Head-body length: 165–260cm (eg Scotland: male *c.* 201cm; female *c.* 180cm; in open range, woodland deer in Scotland as big as English deer). Tail- length: 12–15cm (20cm including hair). Shoulder-height (Scotland): male 122cm; female *c.* 114cm. Condylobasal length (UK): male 300–400mm; female 280–335mm. Weight: regional variation, according to habitat and density: male up to 255kg, (captive, eastern males up to 350kg) female up to 150kg. Dental formula: 0/3, 0–1/1 (appears as an incisiform tooth with the true incisors; diastema present), 3/3, 3/3 = 32–34.

General. Antlers grow in spring and summer, cleaned of velvet in August (varying with age and condition). Antler development depends on forage quality of habitat. Cast March–April. Prime male has 5 or more points (or tines) on each antler. Number of tines on antlers are a poor indicator of age, and instead reflect habitat quality and genetics. A yearling stag in the Scottish highlands may have no antlers (only knobs), whereas an animal the same age in SW England may have 12 points. Stags older than 10 years tend to lose points each year. May have more than 10 points. Economically important harvest. Most closely related to N. American wapiti, now regarded as same species (i.e. *C.e. canadensis*). *C. elaphus* regarded as a super-species, subspecies freely interbreeding to get fertile offspring. Numbers increasing, *c.* 180,000 to 300,000 over 20 years in Scotland. Early hybridisation with Sika, affecting all UK Red Deer except some populations in Scotland. Increasing throughout Europe. Large populations in Germany and Spain.

SIKA DEER *Cervus nippon* **Pl. 41**

Status. Introduced.

Recognition. Summer: chestnut red with rows of whitish spots (clearest either side of spine). Winter: males dark grey or black (see vignette), females light brown or grey (spots indistinct or absent). Hinds with calves keep summer coat longer than those without. Rump (and upper surface of tail) white and outlined in dark brown, also distinctive white metatarsal glands. Antlers: max. 8 points (only one tine on each antler points forward). Moults in spring (May) and autumn (September/November). Droppings: shiny black pellets pointed at one end and indented at the other. Other signs (eg frayed saplings, hair, cast antlers, foot-prints) very similar to, if not indistinguishable from, those of Red Deer.

Habitat. Deciduous or mixed woodland on damp ground with dense undergrowth (eg Rhododendron, Blackthorn, Hazel, Bramble), associated with clearings or fields. Dense conifer thickets.

Habits. Crepuscular. Traditionally active 24 hours/day, but more crepuscular due to human activities. FOOD: Largely grazes on grasses and sedges, but browses too (eg ivy, holly, conifers, heather) and acorns, fungi and bark. Reputedly has better capacity to digest poor quality fibrous food than has Red Deer. HOME RANGE: Little known in Europe. SOCIAL: Male summer territories 2–12ha (dominant male tolerates subordinates in territory). Herds of small family groups of hinds (2–8 normally), and bachelor groups of stags (largest herds in late winter and spring) – particularly in conditions of harsh weather, open environment and poor food availability. COMMUNICATION: Vocalisations of rutting male are high-pitched whistles (hoarse screams later in season), repeated 3–4 times (each call rising and falling in pitch). Hinds bleat and grunt in oestrus (and soft nasal whine to calves).

Breeding. Mating in late August–October. Births in May–June (exceptionally mid-April–December). Sexual maturity 16–18 months (physical maturity 4th year). Gestation *c.* 8 months. Litter size 1 (very rarely twins). Litters per year 1. YOUNG: Weaning age 6–10 months (but begin grazing at 1 month – as do Red and Roe). Parental care by female only.

Lifespan. Max. recorded 15 years.

Measurements. UK – measurements very variable according to habitat. Head-body length: male 138–149cm; female 134–145cm. Tail-length:

< *Sika Deer.* Isolated pockets in continental Europe and Ireland, widespread in Scotland.

> *Fallow Deer.* C Europe, S Sweden and Finland and British Isles, pockets in Italy and Iberia.

12–15cm. Hind-foot length: male 34–38cm; female 32–36cm. Shoulder-height: 80–90cm (eg male *c.* 81cm; female *c.* 73cm). Condylo-basal length: 200–290mm. Weight (UK): male *c.* 64kg; female 41kg (birth 3kg). Dental formula: 0/3, 1/1, 3/3, 3/3 = 34.

General. Causes commercial damage in forests in same way as does Red Deer. All else being equal, antlers increase in size and number of tines until old age, when they decrease, but this rule is rendered unhelpful because diet quality and genetics also cause region variation in antler growth. Bole scoring (vertical incisions in bark caused by thrusting antler tines) apparently only caused by this deer, and only of local importance. Originally introduced to UK from Japan in 1860 and these are the only form known to have become established in the wild, although Manchurian and Formosan Sika have been kept in parks. All forms hybridise readily with Red Deer. Introduction of Sika to Red Deer areas considered irresponsible because of likelihood of hybridisation threatening genetic integrity of Red Deer.

FALLOW DEER *Dama dama* Pl. 41

Name. *Dama* = deer (Lat.) from dam, Welsh *dafad.*

Recognition. Antlers: palmate in older bucks. Most easily confused with Sika, but these never have palmate antlers and their tailtip hangs *c.* 5cm below rump patch. Pelage: varies considerably from fawn to black – certain colours predominate in some regions. Most individuals chestnut brown in summer, more uniformly greyish in winter, and marked with white spots (less distinct in winter). No spots on certain colour varieties: e.g. sandy fawn becomes white adult. Black does have spots – not white but lighter shade of black-brown. Only 'Menil' (very pale, spotted) remains distinctly spotted throughout the winter. Prominent tuft of hairs ('brush', see vignette) on penis sheath – a useful character in young males without palmation. Long tuft *c.* 12cm in females below vulva. Tail relatively long. Black vertebral stripe along dorsal surface of tail of common colour variety (brown stripe in Menil and none in white or black varieties). Most colour varieties have white tail (caudal) patch ringed with black. Moults in May–June and late September–October. Fieldsigns include rutting areas, also thrashed bushes, frayed bark on saplings, marked by strong smelling scrapes (30–300cm diameter) in leaf litter. Droppings: male *c.* 16 × 11mm; female *c.* 15 × 8mm, black, shiny, cylindrical pellets deposited in piles (similar to other deer), usually with one end pointed and the other indented. Playrings – bare ground worn clear of leaf litter (*c.* 3m diameter) by both sexes and all ages running around (often around old tree stump), function unknown. Hairs on barbed wire (tends to go under fence, even where it could easily jump over). Footprints *c.* 6.5cm long (adult), but varying with age, sex and gait.

Habitat. Mature deciduous or mixed woodland, sometimes coniferous, with dense undergrowth and close to open farm- or parkland, often in rather open woodland. Also marsh and meadows (eg Cot Donana, Spain). Patterns of habitat use vary throughout year as seasonal availability of different forage alters. If undisturbed, will lie up in feeding site, or on woodland edge.

Habits. Diurnal if undisturbed, and may lie down to ruminate in open, otherwise crepuscular and nocturnal (older males more nocturnal). FOOD: Preferential grazers, woods used mainly for shelter, emerging to feed in open. Principally grasses and rushes but also browse young leaves from trees. Also acorns, beechmast, berries (eg Bramble, Bilberry, Rose). Also, in winter, heather, holly, bark of felled conifers and other species. Also cereal and root crops. HOME RANGE: Varies with habitat and season: females *c.* 50–90ha, males *c.* 50–250ha (summer in New Forest, UK). Non-territorial, home ranges overlapping. Population density very variable, often controlled by Man culling or feeding. One per 6–8ha increasing by 50% in winter to one per 4ha in E. England. SOCIAL: High degree of sexual segregation (as other deer): adult males in separate herds from those of adult females and 'followers' (usually offspring of current and previous year) in groups of 5–7 in woodland, or temporary aggregations of up to 80 on good feeding area in the open. Aggressive behaviour – most obvious by males competing for display grounds in the rut. Usually resolved by display, but direct fights not uncommon. Young males stay with female herd for first 20 months before joining male groups in (usually) different areas, and return to female area again for rut in August–September. In some populations, however, males remain in female areas longer after rut complete (as late as April/May). In other populations, usually open habitats, mixed herds containing adults of both sexes may persist throughout the year. Female groups loosely hierarchical and led by dominant doe, although individual composition changes frequently. COMMUNICATION: Males in rut give belching groan. Alarm: short bark – especially used by female if young near. Females make a bleating whicker. Various scent glands: suborbital, interdigital, metatarsal and gland associated with penis sheath in rutting males. Colour vision, and can perceive objects up to 60m. Thereafter sight indistinct, but acutely aware of movement. Hearing and scent acute.

Breeding. Mating: strategies vary with habitat, some bucks hold rutting 'stand', others congregate at 'leks' and gather females by calling – herding minimal so do not generally hold them as a harem. Mating in October–November. Births in June–July. Sexual maturity: female 16 months; male 7–14 months, although social hierarchy normally prevents males mating until older. Gestation 230 days. Litter size 1 (twins rare). Litters per year 1. Mammae 4. YOUNG: Fawn similar to adults' summer coat, although white adults begin life as pale sandy fawns. Can follow mother from birth if necessary but normally lies up concealed in vegetation. Weaning age *c.* 12 weeks, but most females still lactating till end of November. Parental care by female only.

Lifespan. Max. 16+ years in wild (but much less normally), 20 years recorded in captivity. Northern populations severely affected by harsh weather (late fawns and exhausted bucks perish). Causes of death include road traffic accidents, and larger carnivores take some fawns. Some males die from rutting injuries. Natural mortality of adults low until old age.

Measurements. Head-body length: 130–170cm. Tail-length: 16–19cm. Hind-foot length: *c.* 43cm – cannon bone 240–260mm in adults. Shoulder-height: 85–110cm. Condylo-basal length: 31–83mm. Weight: max. 130kg (eg UK male (buck) 46–80kg, female (doe) 35–52kg). Birth-weight: mean 4.5kg. Dental formula: permanent 0/3, 0/1, 3/3, 3/3 = 32–34, but sometimes one or two incisiform teeth absent or one or two upper canines present.

General. Extinct in Europe following last glaciation, except for pockets around Mediterranean. Now abundant following widespread reintroductions (Europe) and feral/park animals (UK). Antlers: shed April–May, begin to grow immediately, clean of velvet by August–September. All else being equal, antlers increase in size and number of tines until old age, when they decrease, but this rule is rendered unhelpful because diet quality and genetics also cause region variation in antler growth. Introduced to UK by Romans. Now farmed in many countries. Gait – walk, trot, canter and gallop (stott when alarmed, see vignette), also swim well (eg between mainland and islands).

AXIS DEER *Cervus axis* — Pl. 41

Status. Introduced.

Recognition. White spots present throughout year. White throat patch. No black on tail or rump. Antlers: 3 points on each side, carried on pedicels but only one tine ever points forward (cf. Sika Deer), and antlers are cervine (cf. palmate antlers of Fallow Deer).

Habitat. Woodland.

Habits. Crepuscular and, probably, nocturnal. FOOD: Largely grazing, occasional browsing. HOME RANGE: Female 180 ha, male 500ha (in India). SOCIAL: Gregarious. Herds 5–10 (sometimes aggregations of up to 200) usually led by adult female. Group composition is variable. COMMUNICATION: Harsh bellow in rut.

Breeding. Mating peaks in March–June. Births peak Jan–May. Sexual

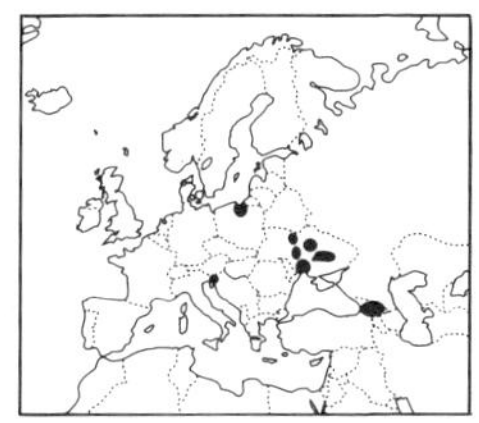

<*Axis Deer.* Isolated introductions to the Italian–Yugoslavian border and N Poland.

>*Elk.* Widespread throughout Scandinavia, with westward extreme in E Germany and S Norway.

maturity: female 14–17 months. Gestation 210–225 days. Litter size 1 (occasionally twins). Litters per year 1. YOUNG: Parental care by female only (young independent after 1 year, but may remain in association with mother for 2 years).

Lifespan. Max. lifespan 20 years.

Measurements. Head-body length: max. 130cm. Tail-length: 20–30cm. Weight: 75–100kg. Dental formula: 0/3, 0–1/1, 3/3, 3/3 = 32–34.

General. An Indian species introduced to Europe (eg Yugoslavia).

ELK *Alces alces* Pl. 45

Name. Moose (USA). *Alces* = elk (Lat.); from *alce* = elk (Gk.); *Elch* = elk (Ger.).

Recognition. Massive, Roman-nosed deer with large pendulous upper lip and hairy nosepad. Pelage: coarse, greyish-brown; legs: greyish-white (in female pale hindlegs extend up as far as tail). Adult male has beard. Two types of antler: palmate or cervine (deerlike). Moults once a year, in spring. Fieldsigns include frayed bushes in rut, wallows and stripped bark.

Habitat. Woodland and associated open country from farmland to mountains. Favours marshland in summer, drier ground in winter. Especially favours river valleys and lakes.

Habits. FOOD: Shoots and twigs of trees, especially pine; bark of aspen, rowan and willow. Summer diet is largely large herbs and leaves, including aquatic plants (for which they wade, swim or even dive), and (in autumn) cereals such as oats. Adult eats 10kg per day wet weight of browse during winter. HOME RANGE: 2.2–18.9 km^2. Some populations are stationary but others in NE Europe migrate up to 150km between summer and winter feeding grounds. SOCIAL: Largely solitary. Summer: solitary females with calves. Winter: herds, including males, under leadership of a female. In rut each male remains with single female at a time, competing with shoving matches. COMMUNICATION: Deep lowing voice. In rut both sexes make nasal, squeaking call. Alarm: muffled cough. During rut, scrapes out a hollow in which it urinates and then wallows until coated in scented mud-pack.

Breeding. Sexual maturity in 2nd year but varies regionally. Gestation

235 days. Litter size in young females is 1, older females have 2–3 calves (1% of births are trios). Litters per year 1 (female still breeding at 20 years old). YOUNG: Reddish-brown. Antlers grow at end of first autumn. Able to follow mother at 2–3 days old. Doubles weight in first month, and grows at 1kg per day thereafter. Calves remain with mother until 10–15 days before birth of next calf, whereupon she drives them off.
Lifespan. Max. recorded 27 years.
Measurements. Head-body length: 200–290cm (female *c.* 25% smaller than male). Tail-length: 7–10cm. Hind- foot length: *c.* 85cm. Shoulder-height: male 180–220cm; female 150–170cm. Condylo-basal length: male *c.* 565mm; female *c.* 512mm. Weight: male 320–800kg; female 275–375kg (birth: 11–16kg). Dental formula: 0/3, 0–1/1, 3/3, 3/3 = 32–34.
General. Swims well and often enters water. Trots very fast and purposefully (see vignette). Can be a hazard to traffic. Antlers grow in April (to *c.* 2cm in first year male), clean of velvet by August–September, shed in December–March. Younger animals keep antlers for longer. *C.e. corsicanus* (from Sardinia and formerly Corsica) is endangered. North American form referred to as Moose.

REINDEER *Rangifer tarandus* Pl. 44

Name. Caribou (USA). *Tarand(r)us* (Lat., Gk.) = reindeer.
Recognition. Both sexes have irregularly branched asymetric antlers (length: male 52–130cm; female 23–50cm) with up to 12 main lines, more secondary lines and prominent brow line in males. Antlers clear of velvet in September, males shed December-January, females shed May. Calf begins to grow antlers within 2 months of birth and keeps them through winter. Summer pelage is dark grey-brown, lighter in winter.
Habitat. Coniferous forest, mountain heath, alpine and arctic tundra (some populations migrate from tundra in summer to forest in winter).
Habits. FOOD: Terrestrial or arboreal lichens at low latitudes, grass and sedges at high latitudes. Spring: leaves, shoots and fungi. Winter: reindeer moss *Cladonia rangiferina*. HOME RANGE: Seasonal shifts between tundra and woodland of more than 1000km. Migrations of 20–150km per day. In contrast, Spitzbergen Reindeer are sendentary. SOCIAL: Highly gregarious. Harem in rut (bulls join female herd just before rut). Disperse into small groups in summer under leadership of old female. Large herds come together for spring migration. Loss of antlers leads to loss of social rank. COMMUNICATION: Grunting.
Breeding. Rut in late September–October, births May–mid-June. Calves can conceive and give birth at 1 yr under very good conditions. Usual age of first mating is at 18 months. Male Reindeer have 1st testicular cycle at 6 months, but obviously not socially mature until later. Gestation 210–240 days. Single young per year. Twins rare. YOUNG: Calf, no spots,

<*Reindeer.* N tundra, including parts of Norwegian and Swedish Lappland.

> *White-tailed Deer.* One isolated population in S Finland.

uniform brown (see vignette). Calves can walk within one hour of birth. Weaned at about one month. Calf stays with mother 1–3 years. Very rich milk (20% fat, 11% protein).

Lifespan. Max. recorded 28 years (females 14–15 yrs; males 9–10 yrs).

Measurements. Head-body length: 185–220cm. Tail-length: 10–15cm. Hind-foot length: *c.* 45cm. Shoulder-height: 82–120cm (regional variation eg Spitzbergen smaller at 82–94cm). Condylo-basal length: *c.* 36cm. Weight: male: 70–150kg; female: 40–100kg (birth: 4–8kg). Dental formula: 0/3, 0–1/1, 3/3, 3/3 = 32–34.

General. Widely protected (eg Norway 1902, Spitzbergen 1925, USSR 1957). Used by Lapps for meat, milk, hide and thread and transport as draft animals. Antlers for carving. Reintroduced to Scotland (Cairngorms). Characteristic 'clicking' sound of reindeer on move caused by small tendon in feet stretching over bony protruberance. Maximum running speed 60km per hour. Hairy nose pad unlike other deer (except Elk) may help to conserve heat.

WHITE-TAILED DEER *Odocoileus virginianus* Pl. 40

Name. *Odocoileus* = hollow tooth (Gk.) from *odous* = tooth, and *koilos* = hollow. *Virginianus* = of Virginia.

Status. Introduced.

Recognition. Between Red and Fallow Deer in size. Rump and underside white. Distinctive pale areas around eyes, muzzle and throat. Pelage: summer reddish-brown; winter greyish. Tail: broad and dark above, white below (largely conceals white rump when held down). Main beam of antlers slides backwards, then abruptly forwards, with erect branches on hind margin. Adults moult twice annually. Spring moult begins mid-March, ends early June. Autumn moult begins August/mid-September and ends by early October. Fawns moult spotted coats at 3–5 months old, acquiring 'adult' coat. Tracks much larger than (though similar to) those of Roe Deer. Droppings same size as hare/Rabbit but slightly oval and deposited in groups, not singly. In wintering areas tree bark s, lichens, mushrooms, reeds and agricultural crops. HOME RANGE: Highly variable depending on habitat quality: 80ha good habitat, 250ha in a poor habitat. In northern parts (eg possibly Finland), summer and winter ranges may be entirely separate. Dispersal is largely by yearling males, whereas females adopt ranges near mother.

Young of both sexes driven away prior to birth of new fawn, only female returning (late summer). SOCIAL: Basic unit is a doe, yearling daughter and two fawns of the year. Adult males solitary or small groups (non-territorial). Male associates with doe until either displaced by a rival or succeeds in mating; mating pairs stay together for a few hours or up to a day. COMMUNICATION: Vocalisations include snorting and blowing. Females call young by low murmur. Young bleat to call mother. Adults can bleat as well, or give hoarse shriek when scared. White underside of tail exposed during almost all alarm and flight situations (hence name). Tail elevated and hairs held erect to make conspicuous white flag which signals danger and prompts group cohesion during flight. Snorting/stamping also alarm signals. Tarsal scent glands produce oily secretion with ammoniacal smell; a tuft of hair covering the gland is raised when deer excited by fear or aggression. By standing with hind legs together and urinating over insides of legs, they saturate the hairs of tarsal glands, the mixture of urine and secretions producing a characteristic odour (probably recognisable individually by other deer). This process is known as 'rub-urination'. Scents convey reproductive information to females, may also terminate seasonal anoestrus and synchronise oestrus to coincide with males' peak condition. Rub-urination also functions in threat between rival males (as do the two signpost types – 'rubs' and 'scrapes'), which serves to establish and maintain dominance hierarchies. Metatarsal scent glands are on outside of each hindleg, about 2.5cm long, bordered by a conspicuous band of white hair. They produce an oily secretion with a pungent, musky odour. Pedal glands between two main toes of each foot have a strong odour and allow deer to retrace their steps (eg to fawn). Small preorbital glands lie in front of each eye – function unknown. Two types of scent marks left by buck in breeding season: (1) young saplings debarked and frayed after thrashing from antlers and rubbed by antlers and (glandular) forehead; (2) scrapes – shallow depression in soil under large trees, often with broken twig above where buck chewed then rubbed twig across forehead as returns to all fours (stands on hind feet to reach twig). May also rub-urinate into scrape. Vision adapted to detect motion rather than resolve detail. Good hearing, excellent sense of smell.

Breeding. Mating in October–January, births in April–September. Sexual maturity 2nd year, except under excellent conditions when females mature at 6 months, males at 1 year. Gestation 200–205 days. Litter size: 1st litter, 1 fawn; subsequent litters, 1–3 fawns, depending on habitat; triplets rare. Litters per year 1. Mammae 4. YOUNG: Fawn spotted, reddish-brown (yellow spotted with white) – cryptic colouring means hard to find when in cover. Weaning age 4 months. Young hidden in dense vegetation for one month, nursed every 4 hours. Female fawns remain with mother for 2 years, return after being driven away prior to birth of new fawn, males for one year. Little parental involvement, and this may be an anti-predator strategy to avoid drawing attention to the fawn.

Lifespan. Max. recorded 15 years (wild) to 25 years (captivity).

Measurements. Head-body length: 180cm. Tail-length: 15–28cm. Hind-foot length: 48–54cm (inc. hoof). Shoulder-height: 90–105cm. Condylo-basal length: male 29–35cm; female 24–29cm. Weight: 50–115kg (birth: 1.5–3.5kg). Dental formula: 0/3, 0–1/1, 3/3, 3/3 = 32–34.
General. Introduced to Europe from N. American and now feral in W. Finland. Carries a cerebral nematode worm that is harmless to White-tailed Deer, but can kill Elk, Reindeer and possibly Red Deer. Since often must travel to find suitable feeding grounds in winter (may shelter in softwood but require mainly hardwood for feeding), snow depth above 0.5m makes travelling uneconomic, and may hence limit populations (snow depth in Finland = 5m for much of year).

ROE DEER *Capreolus capreolus* Pl. 39

Name. *Capreolus* = roe deer (Lat.) (*capra* = goat (Lat.); *capreola* – diminutive i.e. 'small goat').

Recognition. White to buff rump patch (great individual variation and in summer less conspicuous in male than female), black nose and 'moustache', white chin. Appears not to have a tail: distinctive feature of female in winter coat is the anal 'tush', a tuft of white hair which projects backwards between the hind legs, resembling a tail. Pelage: adults lack spots; coat varies from sandy to red-brown in summer, grey-brown to blackish in winter. Antlers: rugose, three points (maximum), grow in winter (most intensively during second half of January and first half of February), shed velvet in March–June, shed antler October–January. Tines point both forward and rearward. Antlers generally less than 25cm in length. Rapid change to winter coat October/November. Change from winter to summer coat in March/June gives moth-eaten appearance. Adults moult before juveniles. Timing dependent on individual condition and social status. Droppings: shiny, black cylindrical pellets pointed at one end (*c.* 14 × 18mm) similar to those of other deer. Rings trodden around tree stump during rut, but difficult to find and recognise. Hairs shed during moult, mainly in spring, are found in scrapes. Male frays young trees and often scrapes at base of same tree (in spring at start of rut); tracks found on soft ground, hair on barbed wire.
Habitat. Woodlands, usually with open ground within, and access to edges of fields. Occasionally moorland with deep heather. In Europe may occur entirely in open in agricultural areas, but the essential feature is usually close proximity of food and cover. Highest numbers in most fertile areas. Scrapes ground before lying in 'bed'. Rests in dense woodland stands (preferably conifer thickets) or dense shrub (bramble, scrub).
Habits. Largely crepuscular and forages throughout night in September–April, but more diurnal if undisturbed, and during May/August due to changes in social behaviour. FOOD: Predominantly browse or forbs (eg bramble, rose, some herbs, conifers, grasses, also young leaves of broad-

leaved trees and bushes in summer), but highly selective – only picks the most nutritious morsels. In autumn, beechmast, acorns, fungi also taken. On moorland: heather, bilberry, cowberry, fungi, top and side shoots of spruce saplings. On agricultural land, eats cereals and broad-leaved weeds among crops. Cycles of feeding and rumination are *c.* 1 hour (summer) to 2 hours (winter) – longer due to high proportion of twigs in diet and morphological changes in the digestive tract. HOME RANGE: 5ha to more than 100ha (eg on fertile coppice land in northern England, territorial males: *c.* 7 ha, retained from year to year and defended April-September; non-territorial males: *c.* 15 ha. Females: *c.* 7ha widely overlapping between genetically related individuals. Population density of 15–25 per km^2 in Central European forests but up to exceptional peaks of 60–70 per km^2 on suitable land. Males forced to disperse after first winter, when adult males become territorial again. Females may disperse but more commonly establish themselves on or adjacent to mother's range. SOCIAL: Both sexes are territorial for the major part of the year. It is not clear whether they are less territorial in winter to save energy when territoriality less necessary, or because food is scarcer in winter and they have to wander far to find it. Male territories are more or less non-overlapping, the limits of their ranges directly associated with the territory boundaries of other males. Female range boundaries also meet neatly, though sometimes mothers let daughters stay on or partly on their territories. There is complete overlap of female territories with male territories, which are slightly bigger. In woodlands, they are relatively solitary and strongly territorial. They prefer dense cover for resting, but will come out into clearings or crop fields for food. Among crops, they will form temporary associations of 7–10 individuals. Roe Deer are opportunistic, simply needing some cover and a supply of high quality food (this being, in woodland, herbs, forbs and browse growing shoots) – a hedgerow may do for this, and in the agricultural areas of southern Czechoslovakia, which has very little shelter, they gather over winter into permanent social groups of 60–70 individuals. Summer: usually solitary or small group of doe and fawns; sometimes also including a buck. Eventually groups form of 2 yearling females or 2–4 yearling and non-territorial males. Yearling female apparently chooses to stay with a territorial buck – during rut 'pair bonds' form for a few days. Winter (on continent): may congregate in herds of up to 30 with linear rank of males (rank changes when antlers cast, reasserted when cleaned of velvet). Dominance of male also diminishes with distance from summer territory. Rank among females less marked. In late winter, even young males (last year's) may be superior to highest ranking female. COMMUNICATION: Alarm bark (both sexes). Courtship: male rasping call, female high-pitched cry. Male rubs head gland (diffusely distributed on front and cheeks) on branches and bushes. Interdigital gland between hindfoot digits marks every step. Scents convey information characteristic of sex, age, dominance, individuality. Also preorbital (see vignette) and preputial glands. Acute sense of smell and hearing. Sight is excellent for detecting moving objects but poor if object motionless.

Breeding. The only Artiodactyl known to show delayed implantation. Mating in July–August, false rut October, implantation late December. Sexual maturity 14 months, but 4 months known in favourable environment. Delayed implantation *c.* 150 days. Post-implantation gestation *c.* 144 days. Litter size typically 75% twins, 20% singles, 5% triplets, but depending very much on environment. Litters per year 1. Mammae 4. YOUNG: Fawn is brownish-black with longitudinal rows of white spots along back and flanks. Black 'moustache' on upper lip very distinctive in fawn. Spots fade *c.* 6 weeks and disappear by October, at the latest. Weaning age 6–10 weeks, though does in good condition give milk until winter. After the mother has mated again, the juveniles continue to nurse. Twin fawns nursed separately (20m apart but may be much more – up to 100 m), 6–10 times daily for a few minutes only (i.e. alone *c.* 23 hours per day). When older, get fed 2–3 times a day for a few seconds only. Young stay with mother till next year's fawn about to be born and then chased off.

Lifespan. Max. 20 years, but exceptional. Normally in wild *c.* 7–8 years male, females slightly older. Age distribution modified by nutritional status of population. 'Exploding' populations have few animals over 6 years old. 'Stagnating' populations have huge fawn mortality and more than 10% of population 7 years and older. Mortality highest up to a few weeks after birth (heavy predation on fawns by foxes, lynx etc. and agricultural machinery), up to 90% die during first winter (starvation or respiratory infections), or spring dispersal due to failure to establish a range.

Measurements. Head-body length: 95–135cm. Tail-length: 2–4cm (visible only during defecation, when raised). Shoulder-height: male 64–67cm; female 63–67cm. Males proportionately larger (cf females) in thriving populations, but little difference or even slightly smaller than females under bad conditions. Condylo-basal length: 150–196mm. Weight: 16–35kg (birth: 1.3–2.3kg). Dental formula: 0/3, 0/1, 3/3, 3/3 (deciduous 0/3, 0/1, 3/3, 0/0 = 32(20)).

General. Occasionally males (perruques) fail to shed antlers (indicative of low androgen levels) and so retain these permanently in velvet, and permanently increasing in size. Occasional antlered females. Many other antler abnormalities. Widespread in central Europe (N. Sea islands to

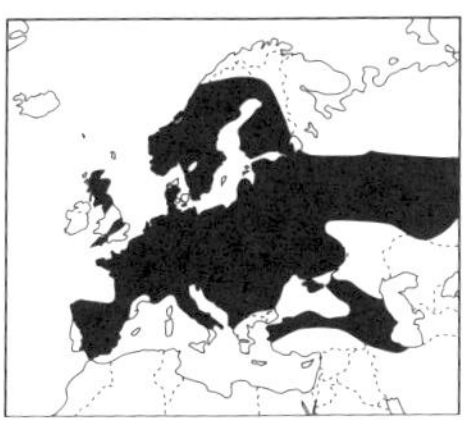

<*Roe Deer.* Throughout Europe, but absent from Ireland, much of Portugal and Greece and large parts of England and Wales.

>*Bison.* Reintroduced in semi-wild conditions in Poland.

above timber line). Western France and UK – local to well distributed. Populations generally increasing. In central Europe, Roe are the deer species with the highest hunting bag (more than 700,000 per year in western Germany) and by far highest income from venison.

BISON *Bison bonasus* Pl. 45

Name. Wisent. *Bison* = *wisant, wisunt* = bison (Lat.); *bonasus* = bison (Lat., Gk.).
Status. Vulnerable
Recognition. Coarse, tangled mane emphasises heavy forequarters especially in male. Short, broad head, stocky frame, humped back. Pelage: dark brown. Horns in both sexes. Moults once a year in spring or early summer. Fieldsigns include conspicuous wallows.
Habitat. Deciduous or predominantly deciduous forest with scattered open glades, and swampy woodlands.
Habits. Nocturnal, and parts of day (rests in afternoon). FOOD: Vegetarian: grasses, buds, leaves of shrubs and trees, small twigs. Acorns (in autumn); heather and evergreens (in winter). HOME RANGE: Population density: 12 per 100ha (Poland), 3–4 per 100ha (Caucasus). Previously, herds of hundreds on migrations and concentrated in favoured feeding grounds. SOCIAL: Herds 10–30, led by old bull. Very old individuals live solitarily. Bulls join matriarchal herds in rut, forming rutting territories (where churn soil with horns). COMMUNICATION: Bellows.
Breeding. Mating in August–September. Births in May–June (females leave herd to calve). Sexual maturity 2–4 years. Gestation 260–270 days. Litter size 1. Calving interval 1–2 years. YOUNG: Weaning age 7 months. Calf stays with mother for 3 years. Female guards calf closely and charges if threatened.
Lifespan. Max. recorded 22 years (captivity: potentially 40 years).
Measurements. Head-body length: 250–270cm. Tail-length: *c.* 80cm. Shoulder-height: 180–195cm. Condylo-basal length: 487–536mm. Weight: male 800–900kg; female 500–600kg (birth: *c.* 30kg). Dental formula: 0/3, 0/1, 3/2–3, 3/3 = 30–32.
General. Extinct in wild in 1919 (in Roman times found in Germany and Belgium, with records of herds of many thousand on migration). Reintroduced to wild from zoos, especially in Bialowiecza forest, Poland. Population now 1500–2000. Similar to N. American buffalo.

CHAMOIS *Rupicapra rupicapra* Pl. 43

Name. *Rupis* = crag (Lat.); *capra* = goat.
Status. The Chartreuse Chamois *R.r. cartusiana* of France is endangered; the Tatra Chamois *R.r. tatra* of Czechoslovakia and Poland is rare. The Apennine Chamois *R. pyrenaica* of the Abruzzo Mountains of Italy is vulnerable.

Recognition. Contrasting marks on either side of head extend from ear over eye to muzzle. Fairly long, upright horns, bent back in shape of hook. Horns in both sexes, but thicker in male and shorter and thicker in Alps than in Appenines. Pelage: summer, pale brown, dark brown legs, dark stripe along back; winter, whole body darker. Moult May–June, grows winter coat September–November. Prints *c.* 6 × 3.5cm, hooves well separated but usually parallel.
Habitat. Rocky, precipitous mountains – up to 3500m in summer (eg Alps, Pyrenees). Summer and autumn: meadows around tree line and above. Winter: in forests to *c.* 800m.
Habits. Diurnal; rests during middle of day – may feed at night if moonlit. FOOD: Grazes on grasses and herbs of alpine meadow; in forests browses on trees (including conifers, bark and needles). HOME RANGE: Variable according to area, sex, age. Most young males wander between the ages of 3–5 years. SOCIAL: Herds up to 100 females and young; males solitary, joining females during rut. Usually groups break up in winter, but sometimes may coalesce into herds of several tens of animals, in relation to local food availability. COMMUNICATION: Sharp, whistling alarm call; goat-like bleating. During rut males mark with scent glands behind horns and display with erect hairs of dorsal stripe. Hearing excellent, smell and sight good.
Breeding. Mating in mid-November–early December, births in May–early June. Sexual maturity 4 years, though 2 year old females may conceive. Gestation 24–26 weeks. Litter size 1 (occasionally 2). Litters per year 1. Mammae 4. YOUNG: Weaning age about 6 months.
Lifespan. Max. 22 years (captive), 15–17 years (wild). Causes of death include avalanches, epidemics (Keratoconjunctivitis), Lynx, Wolves.
Measurements. Head-body length: 100–130cm. Tail-length: 7–8cm. Hind-foot length: *c.* 33cm. Shoulder-height: 70–80cm. Condylo-basal length: 18–20cm. Weight: male 30–60kg; female 25–45kg. Dental formula: 0/3, 0/1, 3/3, 3/3 = 32.
General. Skin used for soft 'chamie' leather. Tends to graze at lower altitudes than Ibex. Fantastic agility. Previously this one species was composed of 8–10 subspecies (opinion differs) but Cantabrican, Pyrenean

< *Chamois.* Mountainous regions, such as Pyrenees and Alps and Appenines (Isard).

> *Musk Ox.* Main population on Greenland, with isolated populations in Sweden and Norway.

and Apennine chamois recently ascribed to a separate species: *Rupicapra pyrenaica*. This species, the Apennine Chamoise or Isard is somewhat smaller in size and has a large whitish patch on throat, neck and shoulder, and rump patches on the dark brown winter coat and pale reddish-brown summer coat.

MUSK OX *Ovibos moschatus* Pl. 44

Name. *Ovis* = sheep, *bos* = ox (Lat.).

Status. Rare in W. Europe. Introduced to Norway from Greenland, some natural movement into Sweden.

Recognition. Heavy coat of long (up to 70cm) guard hairs covers tail and outer ears, and can hang almost to hooves in winter. Horns in both sexes grow from broad bases in centre of crown, somewhat narrower in females (separated by deep, hairless groove in older animals, but by ridge of hair in younger ones). Loses winter pelage April/June, new growth in August.

Habitat. Arctic tundra. Southern limit coincides with July temperatures of 10°C. Winter: above tree line where little snow. Summer: birch forest or upper coniferous forest.

Habits. Diurnal. FOOD: Winter: lichens, vaccinium twigs, grass, sedges. Summer: Willow, Dwarf Birch, herbaceous plants (eg Alpine Lettuce, Mountain Angelica). HOME RANGE: Movement between winter and summer ranges up to 80km. SOCIAL: Gregarious. Summer: old males form harems *c.* 10 animals. Winter: herds of up to 50 males, females and young. Bulls compete with head-on charges at 40km per hour from 50m distance; up to 10 charges before loser gives up. COMMUNICATION: Bulls bellow; calves bleat. Name derives from musky scent of urine sprayed on fringe of abdominal fur.

Breeding. Rut: July and August. Birth: late April–mid June. Sexual maturity: male 5 years; female 2 years. Gestation 8–9 months. Litter size 1 (occasionally 2). Calving interval: usually 2 years (maximum record 9 calves in 13 years). YOUNG: Curly coat (see vignette).

Lifespan. Max. 23 years. Prey to wolves.

Measurements. Head-body length: male 200–250cm; female up to 190cm. Tail-length: 10–12cm. Hind-foot length: 42–51cm. Shoulder-height: male 130–165cm; female max. 110cm. Condylo-basal length: male 42–50.5cm; female 39–45.5cm. Weight: 225–400kg. Dental formula: 0/3, 0/1, 3/2–3, 3/3 = 30–32.

General. Reintroduced in 1940s to Norway. 5 animals wandered to Sweden where more than 30 now. When threatened by Wolf or Man, herd closes in characteristic ring with horns facing outwards and calves in centre.

ALPINE IBEX *Capra ibex* **Pl. 42**

Name. *Capra* = goat (Lat.).
Status. Out of danger.
Recognition. Both sexes have horns; male: curved backwards (maximum 85cm) with relatively flat anterior surface and ribbed with prominent transverse ridges; female: much smaller horns. Pelage: brownish-grey above, pale belly. Shed winter coat May, regrow winter coat October/November. Droppings very similar to domestic goats', spherical/slightly cylindrical, *c.* 1cm diameter, always separate. Leaves little other indication of presence in rocky areas.

Habitat. High mountains (2000–3500m). Only descend below tree line in spring (seeking flush of fresh grasses).
Habits. Largely diurnal. FOOD: Grasses, herbaceous plants, lichens. HOME RANGE: Female *c.* 1km^2; male groups tend to move on higher ground than females in summer. Population density 9 per km^2 (average) but varies according to habitat. SOCIAL: Female and young of up to 3 years of age live in sedentary groups of up to 50 throughout year. Males more mobile, grouping in summer and autumn, more solitary in winter. Adult males join females only in breeding season. COMMUNICATION: Vocalisation is short whistling hiss. Hearing excellent, smell and sight good.
Breeding. Rut: December–January, births late April–early May. Females sexually mature at 3–6 years. Gestation 21–23 weeks. Litter size 1 (occasionally 2). One litter per year. 2 mammae. YOUNG: Weaned at 6–7 months (see vignette of kid). Female protects kid by attacking with horns or decoying intruder.
Lifespan. Max. recorded 18 years. Avalanches are important cause of death.
Measurements. Head-body length: male 130–150cm; female 105–125cm. Tail-length: 12–15cm. Hind-foot length: *c.* 31cm. Shoulder-height 65–90cm. Condylo-basal length: 22–26cm. Weight: male 65–120kg; female 40–70kg. Dental formula: 0/3, 0/1, 3/2–3, 3/3 = 30–32.
General. Formerly Ibex were abundant in the Alpine chain above tree limit. Then reduced by over-hunting to one herd (Gran Paradiso National Park, Italy). Subsequent reintroduction campaign successful (eg more than 40 colonies in Switzerland, total population *c.* 3500 descended from

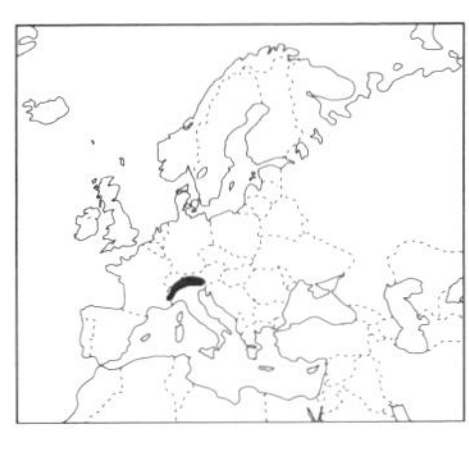

<*Alpine Ibex.* Confined to the Alps.

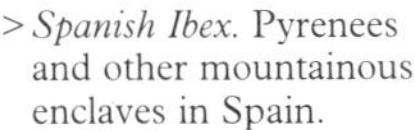

>*Spanish Ibex.* Pyrenees and other mountainous enclaves in Spain.

34). Males, in particular, subject to danger of avalanche because seek richer food on dangerous terrain. Can produce fertile hybrids with Domestic Goat. Siberian form has larger horns (128–143cm) and a conspicuous black beard.

SPANISH IBEX *Capra pyrenaica* **Pl. 42**

Name. Spanish Goat. *Pyrenaica* = of the Pyrenees.
Status. Endangered subspecies *C.p. pyrenaica* (Ordesa National Park, Spain).
Recognition. Horns twisted in slight spiral outwards and upwards with sharp posterior keel, but shape variable. Lacks transverse parallel ridges (cf Alpine Ibex). Pelage: pale with black stripe along back and dark legs, flanks, chest and forehead (extent varying with locality). Male has beard. Moults once a year – April to May. Droppings 1.5–1.8 × 0.1–1.0cm (similar to Domestic Goat). Footprints 5–7 × 3–4cm.
Habitat. Mountainous areas of Pyrenees, central and southern Spain.
Habits. Mainly diurnal with two peaks of activity – dawn and dusk – with a rest period midday. FOOD: Grazes on alpine grasses and herbs, and also browses when food available. HOME RANGE: During spring season, males 4.28 km^2 and females 0.87 km^2, decreasing during winter rutting season to 1.29 km^2 and 0.25 km^2 respectively (these averages for a high population density). At low population density, home range is increased, eg an average of 3.21 km^2 (spring) and 1.94 km^2 (rutting) in females. Dispersal distances average 1.8km. SOCIAL: All ages and sexes in large groups of 50–60 individuals outside breeding season.
Breeding. Rut *c.* 10th November-20th December, births in May. Sexual maturity 2.5 years for females. Male access to females limited by dominant males. Gestation 23–24 weeks. Litter size 1, occasionally 2. One litter per year. 2 mammae. YOUNG: Weaned at 6 months.
Lifespan. Max. recorded 12–16 years. Some juvenile mortality attributable to Wolf and Golden Eagle.
Measurements. Head-body length: female 116–136cm; male 137–153cm. Tail length: 12–13cm. Hind-foot length: female *c.* 30cm; male 29–34.5cm. Shoulder height: female 59–75cm; male 68–77cm. Weight: female 25–35kg; male 60–80kg. Dental formula: 0/3, 0/1, 3/3, 3/3 = 32.
General. Similar to Alpine Ibex (and Domestic Goat, with which some cross-breeding has produced variation in colours). Forms from southern Spain smaller than those in Pyrenees and vary in horn shape and markings. Formerly found in all mountain regions of Spain.

CRETAN WILD GOAT *Capra aegagrus* Pl. 42

Name. *Aigagros* (Gk.) = wild goat: corruption of *aix, aigos* = goat; *agrios* = wild (both Gk.).
Recognition. Similar to Alpine Ibex, but male's curved horns have sharper (wedge-shaped) leading edge, and transverse ridges irregularly spaced and often indistinct (vignette shows young female). Moults end of April until the second half of May (Lefka Ori, Somaria Gorge). NB no wild goat exists in Italy: only recent feral ones on Montechristo Island.
Habitat. Mountainous regions, particularly with shrub covered areas, rocky outcrops with caves, overhangs and thickets, and conifer forests (of *Pinus brutia, Cupressus sempervirens* etc.).
Habits. Active both day and night throughout October–May when it is cool, almost totally eliminating daytime movements during the hot dry summer months of June–September. FOOD: Grasses and herbs, eg *Pistacia lentiscus, Poterium spinosum* and *Juniperus macrocarpus*. HOME RANGE: Size related to altitude, with larger ranges at higher, barren sites in the Samaria Gorge. About 2 km^2 for a family group of females and young from spring to early autumn, increasing in winter due to the scarcity of food. On the Mediterranean islands of Dia, Theodoru and Agnii Pontes they exist in densities of 600 per ha, 100 per ha and 70 per ha respectively. Around the Samaria Gorge region the average densities are 3.2 per ha, or 4.3 for females with young. Young males disperse in the breeding season, probably covering 3–5km. COMMUNICATION: Generally similar to that of Domestic Goat. Alarm is shown by a snort. In the absence of visual contact, mothers and kids communicate by a low bleating at 5 calls per minute. Senses of smell and sight well developed.
Breeding. Rut October–December (eg Lefka Ori: end of October, generally after the first autumn rains, to second half of December. On the islands of Theodorou, Dia and Agnii Pontes the rut begins at the beginning of October. Births April/May. Males probably sexually mature at 1 year old, females 10–11 months old. Gestation 142–153 days. Litter size 1–2. One litter per year. 2 mammae. YOUNG: Weigh around 2kg at birth. Weaned at 3 months on the Mediterranean islands, 4.5 months in the Samaria Gorge area. Parental care as in domestic goats; only the female rears the young.
Lifespan. Males commonly live more than 10 yrs; 14–15 yrs on the Lefka Ori. Mortality higher for females.
Measurements. Head-body length: 100–150cm. Tail-length: 11–16cm. Hind-foot length: 21–25cm. Shoulder-height 58–74cm. Condylo-basal length: *c.* 20cm. Weight 15–80kg (cf W. Crete 15–40kg; Iran 45–80kg). Dental formula 0/3, 0/1, 3/2–3, 3/3 = 30–32.
General. Not native to the Mediterranean islands, as evidenced by ab-

< *Cretan Wild Goat.* Confined to Crete and other Greek islands, and isolated sites in Greece and Turkey.

> *Mouflon.* Originally from Corsica and Sardinia, now widespread on mountainous regions of Europe.

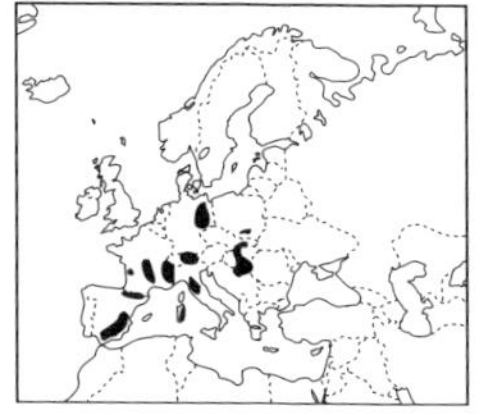

sence of fossils (whereas there are fossils of the rupicaprids). Evidently introduced by Man, becoming feral, around the Neolithic period or perhaps later. High degree of interbreeding with Domestic Goats. Very variable colouring. Much dispute as to whether any pure Cretan Wild Goats left on Crete.

MOUFLON *Ovis orientalis* Pl. 43

Name. *Ovis* = sheep (Lat.); *musimo* (Lat.), *mousmon* (Gk.) = sardinian sheep.
Status. Sardinian Mouflon *O.o. musimon* found in Sardinia and Corsica is Vulnerable; *O.o. ophion*, Cyprus Mouflon also Vulnerable.
Recognition. Small, wild sheep, short-haired coat. Dark tail contrasts with white buttocks. Adult male has white saddle-patch, 'socks' and muzzle, dark neck and shoulders and upper legs. Females, and juvenile males (see vignette), red/brown-light chocolate brown above and white below. Markings much less distinct in summer. Mature male horns curled in almost one full revolution (up to 85cm); females lack horns in Sardinia and Cyprus and have short horny stumps in Corsica. Muzzle and saddle increasingly white with age in Sardinia and Corsica but this does not appear to be the case in Cyprus. Tracks strongly splayed outward.
Habitat. Steep mountainous woods near tree line, but also meadows (migrates to lower altitudes in winter).
Habits. Largely nocturnal/crepuscular, depending largely on Man's activities. Not particularly nocturnal in Cyprus now because no hunting; feed throughout hours of daylight in winter but avoid heat of day in summer. FOOD: Grasses, forbes, young leaves and tender twigs (bark in winter). HOME RANGE: Flocks of ewes and lambs occupy clearly defined home ranges and only joined by rams during rut. Groups include 1–8 females and 1–14 males. Groups small by comparison with most wild sheep, because occupy woodland habitat. COMMUNICATION: Goat-like bleat. Hissing alarm whistle. Hearing excellent, smell and sight good.
Breeding. Rut in October–November, births April–May. Sexual ma-

turity: male first year; female 18 months. Gestation 150–170 days. Litter size 1 (occasionally 2). Litters per year 1. Mammae 2.
Lifespan. Max. recorded 14 years.
Measurements. Head-body length: 110–130cm. Tail-length: 6–10cm (adult). Hind-foot length: *c.* 23cm. Shoulder-height: female: 65cm; male: 75cm. Condylo-basal length: *c.* 225mm. Weight: male 35–50kg; female 30–40kg. Dental formula: 0/3, 0/1, 3/3, 3/3 = 32.
General. Smallest wild sheep, originally from SW Asia. Introduced to Mediterranean islands of Corsica, Sardinia and Cyprus in the Neolithic period. Introduced to Central and South Europe as game. In 1937 only 15 individuals in Cyprus; in 1939 Paphos Forest declared reserve and all goats expelled from the reserve. By 1966 *c.* 200 Mouflon in the reserve.

Order Rodentia – Squirrels, Rats, Mice and Voles

Origins: An insectivore-type ancestor gave rise to the first Rodents, called Paramyids, which arose in the late Palaeocene *c.* 60 mya. Their closest living relative is *Aplodonta*, the American Mountain Beaver. Rodents have some dramatic ancestors: *Castoroides* was a giant beaver, at 200kg and 2.5m long, the size of a bear (but not a very intellectual one, in so far as its brain was the same size as that of a modern beaver).

Main features of the Rodentia: The rodents are classed into three sub--orders, based on the working of the bone-muscle pulley systems that control their gnawing jaws – gnawing being the essence of rodent life. The primitive arrangement is called squirrel-jawed (sciurognathus) in which the chewing (masseter) muscle drops vertically from the cheek (zygomatic) arch of the skull to a bony flange behind and below the teeth on the lower jaw. This flange is similar in the modern squirrels (sub-order Sciuromorpha) and mice and rats (sub-order Myomorpha), but it is angled outward in the more modern porcupine-jawed (hystrognathus) rodents such as porcupines and cavies (sub-order Hystricomorpha). Each of the three sub-orders has opted for a different arrangement of the muscles to add characteristic nuances to their gnawing action – minute and seemingly unimportant differences that in fact underpin one of the greatest mammalian success stories.

The squirrel stock arose about 30 mya in the late Oligocene and is now found on every continent save Antarctica and Australia. A more modern success story is the innovation of the mouse-type jaw-muscle arrangement engineered on the squirrel-model jaw. The mouse-design tends to have fewer molars (three, two or even one) than do other rodents. In the earliest representatives, like *Paracricetodon* from the Oligocene (37–24 mya), the teeth had cusps linked by ridges that foreshadow the modern grinding arrangement characteristic of voles. But it was less than 10 mya, in the late Miocene, that the mouse design really took off, with three explosive waves of speciation: (i) in less than two million years more than 350 species of New World Mice (Sigmodontinae, sometimes called Hesperomyinae) spread from North America to the rest of the New World, while the hamsters (Cricetinae) spread through the Palaearctic. Many retained cuspid teeth. (ii) A niche for diminutive grazers was made possible by the evolution of grasses in the Miocene and was occupied by voles (a subfamily known as Arvicolinae or Microtinae), which arose from the hamster lineage in the late Pliocene (*c.* 2.5 mya). The 100 or so extant species of vole owe their success to teeth on which the enamel forms a pattern of zig-zags aloft wide and high crowns (hypsodont) with parallel sides (prismatic) and roots which remain open throughout their lives, allowing continuous growth. (iii) The true rats and mice (subfamily Murinae) probably arose in SE Asia in the late Miocene and have an omnivorous, but largely vegetarian, diet. Their food is prepared for digestion on the low-crowned cusps of rooted teeth. There fossils are rather uncommon until the late Pleistocene but they were most certainly abundant and diverse in Africa, southern Asia and Australia long before this. In the late Pleistocene, in company with early man, some murines (such as *Rattus* and *Mus*) radiated worldwide to produce the 500 murine species we know today. Human association with commensal Rodents is truly ancient: bones of Rodents of the genera *Mus* and *Rattus* are found alongside the million-year old remains of Man in mid-Pleistocene encampments.

Finally, the porcupine-like Rodents have Old and New World lineages, the former first known from Egypt. They tend to retain four molars in each jaw, each low to medium crowned with 3–5 transverse ridges. Today the biggest Rodent in the world, the 40kg Capybara, belongs to this suborder, but five million years ago its Pliocene relatives in South America were the size of a modern rhino.

The majority of Rodents are seed-eaters, but some are insectivorous and some are very versatile omnivores. Rodent teeth are self-sharpening because they have a thick strip of hard enamel along the surface nearest the lips (labial surface), while the surface nearer the tongue (the lingual surface) is softer and wears away more rapidly. Thus the enamel strip stands out as a sharp ridge. In contrast, since Lagomorph incisors are encircled by enamel, they require a special sharpening mechanism.

Rodentia in Western Europe: Nearly 40% of all living mammalian species are Rodents, making them the largest order of mammals. Together they amount to 30 families, 389 genera and more than 1702 species. A further 12 families and more than 300 genera are known from fossils. There are three sub-orders: Sciuro-, Myo- and Hystricomorpha, all three of which are represented in Europe by a total of seven families.

The Castoridae or beavers are large, social Rodents with huge incisor teeth and aquatic adaptations including flat, scaly tails, webbed hind feet. Their throats are uniquely structured so that they can be blocked by the back of the tongue, and the lips can close behind the incisors so the animal can gnaw and carry sticks underwater without choking. They are well-known for building dams.

The Sciuridae include squirrels, marmots, flying squirrels and sousliks. They typically have a long, cylindrical body, bushy tail, good vision, and occupy a diverse range of habitats. Some are highly arboreal. Squirrels typically have bushy tails (80–90% of the length of their heads and bodies) and five upper and four lower cheek teeth which are low crowned and cusped.

The Muridae is an enormous family of rats, mice, voles, lemmings, and the blind molerats. The voles and lemmings are part of the subfamily Arvicolinae and are adapted to feeding on grass and other tough vegetation, having high crowned cheek teeth. The rats and mice, of the subfamily Murinae are generalists, having low crowned, cusped cheek teeth. Blind molerats have specialised in being fossorial (NB blind molerats, subfamily Spalacinae, are not the same thing as the African molerats, family Bathyergidae). The Muridae have three upper and three lower cheek teeth, and almost naked tails. The tails of the mice and rats are relatively longer (80–120% of head and body length) than those of the voles (25–60% of head and body length).

The Gliridae or dormice are the only Rodents without a caecum. In common with the Capromyidae, they have four upper and lower cheek teeth, but these are low crowned with transverse ridges (those of the Coypu, in contrast, are high crowned). Dormice typically have a long hibernation period (usually 7 months in European species), during which they accumulate fat. They are intermediate, in form and behaviour, between mice and squirrels.

The Hystricidae or Old World porcupines are partly covered in long quills and spines. Their mammary glands are situated on the side of the body and they are generalist herbivores.

The Capromyidae contains only one species, the Coypu, introduced to Europe from South America. It is a large robust rodent, living on river banks, with many aquatic adaptations.

The Zapodidae contains the birch mice and jumping mice, of which only birch mice are found in Europe. They are nocturnal, and travel by jumping. Birch mice can eat large amounts of food at one time and can survive prolonged shortages of food. They hibernate for about half the year, and feed mainly on seeds, berries and insects.

GREY SQUIRREL, *Sciurus carolinensis* Pl. 46

Name. *Skia* = shadow (Gk.); *oura* = tail (Gk.); ie sits in shadow of own tail).

Status. Introduced.

Recognition. Distinguished from Red Squirrel by grey back and tail (flanks sometimes reddish), larger more robust build, and small ear tufts (often unnoticed). Melanism occasional, albinism rare. Body fur moults twice a year, April–July and September–December. Spring moult proceeds from front to back, autumn moult from back to front. Tail hairs (from base to tip) and ear tufts moult once a year, new hairs growing in late summer and autumn. Fieldsigns indistinguishable from Red Squirrel.

Habitat. Common in deciduous and mixed woodland, occurs in hedgerow, trees, parks, gardens. Nest is compact spherical drey, 30–60cm diameter, often built away from main trunk with outer frame of hardwood twigs and leaves, central cavity of dry leaves and grass. Also temporary summer platforms and dens in tree hollows (hole gnawed to 7–10cm diam). Each squirrel may use several dreys.

Habits. Diurnal. Active from before sunrise to after sunset. Typically active for whole day in autumn, for only 1–4 hours in winter (mainly morning), and for 3–8 hours in summer (often with bimodal pattern).

FOOD: Granivore-omnivore. Vegetation eaten includes acorns, beechmast, tree shoots, flowers, samaras, nuts; also fruits, roots, cereals. Strips bark and eats sappy tissue beneath. Occasionally insects and bird eggs. Eats 40–80g per day, caches surplus nuts, mast and cones in scattered sites 2–5cm below soil or in tree hollows (see vignette). HOME RANGE: Males and females normally 2–10ha (UK hardwoods) 3-dimensional range, but both sexes can commute more than 1km for food in winter, and males can cover more than 100ha during courtship. Population densities may reach 9 per ha in summer, following abundant mast crops. Dispersal generally less than 1km but sometimes more than 10km. SOCIAL: Range peripheries overlap extensively, but small core areas may be exclusive to breeding female squirrels. COMMUNICATION: Vocalisations include scolding 'chuck-chuck-cheree' and low 'tuk-tuk' teeth chattering. Also tail flicking and foot stomping. Scent marks using urine – both males and females gnaw patches of bark on trees and spray with urine. Acute vision, hearing and smell.

Breeding. Sexual maturity 10–12 months. Gestation 42–45 days. Litter size 3 (1–8). Usually 1 litter in spring following good mast crops, otherwise in summer–autumn, occasionally a female will produce a litter in each season. Mammae 8. YOUNG: Independent at 10–16 weeks. Weaning age 10 weeks. Parental care by female only; if disturbed carries young to new drey.

Lifespan. Max. recorded 8–9 years (wild); 20 years (captivity). Adult mortality 30–45%, higher in first year. Average life expectancy at birth *c.* 1 year (rising to 2 years if survive to leave drey).

Measurements. Head-body length: 23–30cm. Tail-length: 19.5–25cm. Hind-foot length: 6–7.5cm. Condylo-basal length: up to 56cm. Weight: Extremes of 400–700g but normally 450–650, seasonal variation with peak in winter; birth weight: 14–18g, weaning 180–250g. Dental formula: 1/1, 0/0, 2/1, 3/3 = 22.

General. Introduced from USA to *c.* 30 sites in England and Wales between 1876 and 1929. Common throughout England and Wales, south of Cumbria, locally common in Scotland. The Red Squirrel's disappearance apparently coincided with Grey Squirrels' arrival, although this is not always the case (eg Thetford Forest in Norfolk). Red Squirrel disappearances typically follow Grey Squirrel arrival, although loss of hazels and mature conifers, or disease, sometimes eliminates Reds in advance. Grey Squirrel's, unlike Red, are favoured by good acorn crops. Pest of forestry.

<*Grey Squirrel.* Widespread in England and Wales, with populations also in Scotland, Ireland and, following introductions in 1948, a 2000 km^2 area in the vicinity of Piedmont (SE of Turin) and Liguria (near Genoa) in Po valley (N. Italy).

PERSIAN SQUIRREL *Sciurus anomalus* Pl. 46

Name. *Anomalos* = abnormal (Gr.).

Recognition. Similar to Red Squirrel, but no ear tufts, reddish brown and grey belly. No upper second premolar teeth.

Measurements. Head-body length: 20cm. Tail-length: 13–17cm. Dental formula: 1/1, 0/0, 1/1, 3/3 = 20.

General. Found on island of Lesbos in Aegean Sea, Caucasus Mountains, Anatolia and from Iran throughout Asia Minor.

RED SQUIRREL *Sciurus vulgaris* Pl. 46

Name. *vulgaris* = common (Lat.).

Recognition. Pelage warm reddish-brown (summer) to deep chocolate brown tinged with grey (winter). Tufts on ears in winter (see vignette). Bushy tail. Very variable, over 40 subspecies (some doubtless invalid) and many races dimorphic (i.e. red vs brown phases) or trimorphic (red, brown and black phases). Colour may be very variable in any one locality, ranging from almost black to buff via all normal reds and browns. Melanism less common in UK than on continent. Some islands have populations of only black animals, eg Funen Island (Denmark). Male squirrels in breeding condition recognisable by darkened, hairless skin of scrotum in adult male, stained orange-yellow (subadult males have bright white belly fur); breeding female has swollen, hairless teats. Spring moult takes *c.* 6 weeks. Hind paw prints larger, with 5 toes, placed in front of much smaller forepaw tracks with 4 toes. Food remains include long, spiral twists of stripped bark on conifers, split nut shells, stripped conifer cones and husks like chestnut/acorn etc. The skull distinguished from Grey Squirrels because the condylo-basal length is 2 mm: the cranium is also deeper and postorbital process longer and narrower.

Habitat. Large blocks of conifer forest (50ha) (UK race adapted to native Scots Pine), also parks and gardens. Up to 2000m in the Alps and Pyrenes. Less abundant in small conifer woods, or pure hardwoods (eg beech or oak/hazel). Nest is spherical drey (c. 30cm diam.) usually above 6m; outer frame of twigs; cavity (12–16cm diam.) lined with moss and grass (thicker lining in breeding drey). Drey usually next to main trunk of tree (of any species); also in bushes. May also have a nest in hollow trunk or branch. Each squirrel may use several dreys.

Habits. Diurnal, (emerging within 30 mins of dawn). Major peak of activity 3–4 hrs after dawn; secondary peak 2–3 hrs before dusk in the summer. In winter, single morning peak. Do not hibernate, but remain in nest for several days in harsh weather. The two activity peaks typical in spring may arise because of a seasonal digestive 'bottleneck': at this time of year all the nuts from the previous autumn have been used up so the squirrels eat bulky buds and shoots – the gap between the two peaks of activity may allow their full stomachs to empty in preparation for the next bulky meal. FOOD: Largely vegetarian. Conifer (spruce and pine) seeds, also acorn berries, fungus, bark and sap tissue. Occasionally eats vertebrates (bird remains in 4 of 1600 stomachs sampled), and, very occasionally, eggs. Daily consumption *c.* 5% of body weight. Soil and tree bark eaten for roughage and minerals. Stores food. HOME RANGE: Average size 7ha, or normally 2–10ha (ranges are 3-dimensional eg in Scots Pine, E. Scotland: 470 × 285 × 15 m). Males and females have similar sized ranges (eg in deciduous woodland, England: ranges averaging 4ha, of which 1ha core area heavily used), but males make long excursions in winter looking for females. Seasonal variation (eg Sweden, male, summer: 214m diameter, winter: 122m diameter). Population density: generally 0.2–1.6 squirrels per ha, whether coniferous or broad-leaved forests (but up to 10 per ha reported in Finland), 12 dreys per ha in pinewood (i.e. several per squirrel), 0.8 squirrels per ha (E. Scotland), 0.5 squirrels per ha (E. Anglia, UK). In deciduous woodland autumn Red Squirrel population correlates with that year's hazel crop; spring population depends on overwinter survival rate and thus the hazel crop of the previous autumn. Wet autumns are advantageous because of increased supply of edible fungus, but wet winters lead to chilling. Dispersal is unpredictable: some female juveniles of 10–18 weeks of age moved 1.5km to new wood in one study (UK), with male members of the litter settling near the natal range. SOCIAL: Ranges overlap, especially in winter when ranges larger, but whereas males and males and females overlap at random, females tend to be more spaced out. Sometimes share dens, especially in cold weather. Possible female hierarchy. Low levels of aggression generally. Most common encounters include some chattering and tail flicking. During mating chases dominant males thought to take the lead over subordinates and to monopolise matings. COMMUNICATION: Vocalisations include smacking, chattering calls ('chucking' noise associated with tail-flicking and foot-stamping). Scent glands: 'lip plates' occur (thickened areas of skin associated with secretory glands of the oral region). Scent marks with oral glands and urine. Males and females urine mark on small patches of stripped bark, especially on the underside of branches. Vision excellent with exceptional focussing power; wide angle of vision.

Breeding. Peak of mating in January–March, usually 1 litter in spring (March–May) following good cone crops, otherwise usually in summer (July–September). Sexual maturity 10–12 months (occasionally 6 months, when spring-born males develop scrotal testes in first winter). Gestation 36–42 days. Litter size 3 (1–8). Litters per year 1–2. Female receptive for only one day. Mammae 8. YOUNG: First activity away from nest at

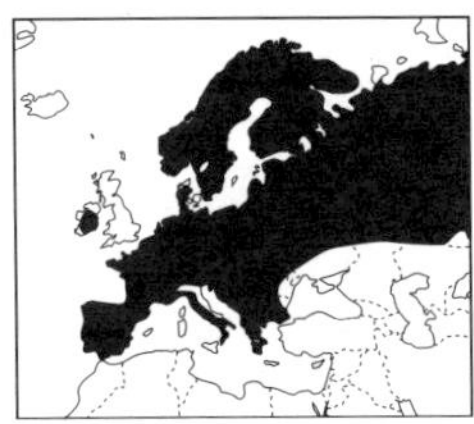

< *Red Squirrel.* Replaced by Grey Squirrel in England and Wales. *Persian Squirrel* is found on the Greek island of Lesbos.

> *Alpine Marmot.* Italian and Swiss Alps and S Germany.

7–8 weeks; independent at 10–16 weeks. Weaning age 7–10 weeks. Parental care involves female only; mother carries young if disturbed. Maternal protective behaviour continues post- weaning for *c.* 2 weeks.

Lifespan. Max. recorded 6–7 years (wild); 10 years (captivity). Causes of death include starvation and road accidents. Predators include martens, raptors, domestic dogs and cats. In one population an estimated 70% mortality before 1 year old, thereafter better survival (eg adult survival 74% per year). In English deciduous woodland higher mortality of summer-born young than in spring-born young, because of food (hazel nut, acorn) shortage in summer.

Measurements. Head-body length: 18–24cm. Tail-length: 14–20cm. Prints of forefoot c. 40 mm long by 20 mm wide, hindfoot c. 50 mm long by 25-35 mm wide. Hind-foot length: 4.9–6.3cm. Condylo-basal length: 44–48mm. Weight: Extremes represented by 200–480g, normally 250–350g (birth: 10–15g). Average weight in English deciduous woods 297g, in Scottish pine forest males average 279g, females 278g; some evidence that weights (especially in winter) higher in coniferous than deciduous woods, eg in Belgian deciduous woodland 305–340g, Belgian coniferous woods 320–350g. Dental formula: 1/1, 0/0, 2/1, 3/3 = 22.

General. Widespread in Europe (and to NE China) and Ireland (and Isle of Wight), but widely replaced by Grey Squirrel in England and Wales, and from Po valley and eastern coast of Italy. Absent from southern Spain and the Mediterranean Is. Adaptations for arboreal leaping: light bones, long curved claws. When jumps fluffed-out tail, spread limbs, and lose skin on flanks help maintain height (see Persian Squirrel vignette). Anti-predator behaviour involves scuttling to far side of tree trunk and pressing flat and motionless to bark. Good swimmer. See comments under Grey Squirrel regarding Red Squirrel's competitive disadvantage except in large blocks of coniferous forest.

ALPINE MARMOT *Marmota marmota* Pl. 47

Name. Mountain mouse, corrupted from *murem montis* (Lat.) (*mus muris* = mouse, *mons montis* = mountain).

Recognition. Heavily built, short legs, large head, rolling gait. Moults once a year, in June, July or August depending on latitude. Well fed animals moult early. In mountainous regions, short, sharp alarm whistle often heard before marmot seen. Conspicuous burrows with cone of displaced earth, flattened vegetation. 4 digits on forefeet, 5 on hindfeet. Prints measure 5 × 4cm. Digital pads evenly spaced about the plantar pad.

Habitat. Alpine meadows, above the tree line (600–3200m above sea level). Steep rocky hillsides, especially south-facing slopes, stabilised rock scree. Deep extensive burrow systems. Nest in deep chamber, lined with grass, entrance blocked up for hibernation.

Habits. Strictly diurnal, but on warm days (above 25°C) may become bimodal as need to withdraw into burrow to cool down. Have poor physiological capacity for heat loss, together with dense fur and a thick subcutaneous fat layer accumulated as an energy store for hibernation. Hence behavioural thermoregulation limits the time available for activity above ground, and this is especially true of the warmer, south-facing slopes. FOOD: Grasses, sedges, herbs (sometimes digging for roots) – unripe fruit and flowers may be taken in summer. HOME RANGE: 0.9–4.6ha, average *c.* 2.5ha. Population density *c.* 1.2 animals per ha. Dispersal starts at 2 years old, but mostly involves 3 year olds. SOCIAL: Family groups in burrow system, group size 2–20. Group typically consists of dominant pair and their offspring from one or more years, all sharing a home-range. In such groups reproduction is limited to dominant female (dominant male does not monopolise reproduction in the same way). Also 'floaters' which lack a well-defined home range. Dispersal after second hibernation at earliest. COMMUNICATION: Alarm is shrill whistle (high pitched piping as run for cover). Screech loudly when fighting. Also rumbling and growling sounds. Scent marking using oral glands. Anal glands not used for marking but for agonistic displays during territorial disputes and other aggressive encounters. Sight and hearing most important.

Breeding. Mating in April–May; births in May–June. Hibernation: October–April. Sexually mature after 2nd hibernation, but usually reproduce in 3rd summer or later. Gestation 34 days. Litter size 2–6. Litters per annum 1. Mammae 8–10. YOUNG: Weaning age 40 days. Parental care continues during first hibernation by warming young.

Lifespan. Max. recorded 15 years. Predators include Golden Eagle, Eagle Owl, Red Fox, with Raven, hawks and Pine Marten taking young animals. Mortality highest among infants (*c.* 20%) and during winter (*c.* 20%).

Measurements. Head-body length: 47–52cm. Tail-length: 15–20cm. Hind-foot length: 7.4–9.9cm for adults (av. 8.6cm). Condylo-basal length: 92–98mm. Weight: adult females 2.8kg (spring), 4.3kg (autumn); adult males 3.0kg (spring), 4.5kg (autumn). Dental formula: 1/1, 0/0, 1–2/1, 3/3 = 16–18.

General. Potentially endangered. Previously killed in large numbers because fat thought to have medicinal and cosmetic properties. Recently successfully introduced to Pyrenees, Carpathians, Black Forest and eastern parts of the Alps. Hibernate in groups, body temperature drops by 5°C. Group members hibernate together in one hibernaculum, which increases infant survival although it may be disadvantageous to the adults.

Hibernation consists of a series of short cycles, the recurrent entrances into and arousals out of which are highly synchronised within groups. Animals whose burrows are on a north- facing slope lose more weight than those on a south-facing slope during winter hibernation. However, the summer microclimate is better (i.e. cooler) on north slopes. Despite the latter, larger and obviously better thriving colonies are found on south-facing slopes since the number of offspring weaned by a mother is in fact more strongly influenced by her weight loss during the last hibernation than by her previous summer's weight gain.

SPOTTED SOUSLIK, *Spermophilus (Citellus) suslicus* Pl. 47

Name. *Sperma* = seed, *philos* = loving (Gk.)
Recognition. Distinguished from European Souslik by marbling on back (creamy white spots), shorter less bushy tail. Moults once a year, end of June to beginning of August.
Habitat. Same as European Souslik, but more frequent on cultivated ground (can survive ploughing).
Habits. Strictly diurnal. Hibernates October–April (SE Poland). FOOD: Mainly grasses and cereals. In spring, all-green plant parts. Also insects, small vertebrates and tubers. SOCIAL: Individuals live in separate burrows within a large 'colony' at population densities of up to 168 per ha. COMMUNICATION: Alarm (see vignette) is high-pitched, melodic whistle. Sebaceous glands at oral angle (mouth corners) and lips of mouth. Oral gland secretions have oily texture and strong 'musky' odour easily detectable by man. Dorsal glandular region larger in male than female (about 60 apocrine glands along back).
Breeding. Mating in April–May, births in May–June. Sexual maturity 1 year. Gestation 23–26 days. Litter size 4–8, max. 11. Litters per annum 1. Mammae 10.
Lifespan. Causes of death include larger birds of prey, storks, gulls, Ravens, Red Fox, cat, dog.
Measurements. Head-body length: 18–25cm. Tail-length: 3–4cm. Hind-foot length: 2.9–3.3cm. Condylo-basal length: 42–45mm. Weight: male: 280g, female: 224g. Dental formula: 1/1, 0/0, 2/1, 3/3 = 22.
General. Large numbers around crops can cause nuisance.

< *Spotted Souslik.* E Europe, with most westerly records in Poland and Romania.

> *European Souslik.* Scattered populations in Austria, Czeckoslovakia, and eastwards to the Black Sea.

EUROPEAN SOUSLIK *Spermophilus (Citellus) citellus* **Pl. 47**

Name. *Citus* = swift (Lat.); *citellus* = little swift beast.
Status. Vulnerable.
Recognition. Almost no outer ears. Short furry tails. Pelage of back somewhat mottled but not spotted. Frequently stands erect in 'begging' posture. Moults once a year between end of June and beginning of August.
Habitat. Plains, meadows, fields. Up to 1300m in Czechoslovakia, 2200m in Yugoslavia. Requires burrowing substrates as in dry open steppes with loam lime-rich soil

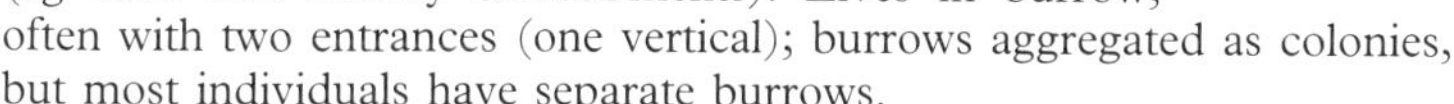

(eg road and railway embankments). Lives in burrow, often with two entrances (one vertical); burrows aggregated as colonies, but most individuals have separate burrows.

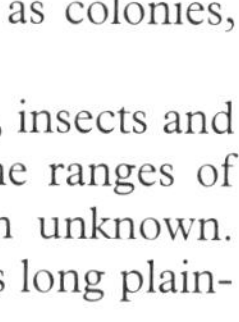

Habits. Diurnal. FOOD: Seeds, also green plant parts, flowers, insects and roots. HOME RANGE: Unknown (related species occupy home ranges of 0.3–4ha). SOCIAL: Lives in colonies, but social organisation unknown. COMMUNICATION: Alarm call is an abrupt whistle. Also makes long plaintive whistle.
Breeding. Mating after emergence in March, births in April–June. Litter size 5–8. Litters per year 1. Mammae 10.
Lifespan. Predators include larger birds of prey, storks, gulls, Raven, Red Fox, cat, dog.
Measurements. Head-body length: 19–22cm. Tail-length: 5.5–7.5cm. Hind-foot length: 3.5–4cm. Condylo-basal length: 41.4–45mm. Weight: 240–340g. Dental formula: 1/1, 0/0, 2/1, 3/3 = 22.
General. Carries food in cheek pouches, stored underground. Builds up fat reserves in autumn (males at end of June, females beginning at end of July), preparation for hibernation from October/November–March.

SIBERIAN CHIPMUNK *Tamias sibiricus* **Pl. 48**

Name. *Tamias* = steward (Gk.).
Status. Introduced.
Recognition. 5 distinct white stripes along back.
Habitat. Woodland with bushy understory (on ground or climbing low shrubs), and found regularly in parks and towns. Nests in burrow or hollow log.
Habits. Diurnal. FOOD: Nuts (lays up stores for winter, transported in cheek pouches; see vignette), seeds, tree buds, mushrooms, berries, cereals. HOME RANGE: 700–3475m^2, larger in autumn than spring, and for females than males. COMMUNICATION: Scent marking with urine (most frequent), also with cheek (oral) glands. Former by both males and females but marking sites differ: males urine-mark in prominent place, e.g. stone; females use constant marking sites on ground, or mark on the male. Scent marking sites are used by all members of the local community.

< *Siberian Chipmunk.* Occasional escapes establish more or less permanent colonies in France, Germany, Austria, Netherlands, Finland.

> *Flying Squirrel.* S Finland and eastwards.

Breeding. Gestation probably *c.* 35–40 days. Litters per year: 1 in Siberia, reputedly 2 elsewhere. Mammae 8.
Measurements. Head-body length: 12–17cm. Tail-length: 8–15cm. Hind-foot length: 31–38mm. Condylo-basal length: 38–42mm. Weight: 50–120g. Dental formula: 1/1, 0/0, 2/1, 3/3 = 22.
General. Middle-European populations (in France, W. Germany, Netherlands and Austria) entirely from escapees.

FLYING SQUIRREL *Pteromys volans* Pl. 46

Name. *Pteron* = wing (Gk.); *mys* = mouse (Gk.); *volans* = flying (Lat.).
Recognition. 'Wing' for gliding formed by membrane of skin stretched out from body between fore and hind legs (inconspicuous when not in use). Very large eyes.
Habitat. Mixed forest (especially mature birches and conifers) of Finland (and Russia and Siberia). Nest in cavity at junction of branch and trunk of conifer or in abandoned woodpecker's hole.
Habits. Nocturnal. FOOD: Buds, shoots, leaves, seeds, nuts, berries, also occasional bird nestlings and eggs. Caches alder and birch catkins in nest holes (sometimes of up to 5 litres capacity). SOCIAL: Groups found in the same tree or nest, reputedly all same sex except during breeding season.
Breeding. Litters born in April–May and in June–July. Litter size 2–4. Litters per year 2.
Measurements. Head-body length: 13.5–20.5cm. Tail-length: 9–14cm. Hind-foot length: 3.2–3.9cm. Condylo-basal length: *c.* 37mm. Weight: 95–200g. Dental formula: 1/1, 0/0, 1–2/1, 3/3 = 20–22.
General. Does not hibernate, but in cold weather may sleep for several days. When landing, raises tail and arches membrane like a parachute to act as brake. Numbers declining (eg in Finland) as modern forestry removes old, hollow trees.

EUROPEAN BEAVER *Castor fiber* Pl. 56

Name. *Castor* = beaver (Gk., Lat.) (possibly from Sanskrit *kasturi* = musk); *fiber* = beaver (Lat.), hence English 'beaver'.

Status. Out of danger.

Recognition. Beavers are the largest European rodents. Tail: broad, horizontally flattened and scaly. Hindfeet: webbed. Nasal and ear passages can be shut while swimming. Ears and eyes small, legs short, head blunt. Northern forms darker than southern. External sexual characteristics not visible – in both sexes the urinogenital tract opens into a common anal cloaca. Large, yellowish-red incisor teeth (upper pair shorter and curved). Paler colour and longer nasal bones than Canadian Beaver, *C. canadensis*. Coat yellowish-brown to almost black, reddish-brown being most common. Presence indicated by lodges, channels, burrows, felled trees, stripped bark, also food stores (in water) and 'refectories' (sites where they eat – accumulate bark etc.). Dams constructed from sticks, tree branches and mud, up to 150 metres long. Defaecate into water – faeces appearing as amorphous mass of shredded wood fibre and other plant material which soon sinks to the bottom. Footprints on soft mud distinctive because of the webbing. Forefoot has 5 toes with pointed claws, though print often shows only 4; 5.5cm long × 4.5cm wide. Hindfoot: 15cm × 10cm.

Habitat. Prefers broad river valleys with mainly floodplains – stands of softwood. Where possible nests in burrow (e.g. banks of Rhône), otherwise a family group of beavers will build a lodge consisting of a large hollow mound of branches with underwater entrances. Inside the lodge the nest is built on a platform just above water level. Beavers construct an elaborate system of canals and dams to preserve the water level in the system, providing safe pathways and facilitating access to feeding areas.

Habits. Largely nocturnal, but more diurnal (especially at dusk) if undisturbed. FOOD: In spring and summer, grasses and leaves and twigs of herbaceous plants (up to 300 spp. recorded), bushes and trees in autumn, and twigs and bark in winter. Tree preferences: willow, aspen and poplar. Fells small trees and drags them into water for storage – cold water temperatures actually help preserve nutritional content of bark. This provides accessible food when river ices over (beavers may then be trapped underwater). HOME RANGE: Highly variable according to habitat configuration – long and narrow if live around streams, round or rectangular for pond or lake beavers that follow shorelines. The home range may be restricted to a small area by physical constraints (eg ponds, suitable slow water along a fast-flowing river) or made up of a series of separated small ponds connected by a small waterway. Range size also depends on quality of vegetation. Densities overall vary from 1.0–1.5 per km^2 to 1.8 per km^2, eg numbers of beavers per distance along a waterway range from 1.2–1.6 to 9.1 per km (rivers), and 2.83 to 22 per ha (ponds and lakes). Along river, one family may use 500m in optimum habitat, 5.5km in poor habitat. Four categories of movement are recognised: (1) movement of an entire family between ponds within a territory; (2) wanderings of yearlings (usually short distance with returns to home lodge); (3) movement of adults who have lost their mates; (4) dispersal of 2 year old beavers from their natal site (the longest movements, with records of over 100 km). SOCIAL: Small family groups, size depending on several variables such as length of time pair have been together, reproductive success etc. Average family is 5–6 beavers: adult pair, yearlings from previous summer and young born during present summer. Young stay at birthplace for 2 years. Monogamous. COMMUNICATION: In water warns of danger by slapping surface with tail. Vocalisations include growling, hissing and screaming. Scent secretion called castoreum used in perfumery (formerly, secretion from one beaver worth as much as a farm hand's annual wages). Several scent glands: paired anal glands, paired castor (musk) glands, glands on soles of feet (all in both male and female). The castor glands (also called preputial glands) open into the urethra, secretion mixing with urine to give 'castoreum' – gives beaver characteristic odour. Size of both anal and castor glands positively correlated with body weight. Mark with both castoreum (sprayed onto earth piles, tree trunks, protruding objects; see vignette) and anal glands (dragged over objects). 'Sign places' or 'scent mounds' of marked earth may be 60cm high and up to 7 per colony, mostly on territorial edge. Used by all colony members, construction dependent on number of neighbours (greatest during times of dispersal). Scent mounds indicate occupancy of territory. Beavers respond to a stranger's castoreum on a scent mound by hissing, tail-slapping, rebuilding of the mound, the depositing of new castoreum, etc. Postures during agonistic encounters include lunging, sham-biting, tail-quivering. Good senses of smell and hearing; poor eyesight.

Breeding. Mating in February, births in June. Sexual maturity at 3 years (occasionally 2 years) for females; successful mating for males unlikely before 3 years old. Gestation 103–108 days. Litter size 1–6, average 2.7. Litters per year 1. Mammae 4. YOUNG: Precocious, fully furred with eyes open at birth, learn to swim in tunnel entrances of lodge within hours. Weaned during first summer.

Lifespan. 7–8 years normally, occasionally up to 25 years. Causes of death include large mammalian predators such as wolves, but also by drowning when caught under ice by sudden rise in water; starvation over winter. Tularemia (disease usually fatally affecting liver, spleen, lungs and lymph nodes) can also cause widespread mortality, as can car accidents.

Measurements. Head-body length: 75–90cm (regional variations, eg Scandinavia: 74–81cm, Elbe: 80–87cm, Rhône: 82–90cm). Tail-length: 28–38cm (eg Scandinavia: 28.5–33.5cm, Elbe: 31–34cm, Rhône: 31–38cm). Hind-foot length: 16–18cm. Condylo-basal length: 125–150mm.

< *The Beavers.* The *European Beaver* is confined to isolated pockets in Scandinavia and eastern Germany, together with scattered sites (triangles) where recently reintroduced.
The *Canadian Beaver* has been introduced to lakes in Finland.

Weight: up to 38kg (eg Scandinavia: 12.5–30kg, Elbe: 14.5–30kg, Rhône: 15–38kg). No difference between sexes. Dental formula: 1/1, 0/0, 1/1, 3/3 = 20.

General. Special lobe of cheek tissue as mechanism for shutting mouth behind incisor teeth in order to gnaw underwater. Grooms with double claw on second toe of hind foot which functions as tweezers. May dam streams and dig canals (up to 150m long) around lodge. Can fell 25cm thick tree in less than 4 hours (and one family can fell 300 small trees in one winter). Dives often last 5–6 minutes (max. 15 minutes). Many subspecies; some consider the west European beaver a distinct species, *C. albicus*, mainly because of skull differences. Formerly, typical holartic distribution throughout Europe and Asia, north of the line Jaffa, Basra, Pamir, Peking and south of line Vorkuta-Ochotsk.

CANADIAN BEAVER *Castor canadensis*

Name. North American Beaver.

Status. Introduced.

Recognition. As for European Beaver, but a darker colour and shorter nasal bones. Colour ranges from black to grey – underfur of western beavers is reddish tipped. Extremes in America range from jet black to pale silver. Moults annually, during summer. Fieldsigns as for European Beaver.

Habitat. Can occupy a wide range; determined by food and water availability. Has been known at altitudes of more than 3400m in N. America. Ideally ponds, small lakes with muddy bottoms, meandering streams. Will occupy artificial ponds, ditches and reservoirs if food is available. Compared to the European beaver, *C. canadensis* has a greater tendency to build dams.

Habits. Nocturnal. During northern winters they remain in lodge under ice for warmth (i.e. in semi-darkness), whereupon beaver 'days' (i.e. their cycle of activity) lengthen to a cycle of 26–29 hours. FOOD: A great number of woody (trees, woody vines) and herbaceous species (including many grasses); this number decreases from the southern range to its arctic or alpine limits. Can use about 30% of the cellulose and 44% of the protein it consumes, by thoroughly chewing its food, through coprophagy and through bacterial activity in the caecum. HOME RANGE: Very variable, according to habitat quality; *c.* 0.6–2.0km of stream. In spring, 2 year olds of both sexes leave the family group. Longest distance recorded 21km (N. America). Site abandonment by a whole family also may occur (reasons not understood). Family movements observed to occur in early spring, and during late summer – early autumn. SOCIAL: Monogamous. Family groups of 3–9. COMMUNICATION: As for *C. fiber*. Good senses of smell and hearing, poor eyesight.

Breeding. Copulation occurs in the water; if mated the female excretes a hard vaginal plug about 2 days later. In northern and montane regions of N. America, mating season is February–March. Sexual maturity in both sexes at 21 months. Gestation 105–110 days. Litter size in N. America 3–4 (but may be as many as 8); varies with food availability, age and population density. Litters per year 1. Mammae 4. YOUNG: As for *C. fiber*. Weaning during first summer.

Lifespan. Causes of death include large mammalian predators (especially wolves, where they occur) and intraspecific strife: fighting beavers try to bite each other's spinal cord. Fights mainly in spring, between males. Victim of the disease tularemia.

Measurements. Head-body length: averages 72cm (55–86cm). Tail-length: 23–32.5cm. Hind-foot length: 15–20cm. Condylo-basal length: 13–15cm. Weight: 14–30kg (exceptionally 45kg). Dental formula: 1/1, 0/0, 1/1, 3/3 = 20.

General. Populations of this species have been introduced from N. America to Finland, Russia and Poland.

COMMON HAMSTER *Cricetus cricetus* Pl. 62

Recognition. Heavily built, fairly large rodent, tail short and furry. Piebald, brown and white, underside mainly black; black mutant occurs.

Habitat. Grassland, rough grass at the edge of fields (not closely grazed pasture), cultivated steppes in loamy and loess soils. Below 500m. Avoids wet habitats. Dens in a system of burrows up to 2m deep with several chambers for food storage and nesting. Nesting chambers lined with soft bedding, grass etc. Tunnels of diameter 6–8cm. Winter nest in deepest parts of burrow.

Habits. Mainly nocturnal or crepuscular. Hibernates October–March/April, may wake occasionally and feed from stores. FOOD: Seeds, roots, etc., including root crops. Insects and other small animals, which are carried in elastic cheek pouches. A store of food may comprise up to 5kg potatoes, 15kg of grain and a total of up to 65kg of food. SOCIAL: Males move onto territory of females for mating; usually driven out subsequently (Kaluga district of Russia – density of about 1 burrow per m^2 has been reported, each burrow with 1 adult occupant). Solitary. COMMUNICATION: Flank glands used to flank-mark object and the substrate. Also a ventral gland, anal and ear glands. Glands on oral angle, lips and soles of feet, preputial glands

<*Common Hamster*. E Europe, with pockets in W Germany, Belgium and NE France.

> *Grey Hamster*. Generally E of the Black Sea; most westerly records from E Greece.

(male) Sex hormones may enhance glandular activity since seasonal and sexual and age differences is found, especially in the flank glands.

Breeding. Mating in April after hibernation (October/November–March/April). Sexual maturity: females at 43 days old (so early females can breed in summer of birth). Gestation 18–20 days. Litter size 3–15. Litters per year 1–3. Mammae 8. YOUNG: Parental care by females only (although in captivity pairs may remain together and raise young). Weaned at 3 weeks.

Measurements. Head-body length: 200–340mm. Ear: 23–31mm. Tail-length: 40–60mm. Hind-foot length: 35–40mm. Condylo-basal length: 43–51mm. Weight: 220–460g. Dental formula: 1/1, 0/0, 0/0, 3/3 = 16.

General. Winter hibernation interrupted every 5–7 days, at which time hamsters eat from cached food. Warren size generally dependent on age of occupying hamster. Formerly in a belt stretching from Russian Steppes, Danube Valley, and the Pannonian Plains through Germany, as far as Southern tip of Netherlands, eastern parts of Belgium and NE of France.

GREY HAMSTER *Cricetulus migratorius* Pl. 62

Name. Migratory Hamster.

Recognition. Vole-sized, with grey fur. Distinguished from voles by larger ears and eyes, and by shorter tail.

Habitat. Cultivated and uncultivated grassland and woodland. Steep wooded slopes in forest-steppe country. Gardens; cornfields. Dens in a system of burrows up to 1.5m deep with several food storage and nesting chambers. Nesting chamber lined with soft bedding, grass etc. and sometimes feathers and sheepswool. 1–5 tunnels leading in.

Habits. Mainly nocturnal or crepuscular. FOOD: Seeds, roots, young shoots, soybean, millet; 37 different plant species recorded in diet; some insects reputedly eaten. SOCIAL: Solitary. COMMUNICATION: Mid-ventral (abdominal) gland which secretes during breeding season.

Breeding. Throughout the year in Armenia. Gestation 17–22 days. Litter size 2–11 (Armenia). Litters per year 3–4. Mammae 8. YOUNG: Parental care by female only.

Measurements. Head-body length: 80–117mm. Tail-length: 22–28mm.

Ear: 13–20mm. Hind-foot length: 15–17mm. Condylo-basal length: 22–28mm. Weight: 33–38g. Dental formula: 1/1, 0/0, 0/3, 3/3 = 16
General. Opinions differ as to whether or not hibernates but thought that activity is simply very much reduced in winter. Stores food. Defends itself vigorously by throwing itself on its back, opening its mouth and exposing its incisors.

ROMANIAN HAMSTER *Mesocricetus newtoni* Pl. 62

Recognition. Almost white underside (any black does not extend behind forelegs); a blackish area on crown continued backwards as median line to between shoulders; an oblique black stripe from lower margin of cheek to side of shoulder.
Habitat. Grassland, cultivated steppes, rough grass at edge of fields, dry rocky steppes or brushy slopes. Uses burrows throughout year; each burrow has several entrance tunnels and compartments for nest, storage etc.
Habits. Largely nocturnal. FOOD: Green vegetation, seeds, fruit and meat. SOCIAL: Lives solitarily (one individual per burrow), fighting intruders. In captivity, social rank of males based on size of flank glands: those with largest glands are dominant. COMMUNICATION: Squeaks; also communicates with ultra-sound (especially males and females during mating). Both sexes have flank glands, those of male larger than of female. Both sexes flank mark by rubbing glands against objects (see vignette) – aggressive encounters increase level of marking in males and marking associated with aggressive behaviour in females. Also glands on side of ears (larger in males, and can be used for sexual discrimination); Harderian glands in eye orbit differ between sexes: only females have black pigmentation and can provide information of reproductive state of females; female has scent glands in oral angle. Vaginal secretions are correlated with sexual behaviour. Olfaction essential for mating behaviour in males. Also used for species identification, and females can determine 'sexuality' of male by his odour. Urine, faeces and preputial gland secretions do not appear to be used for olfactory communication, except that urine from stressed hamsters is avoided and increases flight responses, suggesting release of 'fear odour'.
Breeding. Breeding season in wild restricted to spring and summer, but breed year round in captivity. Sexual maturity *c.* 56–70 days. Gestation *c.* 15 days. Litter size 1–12. Mammae 14 or 16.
Measurements. Head-body length: 130–180mm. Ear: 14–20mm. Tail-length: 12–20mm. Hind-foot length: 19–23mm. Condylo-basal length: 32.8–36.8mm. Weight: 80–115g. Dental formula: 1/1, 0/0, 0/0, 3/3 = 16.
General. Cheek pouches. Very similar to Golden Hamster: Golden 44 chromosomes; Romanian 38. Can interbreed but offspring sterile. Duller

<*Romanian Hamster.* Confined to small area of Romania and Russia north of Black Sea.

>*Greater Mole Rat.* European records confined to Romania.

coat than Golden and dull marks in front of eyes and on nape. Golden originates from Syria – mainly pets but can survive for a while in the wild, even establishing feral populations. Dimensions of Golden as for Romanian (sometimes assigned to same species). Digs very long tunnels and holes close to surface. Scolds and gnashes teeth.

GREATER MOLE RAT *Spalax microphthalmus (polonicus)* Pl. 48

Name. Podolian Mole Rat. *Micros* = small (Gk.); *ophthalmos* = eye (Gk.); *spalax* = mole (Gk.).

Recognition. Tail-less. Eyes covered by membrane of skin. Paws not modified for digging (cf moles). Velvety pelage with row of almost white bristles on each side of head. Prominent incisor teeth used for burrowing. Pelage is often greyish, but variable. Distinguished from Lesser Mole Rat by larger size and smaller ears. A fieldsign is the burrow systems with evidence of plants being pulled in.

Habitat. Fertile steppes and valleys, cultivated, dry valleys and grasslands. Den under mound of soil (pushed up while digging), nest 25–35mm diameter. Complicated burrow system, sometimes near surface, but up to 2m in depth. Burrow contains several food storage chambers.

Habits. Nocturnal and dawn. FOOD: Vegetation: roots in burrows, plants dragged into burrows. SOCIAL: Solitary, except during breeding season. Young travel above ground to disperse. SENSES: The eyes are rudimentary, being completely and permanently hidden under the skin, with atrophied lens cells enclosed in a vesicle and a retinal layer. Retinal cells may be vital for photoperiod reception; although mole rats are totally blind, thermoregulatory responses to changes in photoperiod have been observed.

Breeding. Births in spring. Litter size 4–5. Litters per year 1.

Measurements. Head-body length: 200–310mm. No tail. Hind-foot length: 23–30mm. Ears: 1.2–2.5mm. Condylo-basal length: *c.* 50mm. Weight: 370–570g. Dental formula: 1/1, 0/0, 0/0, 3/3 = 16.

General. Rarely above ground (but presence in owl pellets indicates some nocturnal surface activity). Digs with teeth (or head, in soft soil), pushing soil backwards with paws and feet. Reputedly agricultural pest where abundant. Form in Romania is perhaps specifically distinct (*S. graecys*) from Russian *S. microphtharnius.*

LESSER MOLE RAT *Nanospalax leucodon* Pl. 48

Name. Western Mole Rat. *Leucos* = white (Gk.); *odons, odontos* = tooth (Gk.).

Recognition. Tail-less. Only distinguished reliably from Greater Mole Rat by two small openings at rear of skull. The anterior rostrum of the skull is somewhat narrower than that of the Greater Mole-rat. Length of hindfoot best way to distinguish Greater and Lesser Mole Rats. Fieldsigns: molehills.

Habitat. Open grassland, typically steppe.

Habits. FOOD: Subterranean plant parts, also some stems, leaves and fruit, caches food in burrows. SOCIAL: Solitary.

Breeding. Mating in January–March. Litters per year usually 1. Mammae 6. YOUNG: Weaning age 3 weeks.

Measurements. Head-body length: 150–270mm. No tail. Hind-foot length: 19–30mm. Ears 2.1–3.9mm (cf Greater Mole Rat). Condylo-basal length: *c.* 48mm. Weight: 140–220g. Dental formula: 1/1, 0/0, 0/0, 3/3 = 16.

General. Used to be regarded as an agricultural and forestry pest. Due to urbanisation, intensification and the spread of monoculture, population size and distribution is now much reduced. Totally blind.

WOOD LEMMING *Myopus schisticolor* Pl. 61

Name. *Myops* = short-sighted (Gk.); *schistos* = split (Gk.); *color* = colour (Lat.).

Recognition. Distinguished by dark slate-grey pelage and short tail. Pelage lighter colour in winter. Adults have rusty brown patch absent in sexually immature juveniles. Tunnels and nests become apparent when snow thaws.

Habitat. Coniferous forest, rich in moss and damp with secondary growth. Burrows and runways in moss.

Habits. Largely nocturnal. FOOD: Mosses. HOME RANGE: Long distance migration not recorded. SOCIAL: Less aggressive than Norway Lemming.

Breeding. Breeding season from May–August in E. Finland, but in S. Norway also breeds throughout winter. Sexual maturity *c.* 1 month. Litter size 1–6. Litters at intervals of 25 days (in captivity: one female had 11 litters in 299 days). Mammae 8.

< *Lesser Mole Rat.* SE Europe, eastwards from Yugoslavia.

> *Wood Lemming.* E Scandinavia, S to vicinity of Stockholm.

Measurements. Head-body length: 80–115mm. Tail-length: 10–20mm. Hind-foot length: 15–15.5mm. Condylo-basal length: *c.* 25mm. Weight: 20–45g. Dental formula: 1/1, 0/0, 0/0, 3/3 = 16.

General. Manipulation of sex ratio of offspring (only about 25% of population is male; some females produce only female offspring due to mutation of their sex chromosomes). Population densities vary periodically (cycle *c.* 4 years) but never reach plague proportions (cf Norway Lemming).

NORWAY LEMMING *Lemmus lemmus* Pl. 61

Recognition. Variegated pattern of black, yellow, yellow-brown, highly variable between individuals. Very short tail. Tunnels and nests below snow exposed during thaw.

Habitat. Tundra and alpine zone above tree-line (also birch and pine woodlands). Nest of moss and grass. Intricate tunnel system above ground but below snow in winter, below stones and moss in summer.

Habits. Largely nocturnal. FOOD: Grasses, sedges, dwarf shrubs (mosses important in winter). HOME RANGE: Low density population, 3–5 per ha; high density peaks, 330 per ha. Long distance movements in 'lemming years' when they are bold and aggressive. SOCIAL: Much intra-specific aggression. COMMUNICATION: Vocalisations include whistling and squeaking. Male has preputial scent glands; both sexes have glands in oral angle and lips, ears, soles of feet, Meibomian glands, rump (caudal) glands. Faeces deposited in piles along trails probably serve communicative function.

Breeding. Breeding confined to summer months. Sexual maturity: female, min. 14 days in summer; male, *c.* 21 days. Gestation 16–23 days. Litter size 2–13 (generally 5–8). Litters born every 3–4 weeks, generally max. 6 per annum, but can be more (eg 1 pair had 8 litters in 167 days). Mammae 8.

Lifespan. Max. recorded 2 years.

Measurements. Head-body length: 70–155mm. Tail-length: 10–19mm. Ear: 6–11mm. Hind-foot length: 17–19mm. Condylo-basal length: 22.4–29.4mm. Weight: 10–130g. Dental formula: 1/1, 0/0, 0/0, 3/3 = 16.

General. Population densities vary periodically (cycle *c.* 2–4 years). Mi-

< *Norway Lemming.* Found in Scandinavia, south as far as Stockholm. Range may expand during 'lemming years'

> *Ruddy Vole.* E Scandinaia and eastwards.

grate at high densities, swimming well, climbing poorly, attempting to colonise as they go.

RUDDY VOLE *Clethrionomys rutilus* **Pl. 57**

Name. Northern Red-Backed Vole. *Clethria* = chink, hole (Gk.); *Mys* = mouse (Gk.); *rutilus* = red (Lat.).
Recognition. Upperside reddish brown, cream coloured underneath, ears quite conspicuous. No grey, and overall lighter than the Bank Vole. Tail very short and has a long brush of hairs (10–12mm) at its tip. Cheek teeth become rooted, sometimes when fully adult. On skull, the anterior margin of the bony palate may have gaps (cf Bank Vole and Grey-sided Vole). Moults in March and October. Droppings of *Clethrionomys* voles are rounded in cross-section, 4 times as long as wide, dark brown or black in colour, smaller than those of Wood Mouse, not greenish as those of Field Vole.
Habitat. Pine and birch zones; also above tree-line. Often comes into buildings in winter. Willow scrub and open birch woodland. Makes burrows and nest of grass and moss. Nests between tree roots, stones and shrubs and in tree-holes or tree branches.
Habits. Active both day and night, though more nocturnal in summer. FOOD: Almost entirely herbivorous – tender vegetation, nuts, seeds, bark, lichens, fungus and insects. Food often stored in nest for use when supply short. May forage high above ground level. HOME RANGE: 0.1–0.5ha for females, 1.5–2.0ha for males. COMMUNICATION: Vocalisations include chattering and squeaking. Chirp-like bark, chatter their teeth. Males have flank glands; both sexes have Meibomian glands. Preputial glands as for Bank Vole.
Breeding. Sexual maturity: females 4 months. Gestation 17–20 days. Litter size 1–11. Litters per year 2–3. Mammae 8. YOUNG: Parental care by female only; weaned and independent at 17–21 days. Vignette shows 14 day old Ruddy Vole.
Measurements. Head-body length: 80–110cm. Tail-length: 23–35cm. Hind-foot length: 17–18mm. Condylo-basal length: 19.6–23.9mm. Weight: 15–40g. Dental formula: 1/1, 0/0, 0/0, 3/3 = 16.

GREY-SIDED VOLE *Clethrionomys rufocanus* **Pl. 57**

Name. *rufus* = red (Lat.); *canus* = grey (Lat.).
Recognition. Red fur restricted to band along back. Flanks almost pure grey. Last upper tooth always resembles simple form of Bank Vole, cheek teeth generally more angular and become rooted, sometimes when fully adult. Droppings rounded in cross-section, 4 times as long as wide, dark brown or black in colour, smaller than those of Wood Mouse; cf Field Vole which has greenish droppings. Nests, tunnels and runways in grass. Indistinguishable signs from those of *C. rutilus*.
Habitat. Rocks and mountains with dwarf shrubs; sometimes comes into

<*Grey-sided Vole.* Throughout Scandinavia north of about 60°N.

>*Bank Vole.* Throughout W&C Europe and mainland Britain, and spreading in Ireland.

houses. Globular nests of grass with tunnels radiating from them on ground under snow in winter. Otherwise spherical nests of grasses, mosses, lichens and shredded leaves, usually hidden under roots or stumps, logs or brush piles, in tree-holes or high in tree branches.

Habits. Active day and night. FOOD: Almost entirely herbivorous – shoots, buds of various berry-bearing shrubs, also leaves; bark of dwarf birch; other low-growing trees and bushes. Shoots include those of bilberry and crowberry. HOME RANGE: Females 300–900m^2 depending on density, males 10–15 times larger. SOCIAL: Aggression between adult males but not females. Ranges of males mutually exclusive, but overlap several ranges of females. COMMUNICATION: Chirp-like bark when disturbed; gnash/chatter teeth. Males have flank glands, both sexes have Meibomian glands. Preputial glands as for Bank Vole.

Breeding. Breeding season April–September. Gestation 17–20 days. Litter size 1–11. Litters per year 2–3. YOUNG: Parental care by female only. Weaning age *c.* 14 days.

Measurements. Head-body length: 110–135mm. Tail-length: 25–40mm. Hind-foot length: 18–19mm. Ear: 12–16mm. Condylo-basal length: 22.2–27.4mm. Weight: 15–50g. Dental formula: 1/1, 0/0, 0/0, 3/3 = 16.

BANK VOLE *Clethrionomys glareolus* Pl. 57

Name. *glarea* = gravel (Lat.).

Recognition. Characteristic reddish colour, often with grey on flanks; juveniles much greyer. Cheek teeth become rooted, sometimes when fully adult. Has larger ears and eyes and a longer tail than *Microtus* and *Pitimys* species. Droppings rounded in cross-section, 4 times as long as wide, dark brown or black in colour, smaller than those of Wood Mouse; Field Vole has greenish droppings. As with other species of *Clethrionomys*, fieldsigns include small groups of droppings and hazelnuts with large, clean-edged circular hole not surrounded by clear tooth marks. Characteristic tufts of reddish hair found in carnivores' droppings and raptors' pellets often originate from Bank Voles.

Habitat. Deciduous woods, scrub; also areas with high herb growth, park-land, banks and hedges. Rarely in open areas; prefers cover of scrub/herbage. Pure coniferous forest in Scandinavia; most numerous in conifer stands of 6–30 years old. Mainly in warm and dry areas; sometimes in houses (quite bold). Ball-shaped nest in woodland made of leaves, moss and feathers, in grassland made of grass and moss. Has definite entry/exit hole. Many entrances and nest chambers. Site 2–10cm underground at centre of radiating tunnel system, often under logs or in tree roots. Breeding nests may be in tree holes etc.

Habits. Active both day and night though more nocturnal in summer. FOOD: Almost entirely herbivorous. Fleshy fruits and soft seeds eaten when available, leaves of woody plants preferred to herbs. Dead leaves eaten in winter. Other foods include fungi, moss, roots, grass, insects and worms, buds and snails. Makes food stores in northern regions. HOME RANGE: 0.05–0.73ha. Woodland averages: male 0.2ha; female 0.14ha. Males move more widely than females and more in summer. Population density typically between 10–80 voles per ha. On Skomer Island 475 voles per ha and in Czechoslovakia a study of peak densities of 72 voles per ha indicated that breeding stopped at such high numbers. Both sexes disperse at sexual maturity. SOCIAL: Gregarious. In breeding season females space out in exclusive territories. Males aggressive towards conspecifics at bait. Apparent dominance hierarchy. COMMUNICATION: Chattering and sometimes squeaking. Ultrasound used by mother and young and by males during mating. Males have flank glands, but do not appear to flank-mark (cf Water Vole) and both sexes have Meibomian glands. Glands of oral angle and lips are possibly sexually dimorphic. Male and female of all ages have sweat glands on soles of feet (plantar glands). Preputial glands in male and female: larger in male adults and show seasonal variation in size, and larger in dominant than subordinate males. Urine mark (sexually mature males leave trails, females and immature males leave puddles) – preputial gland secretion deposited simultaneously: speculative functions include sex attractants, dominance ranking, territoriality or aggressive motivation. Males can discriminate between females of a number of subspecies of *Clethrionomys* and prefer odour of oestrous females of own subspecies.

Breeding. Breeding season usually April–September/October, but may continue all year round in favourable conditions (if population low and food abundant – experimental addition of food brings forward start of breeding by 2–3 weeks). Sexual maturity: females 4.5 weeks in laboratory; late season young do not mature until following spring. Sexual maturity of subadults suppressed by presence of adult female. Gestation 16–18 days, may be prolonged to 19–22 days by lactation. Females induced ovulators in laboratory; may become pregnant post-partum; if do not conceive at this time do not become pregnant until after lactation. Litter size 3–5 (average 4.1); up to 8 in lab. Litters per year 4–5. Mammae 8. YOUNG: Parental care by females only. Weaned *c.* 14 days.

Lifespan. Max. 18 months in wild, 40 months in captivity. Common prey of owls, Weasels and Red Foxes, although feeding trials indicate that Bank Voles somewhat less relished by foxes than are Field Voles.

Measurements. Head-body length: in UK late summer generation overwinters at 90mm, and in spring measures 100–110mm. Tail-length: 36–72mm. Hind-foot length: 15–20mm. Condylo- basal length: 20.3–23.3mm. Weight: 14–40g (birth 2g). Dental formula: 1/1, 0/0, 0/0, 3/3 = 16 (pattern of third upper tooth very variable, even within one population).
General. Various island subspecies are larger than mainland form *C.g. glareolus*, eg *C.g. caesarius* from Jersey, *C.g. skomerensis* from Skomer, *C.g. alstoni* from Mull and *C.g. erica* from Raasay (Scotland). Occasional forestry pest on Continent, taking bark off trees up to 5m. Lead content shown to increase near main roads in comparison to individuals from woodland and arable sites. Excellent climber; easier to watch than most small rodents as climbs on fallen trees and among shrubs. In Alps found 2400m above sea level, and to 800m in Britain. Active throughout winter. Considerable periodic fluctuations in numbers, typically high in late summer/autumn, decline in winter; trough in April/May. Can distinguish own and another subspecies from odours or urine, faeces and body. Population trends differ between regions, eg in UK woods and S. Sweden there are no cyclic fluctuations, whereas cyclic population explosions are known in N. Sweden, Norway and Germany.

BALKAN SNOW VOLE *Dinaromys bogdanovi (Dolomys milleri)* **Pl. 61**

Name. Nehring's Snow Vole, Martino's Vole. *dolos* = trick (Gk.); *mys* = mouse (Gk.).
Recognition. Pelage: grey-blue, dense, silky hair above, white below, long tail, large hindfeet. Soles of all feet hairy near 'heel'. Claws on all digits except 'thumb' which has small flattened nail. Upper incisor teeth broad and faintly grooved. Does not climb.
Habitat. Rocky slopes above tree line, 600–2000m. Nests under stones or similar crevices.

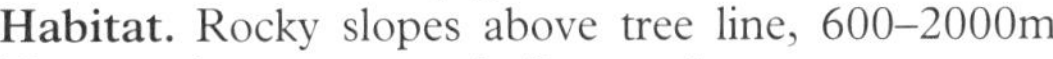

Habits. Nocturnal. FOOD: Grasses (stores food for winter).
Breeding. Breeds twice annually, in March and in June. Gestation 28 days. Litter size 2–3. Litters per year 2. Mammae 6.
Measurements. Head-body length: 100–152mm. Tail-length: 74–119mm (*c.* 71% head-body length). Ear: 13.7–16mm. Hind-foot length: 22.5–26.3mm. Condylo-basal length: 31.6–34.6mm. Weight: *c.* 65g. Den-

<*Balkan Snow Vole.* W Yugoslavia and Albania.

>*Southern Water Vole.* Found throughout most of France and westwards throughout Iberian peninsula.

tal formula: 1/1, 0/0, 0/0,3/3 = 16 (cheek teeth develop roots late in life otherwise similar to *Microtus* species).

SOUTHERN WATER VOLE *Arvicola sapidus* Pl. 55

Name. South-western Water Vole. *Arvum* = field (Lat.); *-cola* = dweller (Lat.); *sapidus* = tasty, savoury (Lat.).

Recognition. Characteristic rounded vole shape. Ears hidden in the brown-grey pelage, slightly lighter and yellowish below. Slightly smaller, lighter in colour and relatively longer tail than Northern Water Vole. The two species of water vole can be distinguished by their skulls. Tooth row usually more than 10mm, nasal bones usually more than 5mm wide. A major source of confusion arises because, although the Northern Water Vole is generally larger and darker than the Southern Water Vole, this situation is reversed where the two species overlap. Thus the most southern forms of the Northern Water Vole are smaller and lighter coloured than the northern forms of Southern Water Voles with which they co-exist. Tracks: Tail does not make a mark. Possible to see the trenches made by the animals as they leave the water regularly at the same spot, leading to feeding areas. Droppings small and cylindrical (c. 5–8cm long), very fine texture, more rounded at the ends than those of the Brown Rat which are quite similar. Deposited in obvious places – on shores, pathways, in tunnels and sometimes in small heaps.

Habitat. Always in close proximity to water, preferably slow-flowing with dense vegetation on the banks. Often in irrigation ditches. Up to 1500m in the Alps, 2000m in the Pyrenees. System of burrows in river bank. Entrance to burrow below water line. Nest of dry grass, larder in the burrow; small spoil heap outside; often has underwater entrance. In marshy terrain, where no suitable banks, makes its nest in the tufts of grass above water.

Habits. Activity periods throughout day and night. FOOD: Aquatic plants, grasses and herbs. Also occasionally insects, crayfish, fish and amphibia. COMMUNICATION: Both sexes have highly developed flank glands – adult male larger than adult female, especially in the breeding season.

Breeding. March–October. Males and females chase each other in the water, emitting small cries; mating takes place close to the edge of the water or even in it. Gestation 3 weeks. Litter size average 6. Mammae 8.

Measurements. Head-body length: 165–220mm. Tail-length: 107–135mm. Hind-foot length: 32–38mm. Ear: 14.3–20.4mm. Condylo-basal length: 39.3–43.7mm. Weight: 165–275g. Dental formula: 1/1, 0/0, 0/0, 3/3 = 16.

NORTHERN WATER VOLE *Arvicola terrestris* Pl. 55

Name. *Terrestris* = earth-dweller (Lat.).

Recognition. Shaggy, brown to blackish (eg N. Scotland) pelage above. Rat-sized but blunt muzzle and shorter tail. Prismatic cheek teeth. Oval scent glands on flanks. Forefeet have 4 toes, hindfeet have 5. Fieldsigns include piles of pith from rushes on water bank and bark-grazing. Faeces are cylindrical, and found at latrine sites. Holes leading to nursery chambers plugged with grass and mud and often surrounded by grazed circle of *c.* 20cm radius. Often seen sitting on haunches while eating grass stalks.

Habitat. Some areas (eg UK) closely associated with fresh water (ditches, slow rivers, lakes). Favours steep river banks, with abundant grass and layered vegetation. Elsewhere (eg Central Europe, especially south of range) less aquatic, favouring pastures and much more fossorial. Sometimes burrowing habit occurs in areas within essentially aquatic part of range (eg Eilean Gamtina; Argyll, Scotland, where distinct form has tooth row of less than 9.2mm). Den is ball of reeds and grass (max. 25cm diameter), usually in burrow. Entrance to burrow in stream bank may be above or below water level. Terrestrial forms make extensive burrows (eg S. Germany), with mounds of earth (with tunnel open) at entrance. If water table is high, nests in tussocks of sedge.

Habits. Active day and night, but predominantly diurnal. FOOD: Largely vegetarian (eg grasses, sedges, roots), including agricultural crops (eg root-crops, bulbs). Occasional animal food (eg fish, see vignette). HOME RANGE: Linear ranges along bank: male *c.* 130m; female *c.* 77m. SOCIAL: Live in pairs. In winter a female, her daughters and unrelated males may occupy a communal nest. Young males disperse at 4 months. COMMUNICATION: Usually silent; alarm call is rasping crick-crick. Faeces concentrated at latrine sites, especially at extremities of home ranges. Other scent sources include Neibomian, preputial and anal glands, glands of the mouth lips and soles of the feet. Both sexes have highly developed flank glands which are larger in males than females when adult, especially in breeding season. Indications of sex, age and population differences in chemical make-up of flank gland secretions. Scent mark by rubbing/scratching hind-feet over flank gland and stamping/drumming feet on ground – 'drum-marking', only by males. Odour of flank gland used in agonistic encounters.

Breeding. Breeding season March/April–September/October (peaks) in mid and western Europe. In western Siberia, May–August. Sexual maturity after first winter in some areas (eg UK), others breed in first year. Gestation 20–22 days. Litter size 4–6. Litters per year 2–5. Mammae 8. YOUNG: Similar to Field Voles (distinguished by longer tail and disproportionately long hind-feet). Naked, with eyes closed, at birth. First emerge from nest at 14 days; may be evicted by mother at 22 days when she produces next litter. Occupy individual home ranges from 3 weeks old. Leave nest when half-grown. Parental care involves both parents.

Lifespan. Max. 5 years (captivity). Mean life expectancy in wild estimated at 5.4 months.

Measurements. Head-body length: 120–235mm (Southern Europe: less than 165mm) (male: *c.* 191mm; female: *c.* 182mm). Tail-length: 40–146mm (55–70% of head-body

<*Northern Water Vole.* Very widespread throughout Europe, but absent from Ireland and most of Iberian peninsula.

>*Muskrat.* Two disjunct distributions, one in E Scandinavia, the other in N, C and W Europe.

length) (north: 80–100mm; south: 40–80mm). Hind-foot length: 27–37mm (male: *c.* 120mm; female: *c.* 114mm). Condylo-basal length: male, 39–44.4mm; female, 37.2–42.5mm. Weight: up to 320g (south: max. = 100g). Dental formula: 1/1, 0/0, 0/0, 3/3 = 16 (tooth row: more than 8mm).

General. Proficient swimmer and diver (often referred to as Water Rat, eg in Kenneth Grahame's famous tale *The Wind in the Willows*). Sometimes absent from suitable riverside habitat because of predation by American Mink, or because suitable patches of habitat are too isolated to maintain adequate immigration.

MUSKRAT *Ondatra zibethicus* Pl. 55

Recognition. Tail laterally compressed (three times deeper than wide). Hindfeet longer than forefeet and fringed with bristles. Toes partially webbed. Much smaller than Beaver or Coypu, larger than Water Vole. Prints resemble stars – 5 claws extend the digits. Forefeet 3.5 × 3cm, hindfeet 7 × 5cm. Usually only 3 digits of forefeet register marks. Leaves 20–30cm runs in vegetation on riverbanks; leaves piles of droppings at entrances and on various projections. Droppings green to black, resemble olive stones. Tail leaves a trail in soft mud. Other fieldsigns include lodge, ventilation shaft, channel.

Habitat. Fresh water, still or flowing, with bankside vegetation (rivers, lakes, etc.). Coastal. Den in summer: burrows in bank (entrance usually below surface); winter: lodge of grass and reeds (up to 1m deep) with two or more chambers and storage chamber (platform with domed roof built in calm inland water).

Habits. Nocturnal and crepuscular. FOOD: Vegetarian (eg sedges, horsetails among 50 plant species recorded in diet). Prefers basal parts of plants (rich in carbohydrate and protein). Occasionally fish and bivalves. HOME RANGE: Population density along shoreline in winter: 3 per km; summer: 10 per km. Longest recorded movement: 13km. Young leave parental lodge and build own *c.* 8m away. COMMUNICATION: Vocalisations include an abrupt whistle. Both sexes have preputial glands modified to form large scent glands. Also Neibomian glands, glands of oral angle, and anal glands; mark with an anal drag.

Breeding. Breeding season March–September. Sexual maturity 12

months. Gestation 25–30 days. Litter size 1–11 (average *c.* 5). Litters per year 2 (1 in north, occasionally 3 in south). Mammae 10. YOUNG: Weaning age 21–28 days. Parental care by female only (male in separate chamber on periphery of lodge when female nursing).
Lifespan. Max. 10 years (captivity), 3 years (wild). Estimated that 80–90% of young die before first winter.
Measurements. Head-body length: 240–400mm. Tail-length: 190–280mm. Hind-foot length: 65–80mm. Condylo-basal length: 59–63.5mm. Weight: 600–1800g (birth: 22g). Dental formula: 1/1, 0/0, 0/0, 3/3 = 16.
General. Proficient swimmer and diver (long distances underwater). Escapees from fur farms spreading in Europe, reaching Finland in 1919, Sweden 1950. Established in UK *c.* 1930, spread *c.* 3km per annum, exterminated 1937. Fur known as musquash.

ALPINE PINE VOLE *Pitymys (Microtus) multiplex* Pl. 60

Name. Fatio's Pine Vole. *Pitys* = pine (Gk.); *mys* = mouse (Gk.); *multiplex* = multitudinous (Lat.).
Recognition. Slightly more yellow, and slightly larger than Common Pine Vole, but otherwise very similar. Pine voles run, but rarely jump or climb.
Habitat. Grassland and open woodland (rarely coniferous forest). Confined to Jura and southern side of Alps, up to 2000m.
Habits. More diurnal than Common Pine Vole. FOOD: Vegetarian (eg roots, rhizomes, bulbs). COMMUNICATION: May use scent to recognise sex of conspecifics, and familiarity.
Breeding. Litter size 2–4. Mammae 4 (all inguinal).
Measurements. Head-body length: 90–115mm. Tail-length: 27–39mm. Hind-foot length: 14.5–17mm. Condylo-basal length: *c.* 25mm. Weight: *c.* 23g. Dental formula: 1/1, 0/0, 0/0, 3/3 = 16.
General. Only reliably distinguished from Common Pine Vole by 48 chromosomes (cf common pine vole, 52–54 chromosomes), but these voles never live in the same area. Other sibling species include Bavarian (*P. bavaricus* probably extinct), Tatra Mountain (*P. tatricus*) and Yugoslavian (*P. lichtensteini*) Pine Voles.

<*Pine Voles.* The Alpine Pine Vole is confined to the Alps in France, Italy and Yugoslavia. The Mediterranean Pine Vole is found in the southern part of the Iberian peninsula. Savi's Pine Vole occurs only in Italy and Sicily, whereas the Common Pine Vole occurs throughout C Europe, south to N Greece and north to S Netherlands.

MEDITERRANEAN PINE VOLE *Pitymys (Microtus) duodecimcostatus* Pl. 60

Name. *Duodecim* = twelve (Lat.); *costa* = rib (Lat.) = 12-ribbed.

Recognition. Dense, velvety coat, yellowish brown to reddish above, silvery to dark reddish grey below. Eyes small. Ears small and partially hidden in fur. Only 3 lobes on inside of third upper molar tooth (cf 4 in Common Pine Vole) and outside ridges very uneven in size. Otherwise very similar to Common Pine Vole and Savi's Pine Vole.

Habitat. Digs underground burrows.

Habits. Largely nocturnal and crepuscular. FOOD: Herbivorous, particularly on legumes and grasses. COMMUNICATION: May use olfactory cues to recognise sex and familiarity.

Breeding. Litter size 1–5. Mammae 4 (all inguinal).

Lifespan. Predators include Barn Owl and Red Fox.

Measurements. Head-body length: 85–107mm. Tail-length: 19–35mm. Hind-foot length: 15–18.5mm. Condylo-basal length: *c.* 25mm. Weight: *c.* 23g (14–28g). Dental formula: 1/1, 0/0, 0/0, 3/3 = 16.

General. Generally similar in all respects to Common Pine Vole. Sibling species: Iberian Pine Vole *P. lusitanicus*, Yugoslavian (and Greek) Pine Vole *P. thomasi*.

SAVI'S PINE VOLE *Pitymys (Microtus) savii* Pl. 60

Recognition. Slightly paler pelage, and tail slightly shorter, than Common Pine Vole and ears even smaller than in other pine voles. Third upper molar only 3 lobes on inside (cf 4 in Common Pine Vole) and ridges on outside of tooth equal in size (cf Mediterranean Pine Vole). Otherwise, very similar to Common Pine Vole.

Habitat. From lowland to mountain terraces. Deciduous, mixed and coniferous forest, meadows, pastures, gardens, agricultural fields and wasteland. Underground burrow.

Habits. Largely nocturnal and crepuscular. FOOD: Herbivorous, feeding on all plant parts. COMMUNICATION: Scent glands include Meibomian (tarsal) and anal glands.

Breeding. Litter size 2–4. Mammae 6 or 8.

Measurements. Head-body length: 72–105mm. Tail-length: 21–35mm. Hind-foot length: 14–16.5mm. Condylo-basal length: *c.* 24mm. Weight: 14–24g. Dental formula: 1/1, 0/0, 0/0, 3/3 = 16.

General. Found in three separate areas (may be subspecies or species): (i) S.W. France and N. Spain (*P.s. pyrenaicus*), (ii) Italy and Sicily (*P.s. savii*), (iii) S. Yugoslavia (*P.s. feltenii*).

COMMON PINE VOLE *Pitymys (Microtus) subterraneus* Pl. 60

Name. *Subterraneus* = living underground (Lat.).
Recognition. Dark greyish pelage. Small eyes. Ears almost hidden in fur. Third upper molar has four enamel lobes. Distinguished from most voles by 4 mammae (cf 8 in others). Distinguished from grass voles by 5 pads on sole of hindfoot (cf 6 in grass voles i.e. *Microtus* species) and shorter tail. Sometimes seals tunnel entrances during rain/ice periods. A more fossorial species than the Field Vole, but its remains are found in raptor pellets so it must spend time above ground.
Habitat. Grassland (especially moist meadows and fields) and woodland (but rarely coniferous forest). Up to 2000m in Alps. Nest of grass, moss and roots usually in extensive burrows just below surface (max. 30cm depth). Holes often stopped against severe weathers (food stores below ground).
Habits. Largely nocturnal. FOOD: Vegetarian (eg roots, rhizomes and bulbs), i.e. a below-ground feeder. SOCIAL: Sociable. COMMUNICATION: Soft squeaks and twitters; hisses with fear. Scent glands include Meibomian (tarsal) and anal glands (enlarged in breeding males).
Breeding. Breeds all year round. Sexual maturity 2–3 weeks. Inhibited in juvenile females by presence of adult female or presence of brother. Gestation 21 days. Litter size 2–3. Mammae 4 (all inguinal).
Measurements. Head-body length: 75–106mm. Tail-length: 25–43mm. Hind-foot length: 13–16mm. Condylo-basal length: *c.* 23mm. Weight: 12–27g. Dental formula: 1/1, 0/0, 0/0, 3/3 = 16.

ROOT VOLE *Microtus oeconomus (ratticeps)* Pl. 58

Name. Tundra Vole, Northern Vole. *Micros* = small; *ous, otos* = ear; *oikos* = house; *nomos* = keeper (all Gk.); *ratticeps* = rat-headed (Lat.).
Status. Rare.
Recognition. Distinguished from Field Vole by slightly larger size, and slightly darker pelage. Inner surface of ear only slightly hairy. Also (cf Field Vole) lacks extra enamel lobe on second upper molar, and first lower molar has three triangular cusps on outside. Fieldsigns include runways in reeds and sedges.
Habitat. Grassland, especially swampy habitats (wetter than those favoured by Field Vole). Nest of moss, dry rushes, blades of grass; above ground in damp areas, otherwise in extensive tunnel system.
Habits. Mainly nocturnal in summer, mainly diurnal in winter. FOOD: Shoots of rushes and other aquatic plants. SOCIAL: During the breeding season, home ranges of males overlap those of several females. Intruding voles are attacked and driven off. COMMUNICATION: Preputial and hip glands on adults of both sexes. As with all *Microtus* species, the size of

<*Root Vole.* C Europe, N Scandinavia eastwards into W Germany, pockets in the Netherlands and S Scandinavia.

>*Field Vole.* Throughout C, W & N Europe, excluding Ireland.

the hip glands is variable in males, but usually largest in older males. Presence in females may be related to geography – those from more northerly climates having them, those from more southerly regions without. Voles are attracted to conspecific odours, especially those of the opposite sex. Little is known of the behavioural function of scent glands among the *Microtus* species, but they can nevertheless be useful for the study of taxonomic relationships.

Breeding. Breeding season from May–September in Northern Scandinavia; winter breeding may occur in Finland north of the Arctic Circle. April–October in Vasa Island, Germany and Austria. Sexual maturity 6 weeks. Gestation 20–23 days. Litter size 2–11. Litters per year 2–5. Mammae 8.

Measurements. Head-body length: 85–161mm. Tail-length: 24–77mm (*c.* 40% head-body length). Ear: 8.16mm. Hind-foot length: 17–22mm. Condylo-basal length: *c.* 25mm. Weight: 25–62g; reputedly up to 103g. Generally larger than Short-tailed Vole. Dental formula: 1/1, 0/0, 0/0, 3/3 = 16.

General. Declining in Netherlands on drained land (whereupon replaced by Common Vole). Swims and dives well. Will climb to reach succulent shoots. Formerly in northern Russia, northern Finland, Poland and West Germany. Relicts in mountains of Scandinavia, western parts of Netherlands, and in parts of Pannonian plains.

FIELD VOLE *Microtus agrestis* — Pl. 58

Name. Short-tailed Vole. *Agrestis* = of the field (Lat.).

Recognition. Slightly larger than the Common Vole. Short tail compared with Bank Vole which is 'redder' or 'chestnut'. Pelage: dark, brownish-grey above, underside, neck and paws grey, flanks sometimes have creamy tinge. Pelage appears longish and shaggy (softer to touch than, eg Bank Vole). Hair grows inside ear and on upper edge. Middle upper molar has small extra enamel lobe on inside at rear. Northern forms tend to be larger. Moults in February to sparse, coarse hairs, interspersed with few fine hairs; in October–November moults to dense fur. Fieldsigns include piles of green droppings, oval when fresh, containing small pieces of cut vegetation. Runways between grass tussocks, leading to underground tunnel entrances. Feeding places with broken off blades of grass.

Habitat. Grasslands (eg meadows, field margins), young forestry plantations, mountain heath, open wood and dunes. Grass sufficiently high to provide cover (i.e. not heavily grazed). Favours damp ground. In Alps, up to 1900m. Female builds spherical nest of finely shredded grass at base of tussock. Underground burrow system, connecting to surface runways.

Habits. Active by day and night, but especially crepuscular. More nocturnal in summer than in winter. Juveniles more nocturnal than adults. FOOD: Grasses and herbaceous plants, some dicots. May gnaw bark in winter up to height of 15cm above ground. Daily consumption up to 30g. Evidence of very occasional animal prey (eg fly larvae). HOME RANGE: 100–1000m^2 (male range twice size of female's). Breeding females have overlapping home ranges; males' ranges more exclusive. Mothers (not young) usually disperse after weaning. SOCIAL: Strongly polygamous and territorial, females aggregate, largely non-territorial. Presence of males or urine accelerates puberty in young females (but less clear effect if male is sexually naive). In captivity: males and pregnant females aggressive to others. COMMUNICATION: Very vocal, loud chattering and chirping calls. Faeces deposited in piles in runways. Scent glands on hip in adult males (very large in breeding season, small at other times). Juveniles, when sniffed, may secrete odour from skin on back of head which stimulates play (see vignette). Olfactory communication plays role in inter- and intraspecific aggression (see Root Vole). Hearing and smell acute.

Breeding. Mostly breeds between April–September, but young may be found throughout year, even in nests under snow. Induced ovulation (as probably are all *Microtus* sp.), commonly conceive post partum. Sexual maturity at 40 days in male, 28 days in female (usually reproduce in same year as born). Gestation 18–20 days. Litter size 4–6, up to 9 in captivity. Litters per year 2–7. Mammae 8. YOUNG: Hairless at first, developing darker fur than adults. Weaning age 14 days. Parental care by female only.

Lifespan. Max. recorded 2 years. Few survive to second autumn (an estimated 2.3% survive more than 15 months) (in captivity, average life expectancy at birth is 7.5 months). Predators include a wide variety of birds and mammals.

Measurements. Head-body length: 78–135mm. Tail-length: 18–49mm (*c.* 30% head-body length). Hind-foot length: 16–20.5mm. Condylo-basal length: 23–29mm. Weight: 14–50g (birth: *c.* 2g). Those born late summer weigh *c.* 17g during winter, increase by June to: males 35–40g, females *c.* 30g. Second-year adults 6–8g heavier than first-year. Dental formula: 1/1, 0/0, 0/0, 3/3 (molar tooth row less than 7mm) = 16.

General. In some areas, 3–5 years periodicity in population density (may damage pastures in peak years). Runs fast, but poor climber. Size and colour variation (eg in UK larger and darker towards north). Occasional colour mutants (eg black, agouti, piebald).

SNOW VOLE *Microtus nivalis* **Pl. 58**

Name. *Nivus* = snow (Lat.); *nivalis* = of the snow.

Recognition. Pale grey (occasionally almost white-tailed) dense pelage and long whiskers (up to 60mm). First lower molar has arrowhead-like cusp pointing forward at front. Distinguished from Balkan Snow Vole by shorter tail. This is the only vole found at high altitudes. Droppings, tunnels, a nest of dry herbs or a small store of food under stones are clues to its presence.

Habitat. Mountain slopes above tree-line (areas where ground cover more than 75% rock), but also in low hills with dry woodland (eg S. France). Over 4000m in Mont Blanc, 120m in Yugoslavia and Adriatic coast. Nest of hay and stalks, also storage chambers, just below surface in tunnel system which has several entrances.

Habits. Diurnal. FOOD: Exclusively vegetarian (eg alpine grasses and shrubs, and fruit, eg bilberry). COMMUNICATION: Chattering in breeding season, and penetrating single whistles. Scent glands include preputial and anal glands, but no hip glands.

Breeding. In SW Alps (1800–2600m) breeding season is early-May–late-August; at lower altitudes (eg 1500m) beginning of May until the beginning of September. Gestation 21 days. Litter size 2–7. Litters per year 1–2. Mammae 8. YOUNG: Weaning age 20 days.

Lifespan. Predators include Eagle Owl, Tawny Owl, Raven, Red Fox, Weasel and Stoat.

Measurements. Head-body length: 110–140mm. Tail-length: 50–75mm (*c.* 50% head-body length). Hind-foot length: 18.5–22mm. Condylo-basal length: *c.* 30mm. Weight: 38–50g. Dental formula: 1/1, 0/0, 0/0, 3/3 = 16.

General. Agile on scree and rocks. Swims well. Basks in sunshine on rocks (see vignette).

CABRERA'S VOLE *Microtus cabrerae* **Pl. 59**

Recognition. Larger than the Common Vole (although the Iberian Common Vole is larger than those further north). Also distinguished from

< *Snow Vole.* Confined to mountainous regions of S Europe, such as Alps and Pyrenees.

> *Cabrera's Vole.* Reported from isolated locations in Iberian peninsula.

Common Vole because of less developed posterior loop of upper tooth row and anterior loop of lower tooth row. Skull almost convex in profile (cf Common Vole almost straight profile). The rows of the upper and lower teeth have, respectively, less developed posterior and anterior loops than those of the Common Vole. Fur on back darker than Common Vole (long guard hairs protrude far on rump). Grey underside more buff than Common Vole. Fieldsigns include small mounds of nibbled grass and leaves. Runways above ground and burrows.

Habitat. Coastal marshes (S. Portugal) to 1500m (Central Spain). Open woodland with shrubby understorey. In Iberia tends to be at lower altitudes or damper ground than Common Vole, but sometimes the two species overlap.

Habits. FOOD: Herbivorous (eg grass, leaves and stems).

Breeding. Mammae 8.

Measurements. Head-body length: 110–130mm. Tail-length: 35–45mm. Ear: 13–16mm. Hind-foot length: 20–23mm. Condylo-basal length: *c.* 27mm. Weight: 40–65g. Dental formula: 1/1, 0/0, 0/0, 3/3 = 16.

General. Sometimes considered as an Iberian form of Gunther's vole.

GUNTHER'S VOLE *Microtus socialis (guentheri)* Pl. 59

Name. *Socialis* = social (Lat.).

Recognition. Tail pale, hindfeet almost white. Similar to Common Vole but distinguished because triangular cusps on grinding surface of molar teeth tend to be well separated from each other. Also, second upper molar sometimes has extra loop on inner, posterior corner. Very similar to Cabrera's Vole.

Habitat. Grasslands (eg pastures and cereals). Spanish varieties in mountainous regions. Nest of soft hay in burrow with 3–8 passages at depths of up to 20cm, including 1–2 storage chambers.

Habits. Largely nocturnal. FOOD: Green plant parts. SOCIAL: Reputedly sociable. COMMUNICATION: Chattering call; alarm, sharp whistle.

Breeding. Mammae 8.

Measurements. Head-body length: 100–120mm. Tail-length: 20–36mm (20–25% head-body length). Hind-foot length: 18–22mm. Condylo-basal length: *c.* 28mm. Weight: 32–68g. Dental formula: 1/1, 0/0, 0/0, 3/3 = 16.

< *Gunther's Vole.* Confined to E Greece.

> *Common Vole.* Europe excluding British Isles, N Mediterranean and Scandinavia west of the Baltic.

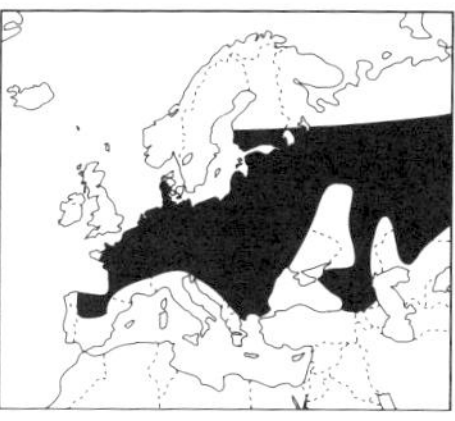

General. Occasional dramatic population explosions. Habits assumed similar to other grass-living voles.

COMMON VOLE *Microtus arvalis* **Pl. 59**

Name. Orkney Vole, Guernsey Vole. *Arva* = field (Lat.); *arvalis* = of the field.
Recognition. Smooth-haired appearance. Tail only faintly darker above than below. Ears almost naked inside, but dense fringe of hairs towards top. Generally distinguished from very similar Field Vole by lighter colour, shorter hair, also Common Vole lacks extra loop on second upper molar (cf Field Vole). Distinguished from Root Vole because first lower molar has four triangular cusps on outer surface. Black individuals common on Orkney; Orkney Voles tends to be darker and have thick coats. Short tail. Fieldsigns include runs in long grass of field margins. Tiny prints hard to find. Droppings in middens, often on path leading from burrow as with Field Voles. Droppings are cylindrical, black or greenish and *c.* 3–4mm long.
Habitat. Dry, short (grazed) meadows, cultivated land, clover, margins of fields. Avoids wet grassland and tall grasses. Nests and storage chambers generally in tunnel system, but occasionally above ground, made of grass and stalks. Tunnels connected by surface runways. Burrows extending up to 6m from nest. Excavates tunnels more than Field Vole and therefore thrives better in grazed areas.
Habits. Largely nocturnal and crepuscular. Alternates periods of activity and rest, each of *c.* 3 hours. FOOD: Similar to Field Vole, preferring *Juncus squamosus* roots. May eat insects. HOME RANGE: Male, up to 1500m^2; female, 25% size of male range. SOCIAL: Occupy nests within 3m of each other. Can be maintained in groups in captivity provided that only one adult male present. In wild monogamous pairs occupy territories during summer. In winter much more overlap. Nursing mothers tolerate other voles inspecting their young. COMMUNICATION: Vocalises with high-pitched squeak. Sources of scent: anal glands, soles of feet, and hips.
Breeding. Breeding season varies with latitude: mid-Germany, March–October; NW France, February–December. On Guernsey mating starts in February and young fully adult by August. Sexual maturity *c.* 30 days, but sometimes females as young as 11–13 days old. Gestation 19–21 days. Litter size 2–12, averaging 5. Litters per year 2–4. Mammae 8. YOUNG: Weaning age 20 days. Male Orkney Vole spends much time with young and will retrieve them when displaced from nest.
Lifespan. Max. recorded 1.6 years.
Measurements. Head-body length: 90–120mm. Tail-length: 30–45mm (*c.* 30% head-body length). Ear: 9–10mm. Hind-foot length: 15–18.5mm. Condylo-basal length: 22–27mm. Weight: 14–40g (Orkney Vole can be 80–100g). Dental formula: 1/1, 0/0, 0/0, 3/3 = 16.
General. Well differentiated subspecies on Sanday (Orkney) Islands

(*M.a. sandayensis*, light dorsal colour, suffused with creamy buff), Orkney mainland (*M.a. orcadensis*, dark dorsal colour suffused with orange buff) and Guernsey (*M. a. sarnius*, light grey pelage). Subfossil evidence suggests present on Orkney for at least 4000 years. Seems to have slower reproduction rate and population turnover than Field Vole. Local migrations common. Often stands up on hind legs. Swims well. Only populations in UK are on islands.

SIBLING VOLE *Microtus rossiaemeridionalis (epiroticus)*

Name. *Epiroticus* = of Epirus (district of N Greece).

Recognition. Identical to Common Vole, except in chromosome number, and characteristics of spermatozoa and genitalia (requiring dissection). However, larger than most subspecies of Common Vole.

Habits. Mainly crepuscular. FOOD: Herbivorous, mainly leaves and plant stems in summer, roots in winter.

Breeding. Breeding season March–October. Litters per year up to 3.

Measurements. Head-body length: 105–128mm. Tail-length: 33–52mm. Ear: 10.4–12.4mm. Hind-foot length: 16–18mm. Condylo-basal length: 24–27.5mm. Weight: 25–46g. Dental formula: 1/1, 0/0, 0/0, 3/3 = 16.

HARVEST MOUSE *Micromys minutus* **Pl. 50**

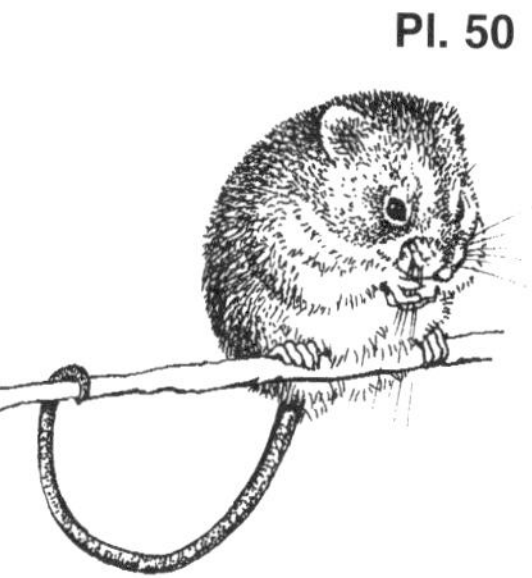

Name. *Micros* = small (Gk.); *mys* = mouse (Gk.); *minutus* = tiny (Lat.).

Recognition. Russet-red above, sharply delineated from white underside. Ears: hairy (vole-like). Blunt muzzle. Tail tip prehensile. The smallest European rodent. During moult appear bi-coloured on back, in the autumn males moult approximately 4 weeks earlier than do breeding females. Tiny droppings *c.* 2mm long. Nests particularly conspicuous in autumn and winter as vegetation dies back.

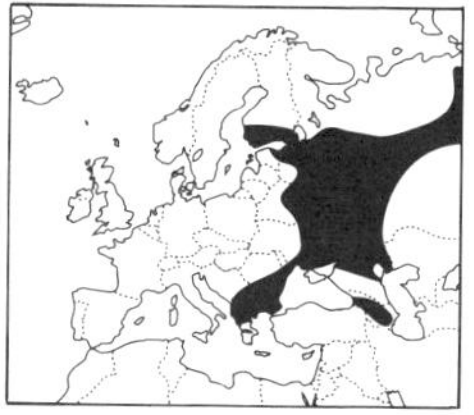

< *Sibling Vole.* Extending into S Finland and countries W of the Black Sea.

> *Harvest Mouse.* Widespread in C&W Europe, excluding Iberia, most of Scandinavia, Ireland and most of Scotland.

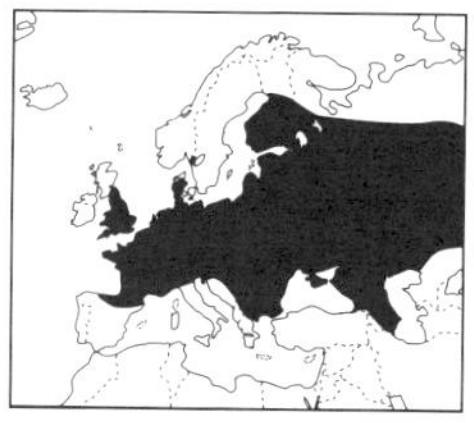

Unlike those of dormice, nest is strongly linked to stems of vegetation for support. Food remains include sickle- shaped slivers of grain.
Habitat. Long grass, dry reedbeds, ungrazed hay meadows, cereal (more in oats and wheat than barley). In summer: ball of shredded grass (8–10cm diameter), 30–60cm above ground on grass stalks or brambles, main walls of grass blades, inner chamber lined with shorter shredded lengths. In winter uses a larger nest, tends to be less spherical and larger and at ground level or in burrow underground. One entrance to nest.
Habits. Largely nocturnal. FOOD: Seeds (cereals), fruits and berries, insects (especially in winter), green shoots in spring when other food scarce. HOME RANGE: 0.04–0.06 ha. Considerable overlap of adjacent ranges. Dispersal recorded over 100m. SOCIAL: Largely solitary.
Breeding. Breeding season May–October (occasionally till December). Gestation 19 days. Litter size 1–8. Litters per year 3–7 in wild (8 in captivity). Mammae 4. YOUNG: Similar grey-brown to House Mouse, but smaller. Weaning age 15 days. Young chased from vicinity of nest at weaning if mother pregnant.
Lifespan. Max. recorded 18 months in wild (5 years in captivity).
Measurements. Head-body length: 50–80mm. Tail-length: 50–70mm. Hind-foot length: 12–16mm. Ear: 8–10mm. Condylo-basal length: 16–18mm. Weight: 5–11g (pregnant: 15g; birth: 0.7g). Dental formula: 1/1, 0/0, 0/0, 3/3 = 16.
General. Cycles of abundance in some regions. The quality of habitat affects body size and weight, those in more marginal habitats (woods and wood edges) are smaller overall than those found in meadows. Climatic factors also affect appearance, areas having high humidity tending to have populations with darker coloured pelage. Spends more time at, and below, ground level in winter. Climbs well.

ROCK MOUSE *Apodemus mystacinus* Pl. 52

Name. Broad-toothed Field Mouse. *mystacinus* = moustache (Gk.); *apodemus* = away from (*apo-*) home (*demus*) (Gk.).
Recognition. Larger and more greyish than other *Apodemus* species and longer whiskers (47mm, cf 40mm in Yellow-Necked Mouse). 185–205 tail rings.
Habitat. Dry woodland, scrub, rocky hillsides, walls along field margins. Den site is under boulders.
Habits. FOOD: Olive and cherry stones, juniper berries, green plant parts, insects.
Breeding. Gestation 23 days. Litter size 2–9. Litters per year 3 max.
Measurements. Head-body length: 100–130mm. Tail-length: 102–140mm. Hind-foot length: 24–28mm. Ear: 17–21mm. Condylo-basal

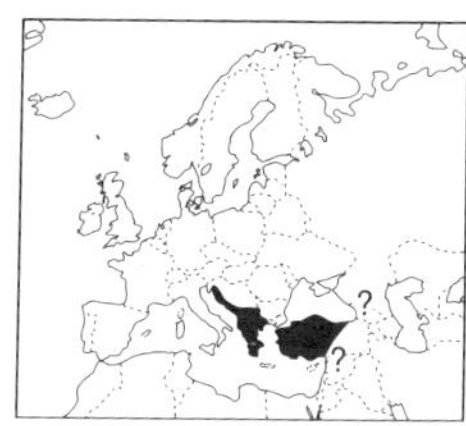

<*Rock Mouse.* Yugoslavian coast, S through Greece and islands, including Crete, Rhodes and Cyprus.

>*Pygmy Field Mouse.* E Europe: Yugoslavia, Romania, Hungary, Bulgaria and E Austria.

length: 25–30mm. Weight: *c.* 42g (28–56g). Dental formula: 1/1, 0/0, 0/0, 3/3 = 16.

General. Debate over whether Kirk Mouse (*A.m. kirkensis*) is a subspecies of Rock Mouse or full species. Kirk has shorter hind-foot length (22–24mm). Only found on Kirk Island in Yugoslavia. However, it seems likely that Kirk Mouse is actually a form of *Apodemus sylvaticus.*

PYGMY FIELD MOUSE *Apodemus microps* Pl. 52

Name. Herb Field Mouse. *micros* = small (Gk.); *ops, opos* = face (Gk.).

Recognition. Chest spot small or absent. Upper incisor teeth lack notch typical of *Mus* spp. Similar to Wood Mouse but smaller, greyer with relatively smaller eyes and ears. Appears snub-nosed.

Habitat. Long grass, scrub, copse (similar to Wood Mouse).

Habits. FOOD: Grain, grasses, herbs, insects, and to a lesser extent, roots, flowers and fruit. COMMUNICATION: Scent glands: male sebaceous caudal gland, larger in larger (older) males than smaller ones.

Breeding. Breeding season March–August. Litter size 3–8, averaging 5.

Measurements. Head-body length: 70–96mm. Tail-length: 65–95mm. Hind-foot length: 17–20.5mm. Ear: 13–15mm. Condylo-basal length: *c.* 21mm. Weight: *c.* 17g. Dental formula: 1/1, 0/0, 0/0, 3/3 = 16.

WOOD MOUSE *Apodemus sylvaticus* Pl. 52

Name. Long-tailed Field Mouse. *sylvaticus* = of the wood, *silva* = wood (Lat.).

Recognition. Distinguished from *Mus* species by larger ears, eyes and hind feet. Distinguished from the very similar Yellow-necked Mouse by smaller size, smaller (or absent) yellow neck (collar longer than broad), and by fewer rings on tail, which is dark above, pale below. Pelage: dark brown above, yellow-brown flanks, greyish-white below, often not clearly demarcated (cf Yellow- necked Mouse). Albinism of tail-tip reputed to reach 3% in some populations. Silver-grey, piebald, semi-hairless and melanistic forms known. Juveniles can look quite like *Mus*. Characteristic musky smell of House Mouse lacking. Juvenile pelage: grey-brown above, dark grey-white below. Moult in France is in spring and autumn, but in UK it has no seasonality, except in Scilly Isles and Cornwall, where restricted to August–January. Regular ventral to dorsal post-juvenile moult at 5–7 weeks. Droppings larger than those of *Mus*, usually 3–5mm in length. Grain manipulated with forepaws, leaving husk. Hazel shells have hole surrounded by tooth marks. Carpels of rose hips eaten but flesh discarded. Snail shells opened by biting through shell away from spire. Food-stores under logs, crevices and underground or disused birds' nests. Footprints: Four foretoes, five hindtoes, slightly larger than those of Bank Vole and House Mouse.

Habitat. Woodland, throughout arable land, gardens, bramble and bracken scrub, sand dunes. Normally not above tree-line. Highly adaptable. Nest of leaves, moss and shredded grass, depending on availability, often underground (deeper in winter). Excavates own burrow, often with 1 entrance plus nest chamber and food storage chamber. Adult males, but not adult females, often share burrow during the breeding season in cereal fields. Larger communal nests frequent in winter, probably involving individuals of both sexes.

Habits. Largely nocturnal, although frequently up to 2 hours before dusk and after dawn during the summer; pregnant and lactating females active for short bursts throughout day in summer. Winter: peaks of activity at dawn and dusk. Summer: one major activity peak after dark. Activity reduced by moonlight, and inhibited by combined wet/cold conditions. Males more active than females. Lactating females return to nest site often. FOOD: An opportunist, proportions of food type taken greatly affected by their availability. Vegetable foods include seeds, seedlings, buds, fruits and nuts, fungi, moss, galls. Animal foods include snails, arthropods, earthworms. Winter: large number of tree seeds. Spring and summer: larval moths and butterflies, centipedes, buds, shoots. Differences exist in diets of different age/sex classes. Specially favoured foods include sweetcorn, wheat, blackberries and wild oats. In cereal fields commonly eat arable weeds. HOME RANGE: Highly variable, but larger in farmland than woodland. Home range sizes differ between habitats and sexes. The recorded average for home ranges of males versus females, respectively, are for woodland 6276m^2 *vs* 1919m^2, for arable farmland 17,657m^2 *vs* 2641m^2 and for sand dunes 36,499m^2 *vs* 15,826m^2. Both sexes can move large distances – some regularly move over 2km every night. Population density: 0.25–100 per ha with predictable intra-annual dynamics (stable summer, rise in autumn, crash over winter and spring). Home range size increases with sexual maturity. Adult males probably disperse throughout year, but females disperse mainly in autumn/winter. SOCIAL: In winter overlapping ranges with communal nesting of both sexes. In summer dominant male with larger range (eg 1.6–2.4ha) containing subordinate males and females. Females defend exclusive breeding territories. Males' home ranges overlap. COMMUNICATION: Vocalisations include squeaks and ultra-sound during aggression. Ultra-sound when mating from male (and from young until 24 days old). Sebaceous (i.e. oily) glands on oral angle and lips. Both male and female have preputial (clitoral) glands. Male sebaceous caudal gland, larger in larger (older) males than smaller males. Smell and hearing acute. Insensitive to infra-red and red light, so easy to watch with red filtered torch.

Breeding. Season March–October (peak: July–August; occasionally throughout year). Reproductive tract usually regressed or undeveloped in winter non-breeding season. Early-born young are sexually mature later in the same year (males 12g; females 15g). Later young mature following year. Gestation 19–20 days, longer during lactation owing to delayed implantation. Litter size 2–9, usually 4–7. Varies with geographical location and possibly also with season. Up to 4 litters per year, usually 1–2. Mammae 4 pairs (3 abdominal, 1 axillary). YOUNG: Born naked and blind, then developing grey fur like House Mouse. Weight at birth: 1–2g. Weaning age 18–22 days. Parental care by female only.

Lifespan. Max. 18–20 months. Few adults survive from one summer to next. High mortality of over-wintered adults in spring, and of juveniles in spring and summer. Predators include: foxes, Weasel, Stoat, Badger, martens, Domestic Cat, Little Owl, Short-eared Owl, Tawny Owl, Barn Owl, Kestrel. Mortality increases in spring, probably as a result of intraspecific strife causing dispersal and increased predation.

Measurements. Head-body length: 97–110mm (eg male, 86–103mm; female, 81–103mm). Tail-length: 69–115mm (130–180 rings on tail). Hind-foot length: 20–24mm. Condylo-basal length: 22–26mm. Weight: 13–27g (female max. 24g, or 30g or more if heavily pregnant). Dental formula: 1/1, 0/0, 0/0, 3/3 = 16.

General. Very agile (less restricted to runways in thick vegetation than voles). Suffers pollution toxicity, eg woodland individuals found to contain 1.14 ppm wet wt lead, roadside 1.66 ppm wet wt lead; susceptible to poisoning from dressed grain: feeding on grain led to 68-fold increase in Dieldrin loading, 11-fold increase in mercury. Homing experiments: 100% returned from 336m, 50% from 663m. Some evidence that Wood Mice orient using magnetic compass. Uses platforms above ground (eg old birds' nests) as feeding stations. Two distinct forms postulated in mainland Britain, in apparent breeding contact but probably of different origins: an eastern type (related to French populations) and a west/north

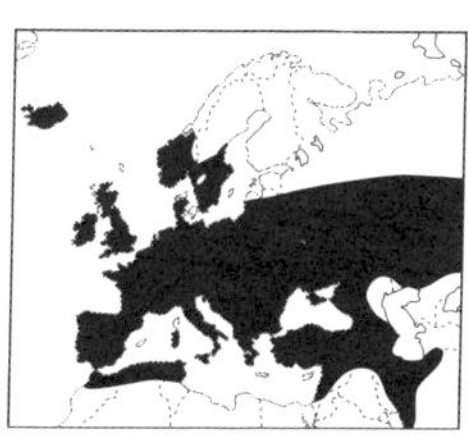

< *Wood Mouse.* Widespread throughout Continental Europe and British Isles and southern Scandinavia.

> *Yellow-necked Mouse.* C Europe as far west as France and S UK.

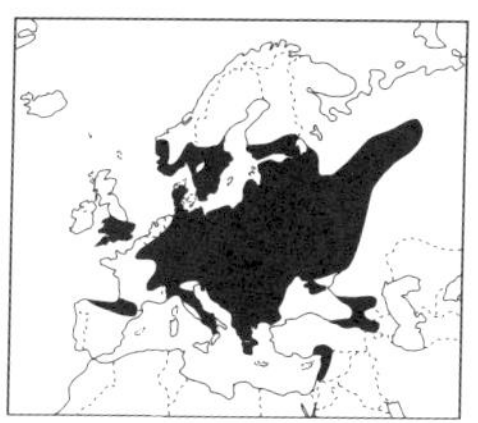

type showing affinities to the island forms; however, minimal evidence to support this speculation.

YELLOW-NECKED MOUSE *Apodemus flavicollis* Pl. 52

Name. *Flavus* = yellow (Lat.); *collis* = neck (Lat.).
Recognition. Distinguished from *Mus* species by larger ears, eyes and hind feet. Distinguished from the very similar Wood Mouse by much larger yellow patch on throat joining upper brown fur (collar is broader than long, but less conspicuous in southern part of range). Richer brown upper parts, purer white underneath (and sharper demarcation) than Wood Mouse. Also much (1.5×) heavier than Wood Mouse and squeals and bites more readily if captured. Fieldsigns include nests, nut shells and seed remains. Carries food to eating place. Droppings larger than *Mus* species.
Habitat. Mainly woodland (often close to arable farmland), hedgerows and field margins, orchards and wooded gardens. More common than Wood Mouse in alpine coniferous forest (and above tree-line), less common than Wood Mouse in open scrub and fields. Reputed to climb in woodland canopy to greater extent than does Wood Mouse. Nest of leaves or shredded grass, under tree-stump or in burrow of moles or voles; occasionally nests in tree crevices or digs own burrow.
Habits. Single nightly activity period. FOOD: Largely seedlings, buds and fruit. Invertebrates estimated to be 10% diet (eg larvae and pupae). HOME RANGE: Assumed to occupy larger home range than Wood Mouse. Population density: usually 1–10 per ha, up to 50 per ha. SOCIAL: Overlapping ranges, but detailed social structure not known. COMMUNICATION: Young mice left in the nest by mother emit squeaks at 56–60kHz which attract breeding females, even if young not their own. Scent glands: sebaceous (oily) glands in oral angle and lips. Both male and female have preputial (clitoral) glands. Male's sebaceous caudal gland larger in larger (older) males than smaller ones. Hearing and sense of smell acute.
Breeding. Breeding season April–October, with occasional births throughout year. Breeding season often shorter than that of Wood Mouse in Great Britain. Early-born females can breed in year of birth. Litter size usually 5 (1–8). Litters per year 3. Mammae 8. YOUNG: Hairless at first, then developing grey fur; yellow collar becomes noticeable at *c.* 2 weeks old.
Lifespan. Max. 2 years. Good juvenile survival in spring and summer, peak density in autumn then steady mortality rate throughout winter.
Measurements. Head-body length: 88–130mm. Tail-length: 90–135mm (165–235 rings on tail). Hind-foot length: 23–27mm. Ear: 16–20mm. Condylo-basal length: 25–29mm. Weight: 10–45g (male *c.* 29 g; female *c.* 27g). Dental formula: 1/1, 0/0, 0/0, 3/3 = 16.
General. Caches food for winter. Climbs and jumps well.

STRIPED FIELD MOUSE *Apodemus agrarius* Pl. 53

Name. *Ager, agri* = field (Lat.).

Recognition. Smaller ears than other *Apodemus* species. Black stripe from nape to rump. Tail length shorter than head-body length. Upper incisors not notched. Relatively short whiskers. Moults in spring and autumn.

Habitat. Fringes of woodland and scrub, associated with damp habitats and river valleys. May enter stables and barns in winter. Digs own burrow but is also found in burrows of other small rodents. Found less often in nests above ground than is Wood Mouse or Yellow-Necked Mouse and it has been suggested that its relatively short whiskers reflect this more subterranean existence.

Habits. FOOD: Plant material such as seedlings, buds, fruits, nuts, and animal food such as insects and their larvae, worms and molluscs. Thought to eat relatively more animal matter than do other field mice. COMMUNICATION: Wide range of aggressive squeaks. Male has sebaceous (i.e. oily) caudal gland, larger in larger (older) males than in smaller ones, and considerably larger than those found in Wood or Yellow-Necked Mice.

Breeding. Litter size 5–6. Mammae 8.

Measurements. Head-body length: 73–123mm. Tail-length: 70–85mm (60% head-body length), 120–140 rings on tail. Hind-foot length: 17–21mm. Ear: 15–19mm. Condylo-basal length: *c.* 23mm. Weight: 16–25g. Dental formula: 1/1, 0/0, 0/0, 3/3 = 16.

General. Always placed in genus *Apodemus*, whereas other field mice sometimes placed in genus *Sylvaemus*.

BROWN RAT *Rattus norvegicus* Pl. 54

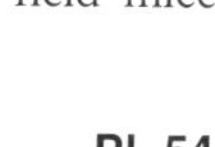

Name. Common rat, Norway Rat. *Rattus* = rat (Lat.); *norvegicus* = of Norway (= incorrect inference of species' origin).

Status. Introduced.

Recognition. Pelage: variations from brown to black; generally brindled brown-grey above, paler grey below (sometimes white patch on chest and light coloured forelegs). Distinguished from Black Rat by shorter, more fleshy, more hairy ears (barely reach corner of eye when folded forward). Tail

<*Striped Field Mouse.* Widespread in E Europe, with western populations reaching Germany and N Italy.

>*Brown Rat.* Ubiquitous throughout W Europe

shorter, thicker and dark above, pale beneath. *R. rattus* tail longer than body and uniformly coloured. Eyes of Brown Rat smaller than Black Rat. Melanism 1–2% of some populations. Tracks: Hindfoot 3.3 × 2.8cm; 5 toes, 5 digital pads, 6 palm pads. One greatly enlarged proximal pad as in *R. rattus*, long heel area. Forefoot 1.8 × 2.5cm, 4 toes and 4 deeply impressed digital pads. 5 palm pads. Trail: When running partial registration of tracks widely spaced about median line, rarely with tail drag; stride 14–15cm. Bounding gait tracks in groups of four about 60cm apart. Walking stride about 15cm. Star-like appearance of footprints with rarely sign of tail drag. Pathways or runs between holes appear as depressions in vegetation or well worn trails of bare compressed earth 50–100mm wide. They are continuous, unlike Rabbits', characteristically along walls, hedgerows and banks close to cover. Smear marks: as rats travel along well defined paths, where their bodies contact solid objects, the fur leaves a dark greasy deposit. Usually seen around entrances, steps and walls. Under beams single loop smear may be left, cf twin loop smear of *R. rattus*. Gnawing may be evident on wooden doors or floors of outbuildings. Droppings: colour varies according to diet, but moist and glistening light brown when fresh, drying within hours to darker matt appearance. Spindle shaped, tapering and often pointed at one end. Deposited in groups – Brown Rat may produce about 40 per day. Size varies but averages 15–20mm long and 5–6mm across. *R. norvegicus* pellets can accurately be separated from *R. rattus* pellets by dividing he diameter of the droppings by their length. If the result is between 0.42–0.56 then they were produced by the Brown Rat *R. norvegicus*, if the result is between 0.31–0.37 they were produced by the Black Rat *R. rattus*. However, the accuracy of this formula may be affected by the rats' diet.

Habitat. Generally associated with people (eg refuse tips, grain-stores, sewerage systems, farmyards) or similar food supplies (eg pheasant-rearing pens). However, some populations spend all or part of year in crops and field margins (favouring dense cover close to water), and some live on uninhabited islands (eg off W. Scotland) and other habitats far from people (eg found 70km from human settlements in parts of Russia). Burrows (65–90mm diameter) often near food source. Earth spoil heaps in front of entrance, unlike Northern Water Vole, which has similar size hole but leaves earth piles away from entrance. Loose earth soon becomes flattened. Dens often in side of bank or tree root, but may be on flat ground. In open, live in system of underground tunnels, extent varies according to conditions, but depth rarely more than 50cm. Several holes to system linked on surface by distinct well trodden runs. Also nests under floorboards and in similar domestic crevices. Also hayricks and agricultural buildings.

Habits. Mainly nocturnal but diurnal if predation at night (eg foxes at rubbish tips) make it too dangerous to be nocturnal. Also, subordinate rats may be diurnal if dominant rats prevent them from feeding undisturbed at night. Peak activity an hour or two after sunset and second peak an hour or two before sunrise. Also some evidence of less dominant animals feeding during daylight hours. May be seen throughout day if relative disturbance low, or population very high. Rats with access to food and shelter eat at fairly regular intervals; longer intervals follow larger meals, but larger meals do not necessarily follow longer intervals. FOOD: Omnivorous, favouring protein and starch-rich foods. Vegetation (eg agricultural cereals and seeds). Animal food includes invertebrates (eg slugs, snails, larvae etc) and vertebrates (eg frogs, birds' eggs, young mammals); also scavenges (meat, bones, etc, including oddities such as soap and candles). Hoards food (especially lactating females and subordinates). Cannibalism symptomatic of protein deficiency. HOME RANGE: 12–400m^2. May travel 3–4km per night, usually keeping close to hedgerows, but will cross open areas for up to 500m. May change home sites every one or two weeks. If living in farm buildings, rats tend not to move far – mean range length 65m. SOCIAL: Colonies develop from pair or single pregnant female. Rat 'infestations' may not be single large social groups but aggregations of smaller social units ('clans') consisting of a dominant male plus associated harem of females and associated subordinate males. When numbers are low members recognize each other and live harmoniously. Intruders are repelled by residents of a territory and may even be killed. This behaviour almost entirely prerogative of males. Females with young pups may defend the nest against all visitors although some colony members allowed to share nest. Dominance structure may limit access to food (subordinates being more diurnal). Large colonies reportedly more tolerant of outsiders than small ones. COMMUNICATION: Makes sounds audible to human ear at 20kHz or less, also ultrasound. Gentle piping or whistling, often by male when threatened or attacked. Similar sound by female defending nest, acts as deterrent to intruder. Scream alarm call by cornered or defensive animal may deter predators and causes other rats to take cover. Tooth chattering by male rats often first actively performed by resident animal in encounter with stranger. Squeaking or squealing by young rats in nest when hungry or cold. Ultra-sounds produced for communication between individuals, highly directional so does not alert predators at a distance; used during copulation by male and during aggressive behaviour: subordinate males have a characteristic call at 30kHz. Newborn pups use ultrasounds which elicit maternal behaviour. Scent of urine gives indication of reproductive condition and individual identity. Leave odour trails of urine and genital secretions and these are followed by other rats. Both sexes (males more than females) use anogenital drag and leg lifting urine marking postures. Female rats can recognize their offspring and distinguish their sex by smell. Scent glands include a nasal gland, anal glands, Harderian, Meibomian and lachrymal glands associated with the eye; sebaceous glands on soles of feet; preputial glands – odour discriminates between sexes, sexually active/inactive animals, mature or immature and kinship (able to recognize related individuals). Rats 'freeze' when they detect cat or mink odour, and avoid fox odour; cow and pig odour tends to be attractive, whereas human odour has little effect on behaviour. Sense of touch very important: use whiskers to orientate in dark. Tactile hairs all over fur. Move in close contact with objects, learning environment and establishing well defined runways – kinaesthetic sense ('memorising' surroundings) means will navigate around a familiar object even when it has been removed. Cautious of new foods (neophobia), taking little at first until possible toxin effect occurs. Rats deficient in certain substances tend to select new foods. Eyes adapted for nocturnal vision – excellent light sensitivity but poor visual acuity. Almost certainly colour blind, although some discrepancy over evidence of their ability to perceive red light. Recognize shapes and movements well. Blind rats able to survive and behaviour little affected. Good sense of hearing extending into ultrasonic range, up to 100kHz, greatest at 40kHz. Produce ultrasound during copulation, to affect aggression and probably for echolocation in the dark.

Breeding. Can breed all year if abundant food and mild weather. Reproduction rate increased if population decreased by trapping or poisoning. In very cold weather, does not breed and males may lack sperm. High altitude decreases reproductive performances. Sexual maturity: female, 8–12 weeks (95% at 136.6–152.1g, median 144.5g). Gestation 20–23 days. Litter size 1–15 – usually 7–9. Litter size varies with body size (eg *c.* 6 at 150g, *c.* 11 at 500g). Average production in one study 24 young per year. Litters per year up to 5 (theoretically 13 at 4- to 5-week intervals). Mammae 6 pairs (3 pectoral, 3 inguinal). YOUNG: Potential number of young per year up to 56 (Germany). Number in litter affect growth rate – fewer in litter, heavier the pups at 14 days and weaning. Maternal food intake and temperature of surroundings affect growth. Born blind and hairless. Eyes open, and fully haired at 7–10 days. Weaning age *c.* 21 days (*c.* 40g, Head-body length 110mm). Parental care by female only. She will retrieve young that stray from nest and defend them from intruders. May move young if nest disturbed. Older pups may continue to suckle in presence of new litter.

Lifespan. Max. 3 years but in wild assume less than 18 months. Females longer than males. In farm populations only 5% survive the year. Evidence that lifespan inherited, diet affects longevity. Estimated *c.* 90% adult mortality. Predators include Tawny Owls, Weasels, Stoats and occasionally Badgers. In Ireland, where Bank Voles absent, Brown Rats are principal prey of Barn Owl. Cats probably one of chief predators in urban areas and farms. Also Red Foxes particularly in rural areas. Aggressive adults may deter the smaller predators.

Measurements. Head-body length: 214–290mm. Tail-length: 170–230mm (80–100% head-body length) (160–190 rings). Hind-foot length:

40–45mm. Ear: 19–22mm. Condylo-basal length: 43–54mm. Weight: usually 275–520g, although 794g is the largest recorded (40g at weaning). Males heavier than females. Growth continues throughout normal life span. Dental formula: 1/1, 0/0, 0/0, 3/3 = 16.

General. Widespread in urban areas throughout most of the world except some areas of mainland tropics and subtropics. Occurs away from human habitation only in temperate regions, or in tropics where there is little competition. In Europe it is introduced and a pest. Successfully controlled by anti-coagulant poisons at first, but then became genetically resistant to first generation of poisons. Recently, resistance to second generation of anticoagulant poisons in several areas eg central southern England, Lower Saxony, Rhine Westfalia. Occurs in most areas except exposed mountain regions and smaller off-shore islands. Digs very well and good swimmer – can stay afloat for up to 72 hours. Can jump up to 77cm vertically and 120cm horizontally. Vector of leptospirosis, causing Weil's disease (bacteria excreted in urine estimated to infect 50–70% of rats in England). May also carry salmonella and toxoplasmosis, but less important vector of plague as generally lives outside houses. Thought to have originated in Asia, possibly China. Arrived in Europe in early part of 18th century and was first recorded in England in 1728. In spite of name, arrived in Norway considerably later, in 1790. Introduced to North America by 1775. Gradually ousted the ship rat *R. rattus* in Europe during the 1700s but was still replacing it in USA in 1950s and in Israel today.

BLACK RAT *Rattus rattus* Pl. 54

Name. Ship Rat, Roof Rat.

Status. Introduced.

Recognition. Pelage: varies from black to brown. Distinguished from Brown Rat by longer, almost hairless ears, larger more prominent eyes, longer and thinner tail. Guard hairs on back and flanks relatively longer than in Brown Rat, giving more shaggy appearance. Black colour often predominant in urban populations. Elsewhere populations may be polymorphic. For example, three morphs, *R.r. rattus* (black), *R.r. frugivorous* and *R.r. alexandricus* (both brown-grey) have been recorded on the Island of Lundy. Tracks: hind foot 2.1cm long, 2cm wide, 5 toes, 5 digital pads, 4 interdigitals and 2 proximals, one elongated. Hand outline common; forefoot 1.5cm long 1.7cm wide, 4 toes, 4 digital pads. Claw marks visible. Tips of toes tend to impress more deeply than those of Brown Rat. Trail: more delicate than Brown Rat. Heavy continuous tail drag may be visible. Stride about 10cm. Droppings: approximately 10–12mm long, 2–3mm wide. Narrow, slightly curved and blunt at both ends (cf Brown Rat). Deposited singly and at random – more widely scattered than those of Brown Rat. See Brown Rat for equations to dis-

tinguish between Brown from Black Rats. Greasemark: distinguished from Brown Rat as smear forms broken loop – 'twin loop' smear.
Habitat. Almost invariably near buildings (especially those with cavity walls, wall panelling, false ceilings), warehouses, supermarkets, food processing plants, ships. Elsewhere replaced by Brown Rat. Lives on rocks and cliffs on Lundy and Channel Islands. Semi-arboreal and fills niche of squirrel in wooded parts of Cyprus, Southern France and Spain. An orchard/plantation pest. Does not dig tunnel systems and in North and Central Europe does not occur in open. Called Roof Rat in tropics where it lives in house roofs and coconut groves. In tropics and subtropics builds nest similar to squirrel drey.
Habits. Nocturnal, with peak activity 2–3hrs after sunset. FOOD: Omnivorous, but greater tendency to vegetable food than Brown Rat (eg fruit, cereals, sugar cane). HOME RANGE: Range appears to be small, about 80–90m, although males may range further. In New Zealand study, mean range length 44m, but related to density. Most recaptures within 50m of original capture point. Population density: 5–12 per ha (Cyprus) up to 52 per ha (Sierra Leone). SOCIAL: This species defends all of its range rather than its pathways as sometimes claimed for Brown Rat. Females more belligerent than those of Brown Rat and threaten other group members, chasing them away from food. Intruders attacked or threatened, males by resident adults of both sexes, females only by other females. Most effective defender is dominant male. Linear order in top males although not in young ones. Two or more females subordinant to dominant male but dominant over all other group members. COMMUNICATION: Similar to Brown Rat. May have specialised scent glands on cheeks and ventral surface of body as 'scent mark' by rubbing these areas onto objects such as tree branches. Senses as for Brown Rat, but sight better.
Breeding. Breeding season mid-March to mid-November. Sexual maturity: females 12–16 weeks (*c.* 90g). Gestation 21 days, but due to delayed implantation, is 23–29 days for lactating females. Litter size 1–16 (average *c.* 7, litter size correlates with body size). Litters per year 3–5. Mammae 5 pairs (3 pairs abdominal, 2 pairs axillary). YOUNG: Potential number of young per year up to 56. Pink, naked with ears/eyes closed at birth. Weight = 4.5g at birth. Weaning age minimum 20 days, weight 30–40g.
Lifespan. Max. probably less than 18 months in wild. Annual mortality estimated at 71–97%. In urban areas Domestic Cat probably chief predator.
Measurements. Head-body length: 150–240mm in old individuals. Grows continuously throughout life. Tail-length: 115–260mm (100–130% head-body length) (200–260 rings). Hind-foot length: 30–38mm. Ear: 24–27mm. Condylo- basal length: 38–43mm. Weight: 145–280g (usually less than 200g). Dental formula: 1/1, 0/0, 0/0, 3/3 = 16.
General. Found worldwide, but in temperate regions ousted by more aggressive Brown Rat. In these areas occurs mainly in vicinity of ports

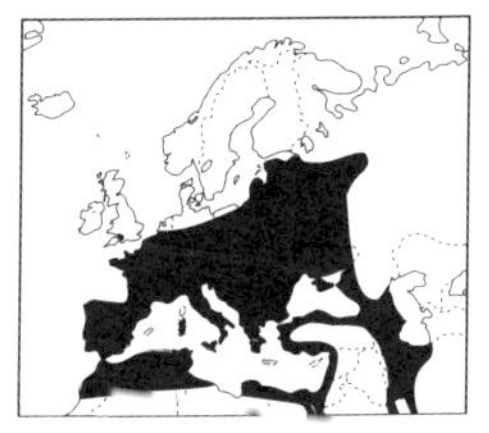

<*Black Rat.* Widespread in continental Europe but now largely absent from British Isles, excepting London dockland and Lundy Island.

and almost exclusively indoors. Now very rare in mainland Britain. Serious pest of agriculture throughout tropical Asia and on islands (notably in Caribbean and Pacific) attacking coconuts, cocoa, sugar cane and most other crops. Probably originated in South East Asia and parts of China. Spread westward through India to the Middle East along trade routes. Date not certain, but Pleistocene caves (*c.* 2 mya) in Crete and middens of German lake-dwelling people contain Black Rat remains. Bones in Roman well in York show it was present in Britain by 5th century AD. By Middle Ages well established in Europe and carried fleas responsible for plague outbreaks. Unlike the Brown Rat, it does not carry *Leptospira interogans* the bacteria causing Weil's Disease. First record in Americas in 1554 in Peru and well established in Virginia, USA, by early 17th century. The classification of the Black Rat is greatly complicated by the different colour morphs. One account states that around the Mediterranean dull-bellied morphs live in houses, but white-bellied forms in olive plantations. Contraction of range in UK, probably due to great susceptibility to rodenticides because of close affiliation with buildings. Also, replaced by Brown Rat (which has greater resistance to cold and adverse conditions). Agile climber – can jump up to 150cm. Gait similar to Brown Rat but appears to move faster. Can swim, but less likely to enter water than Brown Rat. Important vector of typhus and plague. Less neophobic than Brown Rat. As with Brown Rat, can be danger to rare ground-nesting birds: eg nests of the 45 remaining Freira Petrels on Madeira are raided by Black Rats.

STEPPE MOUSE *Mus spicilegus (hortulans)* Pl. 51

Name. *Mus* = mouse (Lat.); *hortulus* = garden (Lat.).

Recognition. Very similar to Algerian Mouse. Tail only slightly shorter than head and body. Similar to House Mouse insofar as incisor teeth are notched, but distinguished by lack of concavity on the inner and front surfaces of the first upper molar. Fieldsigns include burrow system and storage mounds.

Habitat. Grassland and cultivated ground. Burrow system with nest chambers. Food storage mound on surface may contain up to 10kg of grain.

Measurements. Dental formula: 1/1, 0/0, 0/0, 3/3 = 16.
General. Greek Steppe Mice may merit status of separate species, *Mus abbott*. Steppe Mice are surprisingly docile when handled (in comparison to House Mice). Vignette shows one week old mouse. The genetics of genus *Mus* are complicated: Some authorities recognise a separate species of Steepe Mouse *M. spretoides* south of Black Sea and throughout Greece and Albania.

ALGERIAN MOUSE *Mus spretus* Pl. 51

Name. Lataste's Mouse. *spretus* = spurned (Lat.).
Recognition. Clear delineation of brown upper body and greyish white or white underside. Tail shorter than that of House Mouse. Incisors not notched.
Habitat. Cultivated ground, especially damp habitats, scrub, gardens, and open woodland. Avoids human habitations.
Measurements. Head-body length: 70–85mm. Tail-length: 55–70mm. Hind-foot length: 15–17mm. Ear: 11–14mm. Condylo-basal length: 18.7–21.4mm. Weight: 12–18g. Dental formula: 1/1, 0/0, 0/0, 3/3 = 16.
General. Similar to Steppe Mouse or outdoor forms of House Mouse. Does not build food mounds and is less colonial than Steppe Mouse. Vignette shows 14 day old mouse

HOUSE MOUSE *Mus musculus* Pl. 51

Recognition. Pelage uniform greyish- brown above and only slightly lighter below. Eyes and ears relatively smaller than Wood Mouse, underside much less white. Tail as long as body, prominently ringed, and relatively scaly. Similar habitat and colouring to Brown Rat, but much smaller with sharper snout. There are two forms, the eastern form, *M.m. musculus* and the western form, *M.m. domesticus*. Tracks: Hindfoot 18 × 18mm – 5 toes, 4 interdigital pads, 2 proximal pads. Interdigitals and 1 proximal small and round, other proximal pad is large and oval. Forefoot 10 × 13mm – 4 toes, 3 interditigal pads, 2 proximals all small and round. Trail: Walking stride *c.* 5cm, tracks point slightly outwards with undulating tail drag gait. Running trail shows partial over-registration, tail drag

<*Algerian Mouse.* Found in S Iberia. Similar *Steppe Mouse* occurs N of the Black Sea and E through Hungary.

>*House Mouse. M. m. domesticus* in west, *M. m. musculus* in east.

and stride of about 7cm. Jumping mice leave tracks in groups of 4 with about 10cm between each group, but some jumps of up to 50cm. Droppings *c.* 6mm long by 2–2.5mm wide (smaller than Wood Mouse) left in concentrations in favoured places; colour varies with diet. Distinguished from bat droppings of similar size by not crumbling when crushed. Associated with 'urinating pillars' which are accumulations of droppings, dirt, grease and urine up to 4cm high and 1cm wide. Smears: Twin loop smears similar to Black Rat but smaller. Found at roof height around joists. Also identifiable by characteristic 'stale' smell, like acetamide, and 'kibbled' cereal grains (see below).

Habitat. Common in wide range of urban habitats – houses, shops, factories, warehouses and mills. Also recorded in cold stores and coal mines. In rural areas found in farm buildings, granaries, piggeries, refuse tips and hedgerows. Also in open fields, but less abundant than Wood Mice in hedgerows. Commensals are larger than field populations, and less inclined to burrow and store food. In favourable conditions (eg around the Mediterranean), may revert to field existence and live quite independently of humans as do many island populations. Burrow in soil is round in cross-section (cf oval for shrews). with single entrance connected to nest chamber about 20cm deep. Sometimes a food cache over a mound of up to 0.5m may be built. In buildings, nests made from any available material. May live under floor boards or among stored products. Shares communal nests especially when the temperature is low (number of females per nest is proportional to temperature).

Habits. Largely nocturnal. Grooming and inactivity when awake in laboratory occupy equal times in dark and light, but movements, feeding and drinking mainly in dark and sleep in light. FOOD: Omnivorous but with preference for cereals. Eat only what is necessary to maintain health and select components for balanced diet. Average consumption 3.5g per day. Tolerant of acid conditions and can live without drinking water when moisture in food is 15–16%. May use 20–30 different sources of food in one night and if only one source may make 200 visits, taking about 20 mg each time. Pick up grain in forepaws, turn side on, de-husk, and eat about two-thirds in several bites. Remains known as 'kibbled' grain. Fat and protein foods taken, but green foods and fruit of secondary importance. Odd substances such as plaster, soap, tallow and glue taken occasionally. In fields may eat insect larvae, roots, worms and arthropods. HOME RANGE: Individual mice have been recorded moving up to 2400m, but normally much less, particularly when in buildings. May be as little as $4m^2$ in chicken barn whilst in Russia some have home range of $1–2km^2$. Field-dwelling mice may be semi-nomadic. In buildings mice do not move more than 3–10m as a rule. *M. musculus* will rarely cross roads or open spaces. Dispersal largely by subadults (3–10 weeks old) in stable habitats, while adults have to move between temporary habitats. Distance depends on habitat availability. Frequently transported accidentally by man over large distances. Generally populations of two types – those living in areas of plentiful food which form stable groups, and those on periphery which seldom move into main group. Blind mice less successful than sighted mice at homing. All mice better at homing when distances of less than 135m involved. Late autumn and winter movements away from fields and hedgerows into farm buildings. In spring minor movements in reverse direction. SOCIAL: Aggressive behaviour uncommon in small territorial family groups but adults will attack introduced strange mice of either sex. At moderate densities, aggressive behaviour increases and males will live with one or more females in vigorously defended territory. At high densities surplus animals live in subordinate state without territory or breeding. Indoor populations similarly arranged with resident males some 40m apart. Live in groups which increase rapidly. Opportunistic colonisers tolerating high level of inbreeding. COMMUNICATION: High squeak in audible range. Subordinate may rear up and squeal when attacked. Lactating female responds to ultrasonic sounds of infants. Main sounds emitted through mouth, although weaker noises from nose. Unlike the Brown Rat, the House Mouse confines its acoustic communication to mother–infant interactions and heterosexual behaviour. Males do not produce ultrasounds when fighting. Family members and landmarks recognized by scent. Main sources of odour are urine, salivary glands (and various glands around the mouth and eyes), the footpads (plantar glands) and around the genital area (preputial glands of male and female, and vaginal glands of female). Urine spots placed on edges of territory and on conspicuous objects. Long brush-like hairs on tip of male prepuce allows easy marking. Urine marking more frequent in males than females and is influenced by sex hormones and social rank (suppressed in subordinate animals). Frightened animals produce fear/stress odours, especially in the urine, which are avoided by other mice. Use whiskers to orientate in dark. Tactile communication operates between individuals and is important in mother-infant relationships, social and sexual behaviour. Movement orientated by tactile stimulus (thignotaxis), with well developed paths and runways (continue to go around an object after it has been removed). Large spherical lens in eye designed for low light vision rather than high visual acuity. Mainly rods in retina, few cones. Not thought to use visual cues in social communication. Able to hear and produce sounds in ultrasonic and audible range with two peaks of sensitivity at 15 and 50kHz. Leave scent trails using urine and genital secretions. Odour used to distinguish neighbours and strangers, also sex and reproductive status. Female oestrous cycle under olfactory control: introduction of male into an all female group produces synchronous oestrous. Bruce effect: strange male scent blocks female pregnancy. Dominant male urine causes subdominates to keep away but attracts females. **Breeding.** May breed throughout year when food sufficient, producing 5–10 litters of 4–8 young per year. In rural populations breeding is seasonal with peaks in May and June and percentage lactating females negligible in January. Sexual maturity 8–12 weeks (7.5–10g). Gestation

19–20 days. Fertilization may occur at postpartum oestrus and implantation then delayed 2–16 days. Minimal (*c.* 5%) loss of embryos after implantation. Litter size in England averages 5.4. Larger litter sizes at higher latitudes with shorter breeding season. Litters per year 10.2 in hayricks, 5.5 in urban circumstances. Mammae 10; 3 pectoral pairs, 2 inguinal pairs. YOUNG: Born naked, eyes and ears closed, but with short vibrissae. Weight 0.8–1.5g at birth. Fully furred at 14 days with incisors erupted. Excursions from nest at 3 weeks when they take solid food. Weaning age 18–20 days at weight 7–8g. Parental care almost entirely maternal (though males may groom pups). Female makes substantial nest of shredded grass, paper, sacking or any available material. May share communal nest when sites limited or population density high and in these cases the females suckle young indiscriminately. Constant maternal care needed for at least three weeks after birth.

Lifespan. Max. more than 30 months in captivity; in wild generally less than 18 months, with females living longer than males. Heaviest mortality among juveniles (including desertion and infanticide). In adults, causes of death include exposure to cold, wet weather coupled with poor food supply. Frequently poisoned in commensal habitats. Predators include birds of prey, owls (especially Barn Owls), Weasels and Stoats, but these have little effect on well established feral populations. Domestic Cat important in intercepting mice moving in open areas, but little effect on thriving populations. In many habitats House Mouse has no predators except Man and occasionally the Brown Rat.

Measurements. Head-body length: 72–103mm (eg W. Europe, *M.m. domesticus*, 75–103mm, N., E. and S. Europe, *M.m. musculus*, 72–98mm). Tail-length: 70–95mm (equal to head-body length) (*M.m. domesticus*, 150–200 rings; *M.m. musculus*, 140–175 rings). Hind-foot length: 16–19mm. Condylo-basal length: 20–25mm. Temporal ridges on skull absent, or in older individuals only slightly developed. Weight: adult 12–22g. Mean for males 14.6g, females 15.8g. Dental formula: 1/1, 0/0, 0/0, 3/3 = 16. Upper incisors with distinct notch on wearing surface visible from side.

General. Worldwide distribution greater than any other mammal excepting Man. Widespread in British Isles and Ireland wherever there is human habitation, including nearly all inhabited small islands. Excellent sense of balance, running and climbing easily. Jumps and swims well. Of great economic importance as pest of crops and stored products. Although consumption of grain low, up to 10% grain in mills fragmented and rendered unsuitable for milling. Mouse voids over 50 droppings per day, which may contaminate foodstuffs. Also building materials and wiring gnawed and damaged. Can reach staggering densities, eg infested farm buildings and grain stores may support populations of several hundred or even thousands of House Mice. *Mus musculus* relatively unimportant as vector of human diseases, but can carry salmonella in their droppings and Weil's disease in their urine. Some evidence that some populations (eg Birmingham, UK) changing behaviour so less likely to eat conventional rodenticides. Originated in the steppes of Central Asia in the Iranian-Russia border area. A wild form, *M.m. wagnerii*, still lives there. It was here that people first started growing cereals and presumably the mouse became commensal when it moved into granaries where cereals were stored. From the Steppe area, the House Mouse spread worldwide by trade and human settlement. Well known in Ancient Egypt. In Europe two subspecies, *M.m. musculus* and *M.m. brevirostris*. Southern form carried to Latin America and California by early Spanish and Portuguese explorers. British, Dutch and French ships took the northern form to northern states of USA.

SPINY MOUSE *Acomys minous (cahirinus)* Pl. 50

Name. *Mys* = mouse (Gk.), *Minous* = Cretan (Gk., Lat.).
Recognition. Back covered in soft bristles which overlay the fur. Distinct division between buffy-grey upperside and white underside. Tail fragile (reputedly often broken). Feet are white. No chest spot. Whiskers up to 5cm long. Often runs with tail curved forward over back.
Habitat. In Europe found only on Crete. Forages in crevices in dry rocky scrub. Enters houses (especially in winter). May live in rock crevice, burrow or building.
Habits. Nocturnal and crepuscular. FOOD: Seeds, grain, insects. SOCIAL: Small, isolated communities with a hierarchical social structure, dominated by a female. Social grooming occurs. Dominance hierarchy among males during breeding season when fight for females. COMMUNICATION: Some research suggests role of maternal odours.
Breeding. Gestation *c.* 36 days. Litter size 2–3. Mammae 6. YOUNG: Well developed at birth. Reports, in captivity, of females cooperating in care of young.
Measurements. Head-body length: 91–128mm. Tail-length: 89–120mm. Hind-foot length: 18–20mm. Condylo-basal length: *c.* 29mm (25.2–31.4mm). Weight: 40–85g. Dental formula: 1/1, 0/0, 0/0, 3/3 = 16.
General. Found in Israel and Egypt, but only in Crete and possibly Cyprus within Europe (cannot tolerate cold).

MOUSE-TAILED DORMOUSE *Myomimus roachi* Pl. 48

Name. *Mys* = mouse (Gk.); *mimus* = imitator (Lat.).
Recognition. Tail short-haired (more reminiscent of mouse's tail than dormice). Much smaller ears than mice.

< *Spiny Mouse.* A population on Crete and possibly Cyprus.

> *Mouse-tailed Dormouse.* Reported from scattered locations in Bulgaria, Greece and Turkey.

Habitat. Variable: woodland scrub, also more open areas, agricultural land, orchards (especially apple and mulberry).
Habits. FOOD: Seeds mainly.
Breeding. Mammae 14.
Measurements. Head-body length: 70–130mm. Tail-length: 60–100mm. Hind-foot length: *c.* 20mm. Condylo-basal length: *c.* 25mm. Dental formula: 1/1, 0/0, 1/1, 3/3 = 20.
General. Probably more terrestrial than other dormice. In Europe, confined to Bulgaria and Turkey.

EDIBLE DORMOUSE *Glis glis* **Pl. 49**

Name. Fat Dormouse. *Glis, gliris* = dormouse, (Lat.; from Sanskrit *girika* = mouse); *dormio* = I sleep (Lat.).
Recognition. Largest dormouse (twice size of Common Dormouse). Pelage is grey to brown, with hint of dark stripe along spine, and encircling each eye. Bushy tail. Distinguished from young Grey Squirrel by small hind feet. Field-signs include nest and food remains.
Habitat. Mature deciduous woodland, mostly in canopy of trees. Does not require lush shrub layer. Also gardens and orchards (and houses). Up to 2000m in Pyrenes. Summer: nest (mosses and fibres) high in canopy close to tree trunk or in hole. Winter: hibernates at lower levels in hollow tree or underground (to depth of 60cm). Also nests in wall cavities, nest boxes etc. Up to eight individuals may share the same diurnal nest.
Habits. Nocturnal. Hibernates October–April. FOOD: Omnivorous. Vegetable food includes nuts, fruits, fungi, bark; animal foods include insects and occasionally eggs and nestlings. Apples taken from stores in buildings in autumn. HOME RANGE: *c.* 100m diameter. Population density: 0.2–4 per ha. Dispersal distance up to 1km. SOCIAL: Sociable, in loose groups of unknown structure. Record of 69 individuals caught in the same roof. COMMUNICATION: Vocal: noisy grunts and squeaks.
Breeding. Breeding season June–August. Sexual maturity 2 years. Gestation 31 days. Litter size 2–9. Litters per year 1. Mammae 12–20.
Lifespan. Max. 7 years.
Measurements. Head-body length: 130–190mm. Tail-length: 120–150mm. Hind-foot length: 24–34mm. Condylo-basal length: 36–44mm. Weight: 70–200g (up to 300g before hibernation). Dental formula: 1/1, 0/0, 1/1, 3/3 = 20.
General. Absent from Scandinavia, very restricted in UK. Agile climber. Becomes very fat prior to hibernation, during which loses up to 50% of weight. Introduced to UK in 1902. Romans fattened them for the table in jars called *gliraria.*

< *Edible Dormouse.* Throughout C Europe, extending W through N Spain. Introduced to England.

> *Forest Dormouse.* Central Europe E of Austria.

FOREST DORMOUSE *Dryomys nitedula* Pl. 49

Name. *Drys, dryos* = oak (Gk.); *mys* = mouse (Gk.); *nitedula* = dormouse (Lat.).

Recognition. Distinguished from Garden Dormouse because smaller, mask less distinct, and tail longer, bushier and without the black and white tuft. Ears smaller and more rounded than those of garden dormouse. Pelage: light grey to reddish-brown above.

Habitat. Largely in woodland with thick shrub-layer. Up to 1500m in Alps. Nest, with entrance at one side, lined with moss, feathers and hair, built in copses or tree canopy or in holes in tree. In winter nests generally underground.

Habits. Largely nocturnal. Hibernates October–April. FOOD: Largely vegetarian (eg lichen, nuts, seeds and fruits), but also takes insects and larvae. COMMUNICATION: Soft melodic sounds; clicking, snarling, hissing, whistling sounds when excited or threatened.

Breeding. In north of range, mating in May–July, births from late June to beginning of July. In warmer areas, breed throughout March–December. Sexual maturity after first winter. Litter size 1–4 in warmer areas, 3–5 in more northern areas. Litters per year 1 in more northern areas, 2–3 in warmer areas. Mammae 8.

Measurements. Head-body length: 80–130mm. Tail-length: 80–95mm. Hind-foot length: 19–24mm. Condylo-basal length: *c.* 24mm. Weight: 30–60g. Dental formula: 1/1, 0/0, 1/1, 3/3 = 20.

GARDEN DORMOUSE *Eliomys quercinus* Pl. 49

Name. *Mys* = mouse (Gk.); *quercinus* = oak (Lat.).

Recognition. Pelage: reddish-brown above, white underside and paws. Tail dark above, long, bushy, white tip ending in flattened tuft. Black mask joining under ears and around eyes. Large ears.

Habitat. Largely woodland (coniferous or deciduous), especially oak woods. Less arboreal than other dormice; frequently on ground in scrub, among rocks. Often in houses. Up to 2000m in Alps and Pyrenes. Either builds own spherical nest (entrance in side), lined with moss, feathers and fur,

or adapts squirrel drey or bird's nest. Nests in shrubs, ivy, tree holes, or underground burrows. Hibernates in hollow trees, wall cavities, caves.
Habits. Largely nocturnal, but active after dawn. Hibernates September–April. FOOD: Nuts (eg acorns), buds, fruits, insects, snails, nestling birds and mice, eggs. HOME RANGE: Each individual or group occupies *c.* 150m diameter territory; territories remain in same place from one year to the next. COMMUNICATION: Wide range of vocalisation: grunting, teeth gnashing, creaking, shrill whistling and murmuring especially during the breeding season.
Breeding. Mates soon after emergence from hibernation. Births in May-June. Gestation 21–23 days. Litter size 4–5. Litters per year 1–2. Mammae 8.
Measurements. Head-body length: 100–170mm. Tail-length: 90–150mm. Hind-foot length: 22–32mm. Condylo-basal length: 30–33.6mm. Weight: 45–120g (up to 210g prior to hibernation). Dental formula: 1/1, 0/0, 1/1, 3/3 = 20.
General. Reported sometimes to move in mouth-to-tail caravan (like some shrews). Loses skin of tail (and eventually, if this happens, the denuded vertebrae) very easily – seemingly an anti-predator device shared with other members of the family Gliridae. Reputedly seriously threatened by direct competition with the Brown Rat, especially in Corsica. Largely absent from British Isles except for report of six killed by a cat in Dover (SE England) in 1991.

COMMON DORMOUSE *Muscardinus avellanarius* Pl. 50

Name. Hazel Dormouse. Probably from *mus* = mouse; *Corylus avellana* = scientific name of Hazel. But note also: *musca* = fly (Lat.); *avello* = I snatch (Lat.).
Recognition. Short muzzle, long vibrissae (30mm). Distinguished from mice by bushy tail. Distinguished from other dormice by small size and bright orange dorsal fur. 10% have white tip to tail. Fieldsigns include small round gnawed hole in hazelnuts with distinctive tooth-marks (hole has smooth inner surface, chisel marks outside).
Habitat. Deciduous woodland, dense shrubbery and coppices, often associated with secondary growth, hedgerows. Nest is compact, woven ball (often stripped honey-suckle, rarely moss and grass) *c.* 10cm diam. (15cm for breeding nest). Sometimes built on foundation of bird's nest or squir-

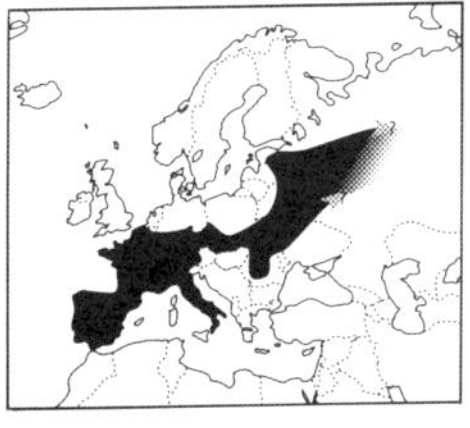

< *Garden Dormouse.* Widespread in W Europe and E, S of E Germany.

> *Common Dormouse.* C & W, excluding Iberia. Present in British Isles, but only in S of England and Wales.

rel drey and lacks distinct entrance. 5m or more above ground but sometimes sited in shrub layer (*c.* 1m off ground). Generally do not burrow.
Habits. Nocturnal. Hibernates October/November–April, probably below ground level. FOOD: Nuts (eg hazelnuts, chestnuts, acorns), berries and fruits. Also occasional insects, birds' eggs and nestlings. HOME RANGE: Arboreal, remaining within small area. Up to 1600m (male) or 700m (female). SOCIAL: Nests often found in small groups. COMMUNICATION: Vocalisations include mewing and purring sounds. Hearing acute, good night vision.
Breeding. Breeding season from May–September, with birth peaks in late-June–early-July and in early-August. Sexual maturity 1 year. Gestation 22–24 days. Litter size 2–7. Litters per year 1–2. Mammae 8. YOUNG: Pelage: more grey and dull. Weaning age probably 6–8 weeks. Young with mother until *c.* 10 weeks old.
Lifespan. Max. recorded 4 years in wild (captivity 6 years). Principal cause of death probably starvation during hibernation.
Measurements. Head-body length: 60–90mm. Tail-length: 55–80mm. Hind-foot length: 15.3–16.4mm. Ear: 10–14mm. Condylo-basal length: 20–23mm. Weight: 15–30g, occasionally heavier (max. 43g). Heaviest prior to hibernation. Dental formula: 1/1, 0/0, 1/1, 3/3 = 20.
General. Four long, slender digits on front feet; hind feet, four digits, plus vestigial stump. Well-developed pads aid climbing. Hibernating animals wheeze when re-awakening.

NORTHERN BIRCH MOUSE *Sicista betulina* Pl. 53

Name. *Betula* = birch tree (Lat. from Gallic word).
Recognition. Distinguished from Southern Birch Mouse and Striped Field Mouse by longer tail. Distinguished from all other mice by black stripe from crown to rump. Upper lip not split. Small premolar tooth in upper jaw. Outermost toe of front and hind feet can be opposed to others for climbing agility.
Habitat. Woodland (especially birch in north) with dense undergrowth; favours wet or marshy areas. In south of range mainly in mountain habitats. Also found in spruce forests in Carpathians. Nest is round ball of grass or moss underground or in tree stump. Digs own burrows.
Habits. Nocturnal. Hibernates October–April/May. FOOD: Largely insects, some fruit and grain. HOME RANGE: 0.4–1.3ha.
Breeding. Breeding season May–June. Sexual maturity in second summer. Litter size 1–11. Litters per year 1 (rarely 2).
Measurements. Head-body length: 50–70mm. Tail-length: 76–110mm (150% head-body length). Hind-foot length: 14–18mm. Condylo-basal length: 17.0–18.8mm. Weight: 6.5–13g. Dental formula: 1/1, 0/0, 0–1/0, 3/3 = 16–18.

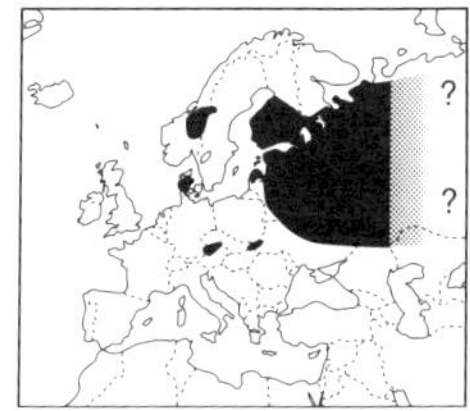

<*Birch Mice.* The Northern Birch Mouse is widespread in S Finland and eastwards, and occurs in pockets in W Scandinavia, Denmark, Austria and Czeckoslovakia. The Southern Birch Mouse is found from Romania eastwards.

General. Very agile. Often runs with tail curved upwards. Weight can drop by 50% during hibernation.

SOUTHERN BIRCH MOUSE *Sicista subtilis* Pl. 53

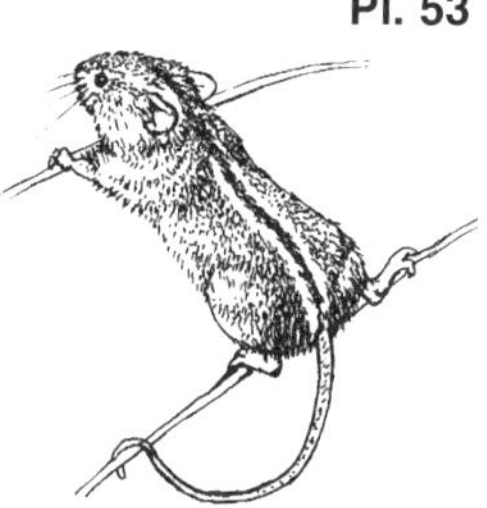

Name. *Subtilis* = delicate (Lat.).
Recognition. Black stripe from crown to rump distinguishes it from all save Northern Birch Mouse. Small premolar in upper jaw. Distinguished from Northern Birch Mouse by shorter tail, slimmer build. Dorsal stripe often interrupted and frequently extends beyond crown of head to broaden on forehead (also, black stripe is edged with pale bands, especially to rear).
Habitat. Dense steppe, rough grasslands, scrub, field margins of lowlands. More open terrain than Northern Birch Mouse. Oval nest of grass and moss, underground in rock fissures, or in shrubs or treeholes (includes storage chambers).
Habits. Nocturnal, but reputedly less so than Northern Birch Mouse. Hibernates October–April. FOOD: Primarily insects and other invertebrates. Also plant parts – mainly seeds and fruits.
Breeding. Breeding season May–June. Litter size 2–8. Litters per year 1. Mammae 8.
Measurements. Head-body length: 55–70mm. Tail-length: 60–90mm (130% head-body length). Hind-foot length: 14–16.5mm. Condylo-basal length: 16.7–18.6mm. Weight: 9.5–14g. Dental formula: 1/1, 0/0, 0–1/0, 3/3 = 16–18.
General. Climbs well using semi-prehensile tail. Outer toes opposable on all feet.

CRESTED PORCUPINE *Hystrix cristata* Pl. 56

Name. *Hystrix* = porcupine (Gk.); *Hys* = pig; *thrix* = hair; *cristata* = crested (Lat.).
Status. Rare, protected in Italy since 1974.
Recognition. One of the largest rodents, long black and (predominantly) white quills (longer quills, 30–40cm, scattered among thicker, shorter ones, covering whole rear of body and tail). Modified hair on upper part of back and neck thick, that on crown and nape generally white and somewhat bristly. Short

quills on rump predominantly black. Distinguished from zoo escapees of other species by crest of long, stiff predominately white spines on head and back (erected when threatened) and wholly dark rump. Also distinguished by relatively small rattle-quills and presence of many blackish rings on quills. Five toes on each paw adapted for digging. Female teats high up sides. Male testes internal so no scrotum. Lumbering gait. The big toe in the forepaw is much reduced. Long vibrissae on the side of the face may reach the shoulders. Fieldsigns include large burrows, food remains, shed quills. Faeces generally 2–4cm long and sometimes form a chain. Alternatively, piled near burrow. Normally no clear imprint of the big toe in the footprint.

Habitat. Dense scrub or open woodland, often near cultivated land. Den in deep burrows or caves, sometimes sharing setts with Badgers and more rarely with Red Foxes. Often digs burrows along banks of canals, but generally where vegetation offers adequate cover.

Habits. Nocturnal. Inactive in poor weather, but does not hibernate. Tends to be less active in open fields during nights with a full moon. FOOD: Roots, tubers, bark, fallen fruit; cultivated root crops, orchards, market gardens. Reported to gnaw bones and eat carrion. HOME RANGE: One adult female was found 15km from a capture site about a year later. SOCIAL: Small family groups. COMMUNICATION: Gruff growling when threatened. Possibly scent communication associated with anal gland secretion. Hearing acute, smell highly developed.

Breeding. Births recorded in most months of year. Sexual maturity about 1 year old. Gestation 110–120 days. The minimum interval between two births in captivity is 91 days. Litter size 1–4, average 1–2. Litters per year 1–2. Mammae 4–6. YOUNG: Young well developed at birth, with eyes open and short, soft spines which quickly harden by 2 weeks old (vignette shows a 4 week old porcupine). Able to feed on solid foods by 40–50 days. Parents rattle quills and defend offspring from intruders, and keep young warm by sleeping on either side of them.

Lifespan. Max. recorded 20 years in captivity. Road accidents and poaching are the main causes of death.

Measurements. Head-body length: 60–80cm. Tail-length: 5–12cm. Hind-foot length: 9–10cm. Shoulder-height: 21–26cm. Condylo-basal length: *c.* 130mm. Weight: 10–20kg. Sexual dimorphism: females are larger and heavier than males. Dental formula: 1/1, 0/0, 1/1, 3/3 = 20.

General. Originated from N. Africa; reputedly introduced by the Romans into Sicily and S. Europe. Also found in Turkey. In Western Europe once confined to Sicily and parts of mainland Italy but gradually extending its range towards the north and eastern parts of the Italian peninsula. When threatened can reverse into adversary, growling, stamping rear feet and rattling spines (those of tail modified to be very noisy). Hunted for meat (often with dogs) and controlled by baiting with poisoned fruit and formerly, by fumigation of burrows.

< *Crested Porcupine.* W & S Italy, and N coast of Africa.

> *Coypu.* Scattered distribution in W & S Europe. Eradicated from SE England in 1990.

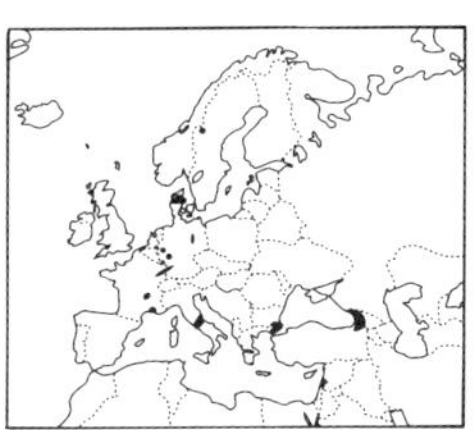

COYPU *Myocastor coypus* Pl. 56

Name. Nutria (refers to fur). *Mys* = mouse (Gk.); *castor* = beaver (Lat.).

Recognition. Large rodent, superficially rat-like; sparsely-furred cylindrical tail, tapering evenly; webs between 4 of 5 toes on hind feet; pelage of glossy brown and yellow-brown guard hairs and grey underfur. Tip of muzzle, chin and most of whiskers white; the top few whiskers are black. Outside surface of incisor teeth orange. Many adaptations to aquatic life: lips can close behind incisors; valvular nostrils; small ears, eyes and nostrils placed high on dorsal surface of head and exposed when swimming. Strongly clawed forefeet. Moult: thick glossy winter coat, thinner summer coat. Fieldsigns include places where they have dug holes for tubers (up to 20cm in depth). 15cm wide runs devoid of vegetation, often leading to regular haul-outs from water. Tail marks (2cm wide) in mud. Paired, crescent-shaped incisor tooth marks on vegetation (17mm wide). Can produce short 'lawns' by intensive grazing next to water. Faeces: long, dark-brown/green, cylindrical, slightly curved with fine longitudinal striations: 2 × 7mm (2 weeks) to 11 × 70mm (adult). Ledges on banks used as grooming stations. Footprints: hindfoot up to 15cm long, imprint of web often visible, up to five claw marks visible.

Habitat. Marshes, fen, also along slow-flowing rivers, estuaries and coastlines. Prefers stagnant (eutrophic) water with thick vegetative cover. Will occupy temporary water surfaces until these dry up and then migrate to other sites, sometimes over long distances and far from water. Burrows (20cm diameter) at edges of ditch or in steep river bank (entrance often half submerged) to depths of up to 6m. Flat nests of dead grasses (*c.* 2–5cm deep and 30cm diameter in dry places (eg UK), but 1m high in wet areas where there is a need for nest to be built up (eg Poland).

Habits. Largely nocturnal and crepuscular (activity peaks at dusk, midnight and before dawn). Diurnal activity during cold winters, and in the absence of diurnal predators and Man. FOOD: Almost exclusively vegetarian: grasses year round. Summer: shoots and basal meristems of sedges and burr-reed and (in autumn) fruits (eg of water lily). Winter: tubers, rhizomes and root crops, and brassicas and sugar beet. Occasionally eats mussels. Food held on ground until small enough pieces broken off to be held up in forepaws. Expert diver and much food gathered underwater. Coprophagous. Selective feeding by Coypu caused massive reduction in reed swamp area in Norfolk Broads (UK). Also altered composition, and all but eliminated certain plants (eg cowbane, great water dock) over large areas when density high (1950s). HOME RANGE: Initially small, increasing with sexual maturity. In fen or swamp, female: 2–4ha; male: 3–5ha (on grazing marsh male ranges of up to 120ha). Can reach densities of up to 20 per ha around favoured parts. Individuals of both sexes disperse over a wide range of population densities. Low densities: females disperse to isolated home ranges. High density: females (daughters) disperse when kin groups meet. Males disperse as young adults, often to neighbouring territories. Migrations of 40–50km have been reported for adults in search of suitable habitat. SOCIAL: Gregarious, polygynous mating system. Kin groups: matriarchal clans with partially overlapping home ranges (female offspring establish ranges affiliated to those of their mothers). Dominant males' range overlap the females', whereas subordinate males pushed to periphery. COMMUNICATION: Loud 'muawk' contact call heard over 200m. Greeting call a soft and repeated grunt. Warning call a continuous low hum. Growls and grinds incisors in aggressive defence. Slaps tail against water surface when disturbed. Does handstands to elevate scent marks. Scent possibly primary mode of communication. Protrudible anal glands in both sexes, larger in males (12g versus females at 4g), who mark more frequently than females. Females scent mark more with urine than with anal glands. Marking often on raised objects or well-used points of entry into water. Prominent white vibrissae (up to 130mm long). Eyesight appears poor, smell and hearing acute.

Breeding. Breeds throughout year. Post-partum oestrus, and oestrous cycle of *c.* 4 weeks. However, Coypu are induced ovulators and so oestrus generally induced by male. Sexual maturity: females 3–8 months, males 4–10 months, depending on season of birth (earliest for animals born in late winter, greatest for autumn births). Gestation 127–138 days. Litter size 2–9 (50–60% prenatal embryo losses through abortion and reabsorption which is common over cold winters and when the female is in poor condition). Maternal manipulation of sex ratio of offspring and investment in young to weaning; young males grow faster than females up to weaning. Mammae 8–10 in two dorso-lateral rows, sometimes teats missing or bitten off. Mother suckles while lying on her underside. YOUNG: Born fully furred, eyes open. Can nurse in water because nipples high on female's sides, though young normally suckled in nests. Birth weights: male: 132–346g, female: 111–327g. Young can swim within a few days. Weaning age 6–10 weeks. Adoptions common in captivity. Lactating females dominant over and aggressive towards males.

Lifespan. Max. 6–8 years in captivity, but less than 0.2% live to 4 years in wild. Predators include Stoats, Dogs and Marsh Harriers. Young taken by various predators (eg owls, hawks, mink, Domestic Cats, Pike). Juvenile mortality from births in autumn/early winter; poor quality juveniles produced when weather conditions severe. Adult mortality probably low, except where caused by Man. Again, severe winters cause large mortality: 80–90% in bad winter. Parasites: a host specific louse can affect over 60% of the population. Ticks rare, fleas and mites very rare. Also carry tapeworms, roundworms, liver fluke, coccidiosis, yersiniasis, ringworm, and a fungus infecting lungs (*Haplosporagium paryum*) normally associated with burrowing animals (moles, Wood Mice). Can catch foot and mouth disease, also pasteurella, salmonella and leptospiriosis.

Measurements. Head-body length: 36–65cm (male: *c.* 60.3cm; female: *c.* 59.3cm). Tail-length: 25–45cm (70–80% head-body length). Hind-foot length: 12.5–14cm. Condylo-basal length: 95–115mm. Weight: adult 4–9kg (average weights: male, *c.* 6.5kg; female, *c.* 6.0kg). Maximum body size reached at 2 years old. Dental formula: 1/1, 0/0, 1/1, 3/3 = 20.

General. Considered a pest in most parts of Europe. In the UK, restricted to E. Anglia (by 1987 an eradication programme had reduced adult female coypus to less than 20. Exterminated by 1989. In France, control measures using poison, trapping and shooting ineffective). Swims by alternating propulsive thrusts of hindlegs and rapid paddling of forelegs. Anti-predator behaviour involves lying immobile under water for several minutes. Webbed feet, waterproof underfur, and valvular nostrils are all adaptations to aquatic life. Probably introduced originally from Argentina (where important source of meat; fur called nutria, prepared from underfur once long guard hairs plucked out), and stocks exported to fur farms in USA, Russia, Africa, Japan, Middle East and Europe in 1920s. First imported to UK in 1929. The fur is still an economic resource in N. America and Russia. In native home of Argentina, Chile and Uruguay, coypus are declining, despite protective legislation. Five subspecies in S. America: *M.c. bonariensis* (Geoffroy) most widespread and the one that the British animals most closely resemble.

Order Lagomorpha – Rabbits and Hares

Origins: The Lagomorphs first appear in the fossil record *c.* 60 million years ago in the late Palaeocene, in Mongolia and China (eg *Eurymylus*). The descent of the Lagomorphs is a puzzle because in spite of the earlier belief that *Euromylus's* ancestors were either Insectivora or ungulates, molecular analysis of pika blood has suggested relationships with tree shrews and carnivores. Yet another school of thought is more influenced by the similarities between rodents and Lagomorphs (in the cranium, foetal membrane and tooth development) and therefore speculate that they should be classified within one superorder, called the Glires.

Main features of the Lagomorpha: Lagomorphs are herbivores. A glance at a Lagomorph's mouth quickly identifies it, because they are unique in having double front incisors. That is, behind the two upper front teeth there are another two, called the peg teeth. All these incisors are completely encircled by enamel. Neither the cheek teeth nor the incisors have roots. Each lagomorph cheek tooth has only two transverse ridges on its grinding surface, and this distinguishes them from rodent teeth, which have up to 5. All Lagomorphs have very short tails, and they all engage in refection, producing a special type of faeces which they eat in order to pass food through the digestive system twice. Rabbits and hares all have slit-like nostrils which can be opened and closed by a fold of skin above.

Lagomorphs in Western Europe: There are only two families of Lagomorph and since no pikas occur in Europe, only Leporidae occur there. They are characterised by long back legs and a flattened body, designed for running at speed, and by their long, mobile ears. The Lagomorphs include 50 extinct genera, and 11 modern genera embracing 58 surviving species. Of these, three species from two genera occur in Europe.

BROWN HARE *Lepus europaeus* **Pl. 63**

Name. Common Hare, European Hare. *Lepus* = hare (Lat.)

Recognition. Distinguished from rabbit by long black-tipped ears (equal to head length), loping gait, also long limbs, larger size. Distinguished from Mountain Hare by longer ears, dark upper surface of tail, and more yellowish pelage. Small Mediterranean race: reddish-brown outer thigh, upperside of neck and back speckled black. Many abnormal colour variants recorded (eg melanic, albino, sandy). Tail held down when running, so black dorsal surface visible. The roots of the upper incisors of Brown Hares reach the suture of the maxilla and premaxilla. The back fur is dense with three different types of hairs, the underfur of 15 mm, the pile hair of 24-27 mm and the guard hair of 32-35 mm. Moults in spring and autumn. Summer coat a bit lighter than redder winter coat. Juvenile moult when *c.* 900g. Obvious fieldsign is the 'form' (see Habitat). Faeces generally lighter colour, larger, more flattened and more fibrous than Rabbit's, but depends on diet and cannot always be reliably distinguished. Trails worn across farmland. Hair in form during spring and autumn moults. Another characteristic sign is gaps or runs through hedge- or fence-rows. Tracks: four toes on each paw – larger than Rabbit's – furry soles often obscure pads (distinguish from dog etc): elongated side by side impressions with hind feet placed ahead of front (latter normally one behind the other), so that the series of prints form a Y.

Habitat. Has colonised most European habitats, but prefers temperate open habitats. Found in most flat country among open grassland and farms. Up to 1500m in Alps, higher in Pyrenes (up to 500m in UK). Traditional ley-farming most suitable, most abundant in arable areas where cereal growing predominates (high densities of livestock on pasture can deter hares). Woods, hedges or shelterbelts frequently used as resting areas during the day. Den is referred to as couch or 'form', and is a shallow depression in open fields or under cover of long grass, scrub, or hedgerow dug out themselves. Hind-quarters in deepest part of form – only back and head visible. Hares re-use forms frequently unless disturbed.

Habits. Predominantly nocturnal, but moderately active by day. Activity extends into mornings/evenings in the summer. FOOD: Mainly herb grasses and agricultural crops (early growth stages of cereals and roots in summer, turnips in winter). Tears bark from saplings (if snow makes grazing difficult). When available, wild grasses/herbs preferred to cultivated forms. Herbs comprise the bulk of the summer diet, whereas grasses predominate in winter. Said to eat animal corpses during bad seasons. HOME RANGE: An adult occupies a range of about 300ha which it shares with other individuals; each individual may concentrate on patches within its home range. Hares move some 1700m daily from day shelter to feeding grounds in NE Scotland. Population densities: Poland 1 per 2.4–5.5 ha, UK 1 per 2–4 ha. No evidence of territorial behaviour and no obvious pattern to distribution of neighbouring home ranges. SOCIAL: Mainly solitary, but pairs or small courtship groups in late winter and spring. Courtship involves 'boxing'. This Mad March hare behaviour actually generally involves unreceptive females chasing off males. 'Chases' occur in which several males pursue one female. Both occur over most of year (long breeding 'season'). Dominant males chase off subordinates. Also temporary feeding aggregations (especially when snow covers ground), organised with dominance hierarchy including both sexes (cf. Rabbit's hierarchy separate for each sex). Sometimes gregarious during foraging, especially in the evening. Hares foraging in groups benefit from each spending less time on the look-out for predators. COMMUNICATION: Screams in distress, but generally silent. Scent produced by anal, inguinal (larger in female; give characteristic odour to hares), and submandibular glands. It is possible that lachrymal and Harder's glands (orbits) may have scent function, for example when secretions are rubbed into forepaws during face washing. Laterally placed eyes allow nearly 360° field of view. Excellent sense of smell enables localisation over great distance, long ears indicate excellent hearing.

Breeding. Breed throughout most of the year (in NE Scotland there is a peak in pregnancies in April–May). About 3 litters (young called leverets, see vignette) produced between February–October (occasionally additional winter pregnancies too). Induced ovulation. Beginning of season may be determined by day length in Britain, although breeding in year of birth may be common elsewhere. Sexual maturity: male 6 months; female 7–8 months. Gestation 41–42 days. Litter size 4, with smaller litters early and late in season; larger litters (up to 10) are exceptional. Litters per year 1–4, average 3. Mammae 6. YOUNG: Born above ground, fully furred and eyes open. Adult weight at *c.* 150 days. Birth weight *c.* 100g. Weaning age variable, but normally in excess of 23 days – last litter of the season may suckle for 3 months. Parental care only by female. Single, brief nursing bout each day; young gather at birth place *c.* 1 hour after sunset and follow female when she arrives. Nursing takes as little as 5 minutes (between which young scatter around place of birth).

Lifespan. Max. 13 years (Poland: 6% over 5.5 years). Pre- and post-implantation losses (i.e. miscarriages) tend to be high at start and middle of breeding season. Animals in first winter may survive less well than older animals, otherwise survival is probably constant with age. Life expectancy (Netherlands) averages 1.04 years. Can live up to 7 years in wild (tagging data) – oldest record from the wild is 12.5 years (Poland). Typically half autumn population is comprised of young animals. Regularly shot as game and pest. Populations in Britain have rapid turnover. Losses due to diseases can be high eg such as coccidiosis (young hares in autumn), yersiniosis (adults in winter). Heavy predation of leverets, particularly by foxes. Agricultural hazards numerous, including machinery and biocides. Agrochemicals have both direct effects (eg Paraquat) or indirect (reducing diversity of food via herbicides). Road casualties may be a significant cause of death in some areas.

Measurements. Head-body length: 480–700mm (Mediterranean race 400–540mm). Tail-length: 70–130mm. Hind-foot length: 115–150mm. Ear: 85–115mm. Condylo-basal length: 83–92mm. Weight: 2500–7000g (male *c.* 5% heavier than female) (birth: *c.* 110g). Dental formula: 2/1, 0/0, 3/2, 3/3 = 28.

General. Game bag records show significant decline in Britain of hare numbers since the 1960s. Modern arable farming methods may be partly responsible. Similar decline in much of Europe, particularly the Netherlands. However, still locally common. Escape predators by out-running them. Inconspicuous when feeding because hold ears flat and body low to ground. Refecate (chew faecal pellets) to recover extra proteins and vitamins. Some authorities consider that the race found in the Iberian peninsula and Mallorcas belongs to the African species, *Lepus capensis.*

The race from the mountains of NE Spain named as *Lepus castroviejoi* in 1979, but its status is equivocal, as is description of a subspecies (*L.e. occidentalis*, de Winton, 1898) for Britain. Extensively introduced into Ireland during last century by hare-coursing clubs, and locally well-established in parts of North-west Ireland. Possibly introduced into Britain by man (no definite record until Roman period). Prized for shooting in much of continental Europe, but in UK hunted mainly with hounds (beagles, harriers) or racing dogs (coursing with greyhounds).

MOUNTAIN HARE *Lepus timidus* Pl. 63

Name. Blue Hare, Tundra Hare, Variable Hare, White Hare, Irish Hare. *Timidus* = fearful (Lat.).

Recognition. Distinguishable from Brown Hare by smaller size, more rounded shape, absence of black top of tail, shorter ears and legs, grey/black coat (summer) and white, or partly white, pelage in winter. Dusky blue underfur shows through on flanks. Moult (i) mid-October–December: brown to white, (ii) mid-Feb–late May: white to brown (see vignette). The suggestion that Mountain Hares have a third moult (early June–mid-September: brown to brown) may arise from the fact that leverets of the year have, in late summer, a slightly different coat colour to older adults. Irish race, *L.t. hibernicus*, is larger than Scottish race *L.t. scoticus* and does not become completely white in winter and is more reddish in summer. In spring, head becomes brown first; females moult earlier than males.

Fieldsigns include the 'form' in long, old heather, with stems of up to 6mm diameter bitten through. The root of the upper incisor does not reach the suture of the maxilla and premaxilla. Trails in heather, normally up and down hill. Following the trails often leads to the form. Greyish-green droppings (10mm diam., impossible to distinguish from *L. europaeus*). Hind- feet heavily furred in winter and therefore leave large tracks.

Habitat. Heather moorland, montane grassland, dry rocky hilltops and occasionally woodland up to snowline. Also, pasture and arable lowlands in absence of Brown Hare. Up to 1300m in Alps. Den is called couch or 'form' under cover of heather or rocky outcrop. Sometimes digs burrow (or takes over Rabbit's) as shelter for leverets, although they are

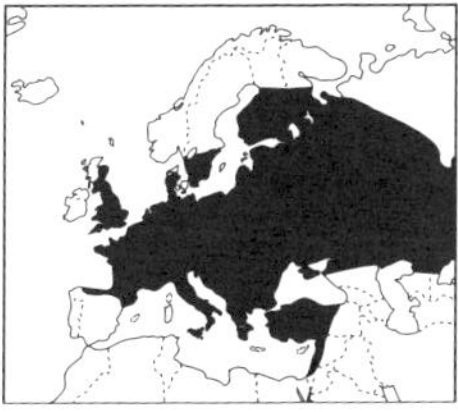

<*Brown Hare.* Widespread in C & W Europe. Absent Ireland, most of Scotland, Sardinia, Balearic Islands and most of Iberia.

>*Mountain Hare.* N Europe, pocket in Alps. Present in Ireland, Scotland, and N England.

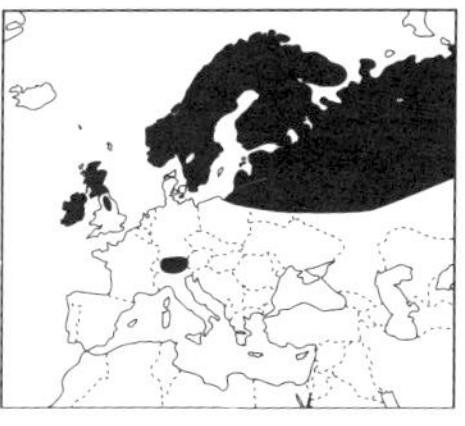

unlikely to have been born there; adult tends to shelter in entrance and flees if disturbed rather than entering burrow. Burrows in snow important in cold weather.

Habits. Crepuscular and nocturnal. Said to graze more in daylight before rain and during snow. More active in daylight during mating season. FOOD: Winter: browses, mainly short young heather, bilberry, with bark and twigs of gorse, willow, juniper when ground covered with snow. Summer: grazes on grasses, legumes, annual herbs and occasional farm crops. HOME RANGE: Variable; around 100ha for males (larger in reproductive season), 85ha for females, with discrete day and night ranges. On hilly ground range long and narrow, over a range of altitudes (tend to move uphill in summer). Great variation in numbers (eg W Scotland: 1 per 80ha, E Scotland: 2.45 per ha). Islands up to 4 per ha, mainland forests 1 per 100 ha. Adults disperse up to a record distance of 300km. SOCIAL: Mainly solitary, but may feed in groups of up to 70 in sheltered places during stormy weather, or at high quality food patches, or perimeter of snow field. Females stay close to leverets during lactation. COMMUNICATION: Screams in distress; female warns leverets of danger with clicking sounds (probably using teeth). Both sexes fight using hind feet (kicking) or forefeet (boxing). Hearing excellent, scent good, sight adapted to detecting movements.

Breeding. Breeding season from February–August (occasionally September). Sexually mature in second year. Gestation 50 days. Litter size 1–5 (max. 8), increasing with decreasing annual temperature. Litters per annum 1–3 (post-partum oestrus) (and circumstantial evidence for superfoetation, i.e. mating before giving birth). Mammae 8. YOUNG: Born fully furred and eyes open. Leave place of birth after one week, but return to suckle. Weaning age 3 weeks. Milk very nutritious. Mother typically nurses 1 hour after sunset, but may do so more than once a day. Occasionally defends young from threat.

Lifespan. Max. 9 years. Big fluctuations in population density, including the death of 75% of adults annually during population declines. Juvenile mortality in first year typically 80%. Average adult mortality 58%. Red Fox an important predator, and Stoats take significant numbers of leverets.

Measurements. Head-body length: 457–610mm (nb distinct subspecies, eg Scotland: *L.t. scoticus* 457–545mm; Ireland: *L.t. hibernicus* 521–559mm; Scandinavia: *L.t.timidus* 520–600mm). Tail-length: 40–80mm. Hind-foot length: 120–165mm (Scotland: 127–155mm, Ireland 144–168mm, Sweden: 150–165mm). Condylo-basal length: Scotland: 63–80mm, Ireland: 69–81mm, Sweden: 87–92mm. Weight: 2000–5800g (female *c.* 13% heavier than male). Dental formula: 2/1, 0/0, 3/2, 3/3 = 28.

General. Circumpolar species if include *L. arcticus* (N. America) and *L. othus* (Greenland). Can run at 64km per hour. Records of sterile hybrids with Brown Hare, but these are rare. Like Brown Hare, refecates (hence wrongly classified as ruminant in the Bible). Important symbol in ancient art and myths.

RABBIT *Oryctolagus cuniculus* **Pl. 63**

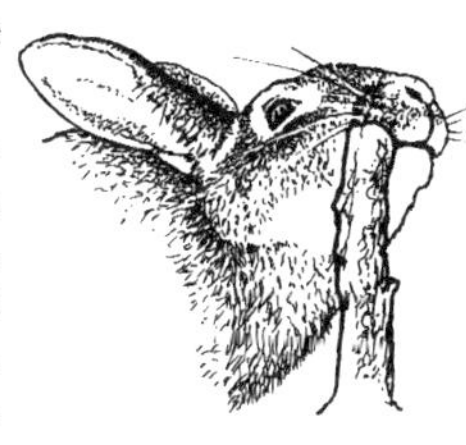

Name. *Oryctos* = digging (Gk.); *lagos* = hare (Gk.); *cuniculus* = rabbit (Lat.).

Recognition. Smaller and less rangy than hares. Ears shorter than head length. Ear tips brown, upper surface of tail blackish-brown (white below, flashes when fleeing). Rufus coloured patch on the back of the neck, and blue/grey fur on belly. Males moult between September and October. Faecal pellets often most obvious sign; Rabbits use them as conspicuous scent stations on dung hills. Droppings are black–dark brown, and about the size and shape of large pea (0.7–1.2cm diam.). Two types of pellets produced: marking and non-marking (former covered with secretion from anal glands). Soft, non-marking pellets re-ingested for nutritional benefits. Pawscrapes are shallow holes 3–10cm deep and 5–15cm long, often with several pellets deposited nearby (often on the edge of the scrape rather than in it). Pawscrapes often found along boundaries of territories. Burrow system is termed a warren. Grass grazed to close-cropped lawn in vicinity of warren, marked with divots and scrapes where Rabbits scratched for roots. The burrows are especially located on slopes or well drained sites; entrances vary from 10-50 cm in diameter. Tooth marks (in winter) indicate where Rabbits have eaten bark of saplings. Tend to use regular trails along which pellets are scattered. Possible to identify sexes in the field: adult females have longer, thinner, finer heads than adult males, which have wider cheeks. At the extreme of the Rabbit's range (i.e. high ground in NE Scotland) the Rabbit behaves more like a hare, creating forms rather than using a burrow.

Habitat. Heathland, open meadow and grassland (including grassy cliffs up to 500m), fringes of agricultural land, woodland, and dry sandy soil (including sand dunes). Avoids coniferous forest. The warren is a system of underground tunnels built up by a large group of Rabbits. At low population densities single burrows occur. Burrow and warren characteristics depend greatly on habitat and on population density. Breeding burrow has a tunnel of 1–2m length, leading to a nest of grass and moss, lined with belly fur. Some females, due to low status or poor terrain, cannot get access to a good warren system and therefore use separate breeding burrows called 'stops'.

Habits. Usually crepuscular and nocturnal, but more diurnal where no human interference. FOOD: Selects leaves of nutritious species from wide range of vegetation (including agricultural crops eg germinating cereals, young trees, cabbages). In winter eats grasses, bulbs and bark. HOME RANGE: Males range further than females, but ranges average 0.4–2ha.

Individuals forage 150–400m from warren (further where population density is low or habitat sparse). Territories of males may include those of several breeding females. Home range directly related to population density (as latter increases, former decreases), which varies dramatically with habitat. Low density populations at *c.* 0.6 Rabbits per ha, high densities *c.* 10 per ha. Large scale reproductive suppression seen when population reaches 25–100 per ha; above this no further density effects seen. Dispersal distance varies with population density; where numbers low, dispersal as much as 800m; where population high, dispersal less than 200m. Enforced dispersal of sexually maturing males in response to increased hostilities from adult males in the group, otherwise adults are very sedentary. SOCIAL: Form pairs at low densities. At higher densities Rabbits form discrete groups, ratio of 1–5 males:1–6 females in stable groups of up to 20 adults. Colonies centred on warrens (dominant individuals breed in best nests near centre). Offspring of dominant does tend to attain high rank as adults. All defend core area in immediate vicinity of warren, but several groups may share 'communal' grazing. Intra-sexual hierarchies exist where females compete for burrows. Satellite individuals of either sex live on edge of community due to limited burrow distribution. COMMUNICATION: Squealing distress call. Thump feet when alarmed (possibly signalling to underground nestlings and/or Rabbits above ground). Olfactory communication very important. Can assess group identity, sex, age, social and reproductive status, and ownership of territory by scent. Harderian, inguinal, submandibular, and anal glands found to be sexually dimorphic for size and secretory activity. Territories marked with urine and faecal pellets (covered with a secretion from the anal glands) and a secretion from the submandibular (chin) glands (see vignette). Females also use urine to mark the sealed entrances to their breeding stops. Males known to squirt urine over conspecifics. The dominant males posses the largest and most active of these scent glands and are responsible for the majority of the scent-marking performed within the group. Scent-marking activity increases during breeding season. Sense of smell very good.

Breeding. Mating occurs throughout the year, with most litters born between February–August. Peak number of pregnancies April–June. Females' vulva engorged and dark red/purple colour when in oestrus, and pale pink and narrow when anoestrous (in autumn). Induced ovulators, and post-partum breeders. Females born early in season can breed same year. Young born early in season more likely to survive winter. Dominant females have greater reproductive success than younger, subordinate females. Resorb foetuses in adverse conditions (social and physical). Population numbers increase from March/April towards peak in September/October. Sexual maturity: females 3.5 months; males 4 months. Gestation 28–33 days. Litter size 3–12, average 5. Litters per annum 3–7 (minimum interval 30 days; post-partum oestrus). Mammae 6. YOUNG: Naked and blind at birth; eyes open 10 days, weaning age 28 days (mother may visit young only once per day for *c.* 5 mins as anti-predator strategy). The demands of lactation upon the mother may delay weaning of the young and the mother's return to oestrus. Female usually gives birth in underground nesting chambers situated in pre-existing burrow systems. Females (low ranking or satellite) dig 'stops' – small burrow where young born and where she conceals entrance with soil between visits. Male protects young, irrespective of paternity, from aggression by females who may attack and even kill any strange youngsters (this defence is the only known paternal care by any leporid species).

Lifespan. Max. 9 years. In one stable island population the annual mortality was 30%.

Measurements. Head-body length: 340–500mm. Tail-length: 40–80mm. Hind-foot length: 75–95mm. Condylo-basal length: 68–75mm. Ear length: 65–70mm. Weight: 1200–2500g (birth: 30–35g). Dental formula: 2/1, 0/0, 3/2, 3/3 = 28.

General. Field of vision: 360°; UK population crashed due to myxomatosis in 1953. Population currently increasing due to attentuation of the myxoma virus and increased immunity of the Rabbits. Die within 3 weeks if prevented from copophragy. Burrowing habit said to have originated in Iberian Peninsula to cope with high predation pressure, but opened the door to exploitation of habitats with only limited cover. The Eastern Cotton-tail Rabbit *Sylvilagus floridanus* is immune to myxomatosis and was introduced by sportsmen into Italy, northern Spain and every French department after the disease had depleted the indigenous rabbit populations. The nape is a rusty colour, the feet pale, the underside of the tail white and there is often a white patch on the forehead. They weigh from 900–1800g. It does not dig its own burrows.

<*Rabbit.* Widespread in W Europe and British Isles, and Balearic Islands, Corsica, Sardinia, Sicily.

>*Cotton-tail Rabbit.* Introduced to Italy, N Spain and whole of France.

Glossary

Where it has proven necessary to use technical terms in this text I have tried to explain them in context. However, this glossary explains some of the common words in a naturalist's vocabulary. Many of these definitions are adapted from the more elaborate glossary to my Encyclopaedia of Mammals (edited by D.W. Macdonald, published in the UK by Harper Collins) and that encyclopaedia should be consulted for fuller accounts.

Adaptation any characteristic that improves the chances of an organism transmitting genes to the next generation (ie producing offspring). Such beneficial changes are genetically controlled and are distinct from alterations occurring within one generation which may not lead to genetic change. Adaptations can affect any level of organization, from cells to whole organisms, and their behaviour. Adaptations are favoured by the process of NATURAL SELECTION.

Adaptive radiation the pattern in which different species develop from a common ancestor (as distinct from CONVERGENT EVOLUTION, a process whereby species from different origins became similar in response to the same SELECTIVE PRESSURES.

Adult a fully developed mature individual, capable of breeding, but not necessarily doing so until social and/or ecological conditions allow.

Allele* One of a pair of genes, or of multiple forms of a gene, located at the same locus of homologous chromosome. Also known as allelomorph.

Amniotic fluid found within the sac which surrounds an embryo. This fluid provides a liquid environment for the embryo, necessary for animals that reproduce on land, and may also cushion the mammalian embryo against distortion by maternal organs pressing on it.

Ancestral stock a group of animals, usually showing primitive characteristics, which is believed to have given rise to later, more specialised forms.

Arboreal living in trees.

Baculum (os penis or penis bone) an elongate bone present in the penis of certain mammals.

Blow (spout) the cloud of spray which accompanies a whale's exhalation at the water surface. This spray may be the water that was lying near the blowhole as the whale exhaled, it may be mucus from the windpipe, or it may occur as a result of the condensation of the water vapour in the whale's breath as it meets cold air.

Brachydont* a type of short-crowned teeth whose growth ceases when the mammal is full grown, whereupon the pulp cavity in the root closes. Typical of most mammals, but contrast the HYPSODONT teeth of many herbivores.

Caecum* a blind sac in the digestive tract, opening out from the junction between the small and large intestines. In herbivorous mammals it is often very large; it is the site of bacterial action on cellulose. The end of the caecum is the appendix; in species with reduced caeca the appendix may retain an antibacterial function.

Calcar a rod of cartilage or bone, which arises from the ankle of a bat and supports the edge of the tail membrane.

Carnassial in the order Carnivora, modified PREMOLAR and MOLAR teeth (the lower first molar and the upper last premolar) characterised by a scissor-like shearing action used for cutting flesh.

Caudal **concerning the tail**

Cellulose* the fundamental constituent of the cell walls of all green plants, and some algae and fungi. It is very tough and fibrous, and can be digested only by the intestinal flora in mammalian guts.

Chorioallantoic placentation* a system whereby foetal mammals are nourished by the blood supply of the mother. The chorion is a superficial layer enclosing all the embryonic stuctures of the foetus, and is in close contact with the maternal blood supply at the placenta. The union of the chorion (with its vascularized ALLANTOIC STALK and YOLK SAC) with the placenta facilitates the exchange of food substances and gases, and hence the nutrition of the growing fetus.

Chromosomes* threadlike structures seen in animal and plant cell nuclei which carry the linearly arranged genetic units.

Class taxonomic category subordinate to a PHYLUM and superior to an ORDER.

Cloaca terminal part of the gut into which the alimentary, urinary, and reproductive systems open, leading to a single aperture in the body.

Colostrum* the first milk secreted by the mammary gland during the first days following parturition.

Competition* the inter- or intraspecific interaction resulting when several individuals share an environmental necessity.

Condylo-basal length the length of a skull, measured from the anterior points of the premaxillary to the occipital condyles.

Conspecific member of the same species.

Convergent evolution the independent acquisition of similar characters in evolution, as opposed to possession of similarities by virtue of descent from a common ancestor.

Coprophagy the eating of faeces. See REFECTION.

Crepuscular active in twilight.

Cretaceous third of the three periods included in the Mesozoic Era. It began 135 million years ago and ended 65 million years ago.

Crustaceans members of a class within the phylum Arthropoda typified by five pairs of legs, two pairs of antennae, head and thorax joined, and calcareous deposits in the exoskeleton, eg crayfish, crabs, shrimps.

Cryptic (coloration or locomotion) protecting through concealment.

Cursorial being adapted for running.

Cusp a prominence on a cheek-tooth (premolars or molars).

Deciduous teeth* teeth of a young mammal which are shed and replaced by permanent teeth. Also known as milk teeth.

Delayed implantation see IMPLANTATION.

Dentine* a bonelike tissue composing the bulk of a vertebrate tooth; consists of 70% inorganic materials and 30% water and organic matter.

Dew claws rudimentary inner (fifth) toe of forefeet in dogs and cats

Dimorphism the presence of one or more morphological differences that divide a species into two groups. Many examples come from sexual differences of particular traits, such as body size (males are often larger than females), and occur because the gene coding for the trait is situated on one of the sex chromosomes. However other dimorphisms, or polymorphisms (where a feature takes several different forms) may be maintained in a population by other mechanisms (e.g. coat colour morphs of Arctic Foxes).

Dioestrus period between two oestrous cycles in a female mammal.

Diphyodont* having two successive sets of teeth, deciduous followed by permanent, as in people.

Dispersal the movements of animals, often as they reach maturity, away from their previous home range (equivalent to emigration). Distinct from dispersion, that is, the pattern in which things (perhaps animals, food supplies, nest sites) are distributed or scattered.

Diurnal active in daytime.

Dorsal on the upper or top side or surface (eg dorsal stripe).

Echolocation the process of perception, often direction finding, based upon the pattern of reflected sound waves (echoes).

Ectotherm* an animal that obtains most of its heat from the environment and therefore has a body temperature very close to that of its environment.

Embryonic diapause the temporary cessation of development of an embryo (eg in some bats and kangaroos).

Enamel* hard glossy natural coating of teeth.

Endemism situation in which a species or other taxonomic group is restricted to a particular geographic region, due to factors such as isolation or response to soil or climatic conditions. Such a taxon is said to be endemic to that region.

Enzyme* any group of catalytic proteins that are produced by living cells and that mediate and promote the chemical processes of life without themselves being altered or destroyed.

Eocene tertiary epoch which began at the end of the Palaeocene approximately 54 million years ago, and ended at the beginning of the Oligocene 17.5 million years later.

Eutherian any placental mammal of the subclass Eutheria, characterized by the embryo developing in the uterus of the female where it obtains food and exchanges gases through the attached PLACENTA.

Evolution an explanation of the way in which present-day organisms have been produced, involving changes taking place in the genetic make-up of individuals that have been passed on to successive generations. According to Darwinism, evolutionary mutations have given rise to changes that have, through natural selection, either survived in better adapted organisms, or died out. Evolution is now generally accepted as the means which gives rise to new species.

Extant still existing.

Family a taxonomic division subordinate to an order and superior to a genus.

Fecundity the numbers of young produced by an organism during the course of its life.

Feral living in the wild (of domesticated animals, eg cat, dog).

Fertilisation union of egg and sperm to form a zygote, which then develops into a new individual.

Fitness* a measure of the ability of an animal or plant (with one genotype or genetic make-up) to leave viable offspring in comparison to other individuals (with different genotypes). The process of natural selection, often called survival of the fittest, determines which characteristics have the greatest fitness, ie are most likely to enable their bearers to survive and rear young which will in turn bear those characteristics.

Follicle* a small sac, therefore (a) a mass of ovarian cells that produces an ovum, (b) an indentation in the skin from which hair grows.

Formen anatomical term meaning opening, eg. lachrymal foramen see p.000.

Fossil record the remains of organisms, or traces of their existence (such as footprints), present in the rock strata and forming a history of the development of life from its origins on earth. The record is very largely incomplete due to the comparative rarity with which fossils are formed, but provides evidence of evolution having taken place, particularly where long series of a particular form can be traced over an extended period of time.

Fossorial burrowing.

Frugivore an animal eating mainly fruits.

Gene* the basic unit of heredity; a portion of the DNA molecule coding for a given trait and passed, through replication at reproduction, from generation to generation. Genes are expressed as adaptations and consequently are the most fundamental units (more so than individuals) on which NATURAL SELECTION acts.

Genotype the genetic constitution of an organism, determining all aspects of its appearance, structure and function.

Genus (plural genera) a taxonomic division superior to species and subordinate to family.

Gestation the period of development within the uterus; the process of DELAYED IMPLANTATION can result in the period pregnancy being longer than the period during which the embryo is actually developing (See also IMPLANTATION).

Harem group a social group consisting of a single adult male, at least two adult females and immature animals: a common pattern of social organization among mammals.

Herbivore an animal eating mainly plants or parts of plants.

Hibernaculum (pl. hibernacula) from the Latin, which means winter quarters.

Hibernation a period of winter inactivity during which the normal physiological process is greatly reduced and thus during which the energy requirements of the animal are lowered.

Home range the area in which an animal normally lives (generally excluding rare excursions or migrations), irrespective of whether or not the area is defended from other animals (cf. TERRITORY).

Homoiotherm* an animal which maintains a constant internal temperature which is often higher than that of the environment; common among birds and mammals. Also known as warm- blooded.

Hormone regulatory substance, active at low concentrations, that is produced in specialized cells but that exerts its effect either on distant cells or on all cells in the organism to which it is conveyed via tissue fluids such as blood.

Hybrid the offspring of parents of different species.

Hybrid/hybridisation the offspring of parents of different species.

Hypsodont* high-crowned teeth, which continue to grow when full sized and whose pulp cavity remains open; typical of herbivorous mammals see BRACHYDONT.

Implantation the process whereby the free-floating blastocyst (early embryo) becomes attached to the uterine wall in mammals. At the point of implantation a complex network of blood vessels develops to link mother and embryo (the placenta). In **delayed implantation** the blastocyst remains dormant in the uterus for periods varying between species, from 12 days to 11 months. Delayed implantation may be obligatory of facultative and is known for some members of the Carnivora and Pinnipedia and others.

Induced ovulation see OESTRUS.

Insectivore an animal eating mainly arthropods (insects, spiders).

Introduced of a species which has been brought, by people, from lands where it occurs naturally to lands where it has not previously occurred. Some introductions are accidental (eg rats which have travelled unseen on ships), but some are made on purpose for biological control, farming or other economic reasons (eg the Common Brush-Tail Possum, which was introduced to New Zealand from Australia to establish a fur industry).

Invertebrate any animal that does not possess a backbone.

IUCN World Conservation Union.

Juvenile no longer possessing the characteristics of an infant, but not yet fully adult.

Karyotype the charateristics and numbers of the chromosomes in a cell. These can be used to distinguish between very similar speces.

Kin selection* a facet of NATURAL SELECTION whereby an animal's fitness is affected by the survival of its relatives or kin. Kin selection may be the process whereby some alloparental behaviour evolved: an individual behaving in a way which promotes the survival of its kin increases its own INCLUSIVE FITNESS, despite the apparent selflessness of its behaviour.

Kingdom* One of the primary divisions that include all living organisms: most authorities recognize two, the animal kingdom and the plant kingdom, while others recognize three or more, such as Protista, Plantae, Animalia, and Mychota.

Lactation (verb: lactate) the secretion of milk, from MAMMARY GLANDS.

Lanugo prenatal hair, eg in seals where, in some species, this hair is retained for a time after birth.

Latrine a place where faeces are regularly left (often together with other SCENT MARKS); associated with olfactory communication.

Lineage line of descent of a related group of animals from their ancestral group. A lineage ultimately extends back through the various taxonomic levels, from the SPECIES to the GENUS, from the

GENUS to the FAMILY, from the FAMILY to the ORDER and so on.

Mamma (pl.mammae) mammary glands the milk-secreting organ of female mammals, probably evolved from sweat glands.

Mammal a member of the CLASS of VERTEBRATE animals having MAMMARY GLANDS which produce milk with which they nurse their young (properly: Mammalia).

Marsupial any member of the subclass Marsupialia containing mammals, characterized by the presence of a pouch to which the young, born in an undeveloped state, migrate during early development. The pouch contains the mammary glands, which vary in number between species, and the young complete their development here. The group was at one time widespread, but now is restricted to Australasia and South America. In Australasia, marsupials, free from competition from EUTHERIAN (placental) mammals, have radiated to occupy most niches elsewhere occupied by placental forms.

Maxilla one of the bones of the upper jaw, which bears all the upper teeth except the incisors

Maxillary teeth all upper teeth other than the incisors, ie those teeth borne by the **maxilla** which is the posterior bone of the upper jaw.

Metatheria* an infra-class of therian mammals including a single order, the Marsupialia; distinguished by a small braincase, a maximum total of 50 teeth, the inflected angular process of the mandible, and a pair of marsupial bones articulating with the pelvis.

Migration movement, usually seasonal, from one region or climate to another for purposes of feeding or breeding.

Monoestrus* having a single oestrus cycle per year.

Monogamy a mating system in which individuals only have one mate per breeding system.

Monotreme* a mammal of the subclass Monotremata, which comprises the platypus and echidnas. The only egg-laying mammals.

Mutation* a structural change in a gene which can give rise to a new heritable characteristic.

Mutualism* mutual interactions between two species that are beneficial to both species.

Natural selection the mechanism, proposed by Charles Darwin, by which gradual evolutionary changes take place. Organisms which are better adapted to the environment in which they live produce more viable young, so increasing their proportion in the population and thus being 'selected'. Such a mechanism depends on the variability of individuals within the population. Such variability arises through mutation, the beneficial mutants being preserved by natural selection.

Neophobia fear of new things.

Niche the role of a species within the community, defined in terms of all aspects of its lifestyle (eg food, competitors, predators, and other resource requirements). Also, the limits for all important environmental features, within which individuals of a species can survive, grow and reproduce.

Nocturnal active at night time.

Oestrus the period in the oestrous cycle of female mammals at which they are often attractive to males and receptive to mating. The period coincides with the maturation of eggs and ovulation (the release of mature eggs from the ovaries). Animals in oestrus are often said to be 'on heat' or 'in heat'. In primates, if the egg is not fertilized the subsequent degeneration of uterine walls (endometrium) leads to menstrual bleeding. In some species ovulation is triggered by copulation and this is called induced ovulation, as distinct from spontaneous ovulation.

Olfaction, olfactory the olfactory sense is the sense of smell, depending on receptors located in the epithelium (surface membrane) lining the nasal cavity.

Omnivore an animal that feeds on both plants and animals.

Order a taxonomic division subordinate to class and superior to family.

Ovulation* see oestrus.

Pair-bond an association between a male and female, which lasts from courtship at least until mating is completed, and in some species, until the death of one partner.

Parallel evolution similar evolutionary development that occurs in lineages of common ancestry. Thus the descendants are as alike as were their ancestors. The nature of the ancestry imposes or directly influences the development of the parallelism.

Parasitism interaction of species in which one, typically small, organism (the parasite) lives in or on another (the host), from which it obtains food, shelter, or other requirements.

Parturition the process of giving birth (hence post partum - after birth).

Peduncle stalk (eg of tail)

Pheromone secretions whose odours act as chemical messengers in animal communication, and which prompt a specific response on behalf of the animal receiving the message (see SCENT MARKING).

Photoperiod day length.

Phylum taxonomic category subordinate to a kingdom and superior to classes and all lower taxa.

Pleistocene the first of two epochs of the Quaternary, preceded by the Pliocene. It is held conventionally to have lasted from approximately 1.8 million years ago until the beginning of the Holocene, about 10 000 years ago, but recent evidence from deep-sea cores may necessitate a revision of the earlier date.

Polygamous a mating system wherein an individual has more than one mate per breeding season.

Polygynous a mating system in which a male mates with several females during one breeding season (as opposed to polyandrous, where one female mates with several males).

Polygynous* a mating system in which a male mates with several females during one breeding season (as opposed to polyandrous, where one female mates with several males.)

Population* a more or less separate (discrete) group of animals of the same species within a given BIOTIC COMMUNITY.

Post-calcarial lobe a lobe of skin found in some bats, attaching to the outer side of the CALCAR.

Post-partum oestrus ovulation and an increase in the sexual receptivity of female mammals, hours or days after the birth of a litter (see OESTRUS).

Predator an animal which forages for live prey; hence "anti-predator behaviour" describes the evasive actions of the prey.

Prehensile capable of grasping.

Refection process in which food is excreted and then reingested a second time to ensure complete digestion, as in the Common Shrew.

Rumen* first chamber of the ruminant artiodactyl four-chambered stomach. In the rumen the food is liquified, kneaded by muscular walls and subjected to fermentation by bacteria. The product, cud, is regurgitated for further chewing; when it is swallowed again it bypasses the RUMEN and reticulum and enters the omasum.

Ruminant a mammal with a specialised digestive system typified by the behaviour of chewing the cud. Their stomach is modified so that vegetation is stored, regurgitated for further maceration, then broken down by symbiotic bacteria. The process of rumination is an adaptation to digesting the cellulose walls of plant cells.

Rut a period of sexual excitement; the mating season.

Scent gland an organ secreting odorous material with communicative properties; see SCENT MARK.

Scent mark a site where the secretions of scent glands, or urine or FAECES, are deposited and which has communicative significance. Often left regularly at traditional sites which are also visually conspicuous. Also the "chemical message" left by this means; and (verb) to leave such a deposit.

Selective pressure a factor affecting the reproductive success of individuals (whose success will depend on their FITNESS, ie the extent to which they are adapted to thrive under that selective pressure).

Sibling species different species which are morphological identical, eg behaviourial differences can lead to reproductive isolation.

Siblings individuals who share one or both parents. An individual's siblings are its brothers and sisters, regardless of their sex.

Speciation the process by which new species arise in evolution. It is widely accepted that it occurs when a single-species population is divided by some geographical barrier.

Species a taxonomic division subordinate to genus and superior to subspecies. In general a species is a group of animals similar in structure and which are able to breed and produce viable offspring.

Subfamily a division of a FAMILY.

Suborder a subdivision of an ORDER.

Subspecies a recognisable subpopulation of a single specie, typically with a distinct geographical distribution.

Subspecies a recognisable subpopulation of a single SPECIES, typically with a distinct geographical distribution.

Suture seam (eg between bones of skull).

Sweat gland* a coiled tubular gland of skin which secretes sweat.

Systematics* see TAXONOMY

T-piece T-shaped cartilage attached to POST-CALCARIAL LOBE in bats, may be used as a feature to distinguish between species.

Tapetum a reflecting layer which lies outside the receptor layer of the retina in the eyes of certain mammals, birds and fishes. It gives a shining appearance to the eyes (eg in cats) and aids vision in dim illumination; by reflecting back any light that is not absorbed by the receptor layer, it provides a second chance for the receptors to be stimulated. (Tapetum has a number of other definitions in other fields of biology).

Taxonomy (systematics) the science of the classification of living organisms.

Territory an area defended from intruders by individuals or group. Originally the term was used where ranges were exclusive and obviously defended at their borders. A more general definition of territoriality allows some overlap between neighbours by defining territoriality as a system of spacing wherein home ranges do not overlap randomly - that is, the location of one individual's, or group's home range influences those of others.

Theria* a subclass of the class Mammalia including all mammals except the Monotremes.

Thermoregulation the regulation and maintenance of a constant internal body temperature in mammals.

Tragus a flap, sometimes moveable, situated in front of

the opening of the outer ear in bats.

Ultrasound sound of frequency above the upper limit of the normal range of human hearing, ie above 20 kHz.

Ungulate a member of the orders Artiodactyla (even-toed ungulates), Perissodactyla (odd-toed ungulates), Proboscidea (elephants), Hyracoidea (hyraxes) and Tubulidentata (aardvark), all of which have their feet modified as hooves of various types (hence the alternative name, hoofed mammals). Most are large and totally herbivorous, eg deer, cattle, gazelles, horses.

Unicuspid tooth with one cusp.

Uterus* the organ of gestation in mammals which receives and retains the fertilized ovum, holds the foetus during development, and becomes the principle agent of its expulsion at birth.

Velvet furry skin covering a growing antler.

Ventral on the lower or bottom side or surface; thus ventral or abdominal glands occur on the underside of the abdomen.

Vertebrate any member of the subphylum Vertebrate in the phylum Chordata, including all those organisms that possess a backbone, such as fish, amphibians, reptiles, birds and mammals. In addition, they are characterized by having a skull which surrounds a well-developed brain and a bony or cartilaginous skeleton.

Vertebrates an animal with a backbone: a division of the phylum Chordata which includes animals with notochords (as distinct from invertebrates).

Vibrissae stiff, coarse hairs richly supplied with nerves, found especially around the snout, and with a sensory (tactile) function (colloquially whiskers).

Zygomatic arch (zygoma) the bone at the side of the skull which forms an arch from beneath the eye socket to a position toward the back of the head. The main chewing (masseter) muscle is attached to the zygomatic arch.

BIBLIOGRAPHY

Much information about mammals can only be found in technical papers published in scientific journals, and these are beyond the scope of this bibliography. However, the following good books provide a start for those wanting to explore details mentioned in this book. Readers of the FIELD GUIDE should refer first to the companion volume, EUROPEAN MAMMALS: EVOLUTION AND BEHAVIOUR, in which species and subject indexes will help you track down details of the mammal you have seen.

General

Readable, non-technical but informative accounts of mammals, along with many illustrations, can be found in:

Macdonald, D.W. (ed) (1984) The Encyclopaedia of Mammals, 2 vols. George, Allen and Unwin, London. Published in one volume by Harper Collins, London.

Similarly general books, but organised species by species, include:

Boyle, C. (1981) The RSPCA Book of Mammals. Collins, London.

Bjarvall, A., and Ullström, S. (1986) The Mammals of Britain and Europe. Croom Helm, London.

Earlier field guides and books on field craft include:

Corbet, C., and Ovenden, D. (1980) The Mammals of Britain and Europe. Collins, London.

van den Brink, F.H. (1967) A Field Guide to the Mammals of Britain and Europe. Collins, London.

Bang, P., and Dahlstrom, P. (1974) Collins Guide to Animal Tracks and Signs. Collins, London.

Lawrence, M.J., and Brown, R.W. (1973) Mammals of Britain: Their Tracks, Trails and Signs (revised edn). Blandford Press, London.

Clark, M. (1981) Mammal Watching. Severn House, London.

More technical books include the excellent:

Corbet, G.B., and Harris, S. (eds)(1990) The Handbook of British Mammals (3rd edn), Blackwell Scientific Publications.

and its four volume continental equivalent:

Niethammer, J., and Krapp, F. (1978, 1982) Handbuch der Säugetiere Europas. Akademische Verlagsgesellschaft, Wiesbaden.

Definitive lists of mammals can be found in:

Nowak, R.M., and Paradiso, J.L. (eds) (1983) Walker's Mammals of the World (4th edn) 2 vols. Johns Hopkins University Press, Baltimore and London.

Honacki, J.H., Kinman, K.E., and Koeppl, J.W. (1982) Mammal species of the world: a taxonomic and geographical reference. Allen Press, Inc. and Association of Systematics Collections, Laurence, Kansas, USA.

Corbet, G.B., and Hill, J.E. (1991) A World List of Mammalian Species

(3rd edn), Natural History Museum Publications/Oxford University Press, London & Oxford.

Three books will provide the enthusiast with first class, clear yet scholarly accounts of the crucial topics of evolution, ecology and behaviour:

Dawkins, R. (1976) The Selfish Gene. Oxford University Press, Oxford.

Begon, M., Harper, J.L. & Townsend, C.R. (1990) Ecology: Individuals, Populations and Communities. Blackwell Scientific Publications, Oxford.

Krebs, J.R. & Davies, N.B. (1987) An Introduction to Behavioural Ecology (2nd edn). Blackwell Scientific Publications, Oxford.

The following books are concerned specifically with selected groups of mammals:

Carnivores

A colourfully illustrated natural history of Carnivores, past and present, is given in:

Macdonald, D.W. (1992) The Velvet Claw: A Natural History of the Carnivores. BBC Books, London.

A classic general account of the Carnivora is provided by:

Ewer, R.F. (1973) The Carnivores. Comstock Publishing Associates, Cornell University Press.

A technical account of Carnivore biology is given in:

Gittleman, J. (1989) Carnivore Behavior, Ecology and Evolution. Chapman and Hall, London.

Monographs on otters include:

Mason, C.F. and Macdonald, S.M.(1986) Otters: ecology and conservation. Cambridge University Press.

Chanin, P. (1985) The Natural History of Otters. Croom Helm, London.

An outstanding account of badger behaviour is presented by:

Kruuk, H. (1989) The Social Badger. Oxford University Press, Oxford.

Other useful, and more general accounts of badger natural history are provided by:

Clark, M. (1988) Badgers. Whittet Books, London.

Neal, E. (1986) The Natural History of Badgers. Croom Helm, London.

A wide review of aspects of the lives of weasels and stoats is published in:

King, C. (1989) The Natural History of Weasels and Stoats. Christopher Helm, London.

A general account of Red Foxes, both rural and urban, is provided by:

Macdonald, D.W. (1987) Running with the Fox. Unwin Hyman, London.

Lloyd, H.G. (1980) The Red Fox, Batsford Press.

A short, entertaining account of urban foxes is given by:

Harris, S. (1986) Urban Foxes, Whittet Books.

Wolves in Europe are covered in:

Zimen, E. (1981) The Wolf, Souvenir Press.

Cetaceans

Informative general accounts are provided by:

Evans, P.G.H. (1987) The Natural History of Whales & Dolphins. Christopher Helm, London.

Martin, A.R. (1990) Whales and Dolphins. Salamander Books Ltd.

Rodents

There is no general account of rodent biology specifically for the layman (see the entry in Macdonald's Encyclopaedia of Mammals), but a good case study is provided by:

Gurnell, J. (1987) The Natural History of Squirrels. Christopher Helm, London.

Artiodactyla

A general account of deer is provided by:

Putman, R.J. (1988) The Natural History of Deer. Christopher Helm, London.

Single species accounts include:

Chapman, N. (1984) Fallow Deer. Anthony Nelson, Oswestry.

Geist, V. (1971) Mountain Sheep: a Study in Behavior and Evolution. University of Chicago Press, Chicago.

One of the best field studies of recent times is summarised in a technical, but readable, account in:

Clutton-Brock, T.H., Guinness, F.E., and Albon, S.D. (1982) Red Deer, Behavior and Ecology of Two Sexes. Edinburgh University Press, Edinburgh.

Aspects of the management of wild ungulates are illustrated in:

Clutton-Brock, T.H., and Albon, S.D. (1989) BSP Professional Books, Oxford.

Insectivores

The European insectivores are covered in three general books:

Churchfield, S. (1990) The Natural History of Shrews. Christopher Helm, London.

Gorman, M.L. & Stone, D. (1990) The Natural History of Moles. Christopher Helm, London.

Morris, P. (1983) Hedgehogs. Whittet Books, London.

Chiroptera

A classic account of bats is given by:

Griffin, D.R. (1958) Listening in the Dark. Yale University Press, New Haven.

Recent accounts of bats include:

Ransome, R.D. (1990) The Natural History of Hibernating Bats. Christopher Helm, London.

Stebbings, R.E. (1988) Conservation of European Bats. Christopher Helm, London.

Lagomorphs

An entertaining account of lagomorphs is given by:

McBride, A. (1988) Rabbits & Hares. Whittet Books, London.

Pinnipeds

An excellent general account of the biology of pinnipeds is presented by:

Bonner, W. N. (1990) The Natural History of Seals. Christopher Helm, London.

A rather more technical account is found in:

Riedman, M. (1990) The Pinnipeds: Seals, Sea Lions and Walruses. Berkeley: University of California Press.

Extinct Mammals

General, highly readable accounts of extinct mammals are given in:

Sutcliffe, A.J. (1985) On the track of ice age mammals. British Museum (Natural History), London.

Savage, R.J.G. and Long, M.R. (1986) Mammalian Evolution. British Museum (Natural History).

Index

English names are printed in Roman type, scientific names are in *italics*. The numbers in Roman type refer to the text pages, those in bold refer to the colour plate in the centre of the book on which that species is illustrated and identification features are described. In all cases, all commonly used English names have been included in this index, but only the generally accepted English name is indexed to both the plate and the text entry. The remaining entries direct the reader to the text description only. Distribution maps are included at the beginning or end of the text description.

NOTES

NOTES

NOTES

NOTES

NOTES

NOTES

NOTES

NOTES